T0252232

# PRIME SUSPECTS

## The Anatomy of Integers and Permutations

Andrew Granville

Jennifer Granville

Illustrated by
Robert J. Lewis

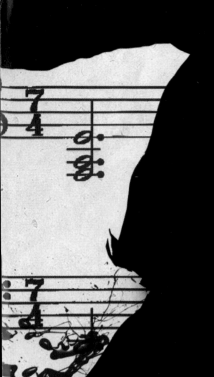

PRINCETON UNIVERSITY PRESS
PRINCETON AND OXFORD

Copyright © 2019 by Princeton University Press

Published by Princeton University Press
41 William Street, Princeton, New Jersey 08540
6 Oxford Street, Woodstock, Oxfordshire OX20 1TR

press.princeton.edu

All Rights Reserved

Library of Congress Control Number: 2018966675
ISBN (pbk.) 978-0-691-14915-8

British Library Cataloging-in-Publication Data is available

Editorial: Vickie Kearn, Susannah Shoemaker, and Lauren Bucca
Production Editorial: Mark Bellis
Illustration Management: Dimitri Karetnikov
Illustration Assistance: Meghan Kanabay
Production: Erin Suydam
Publicity: Sara Henning-Stout and Kathryn Stevens
Copyeditors: Natalie Baan and Cyd Westmoreland

Cover Credit: Robert J. Lewis and Gabriel Cassata
Inside cover ads by Andrew and Jennifer Granville and Tanya Bub

This book has been composed in Adobe Jenson Pro,
CCMeanwhileUncial & Kingthings Typewriter

Printed on acid-free paper. ∞

Printed in China
1 3 5 7 9 10 8 6 4 2

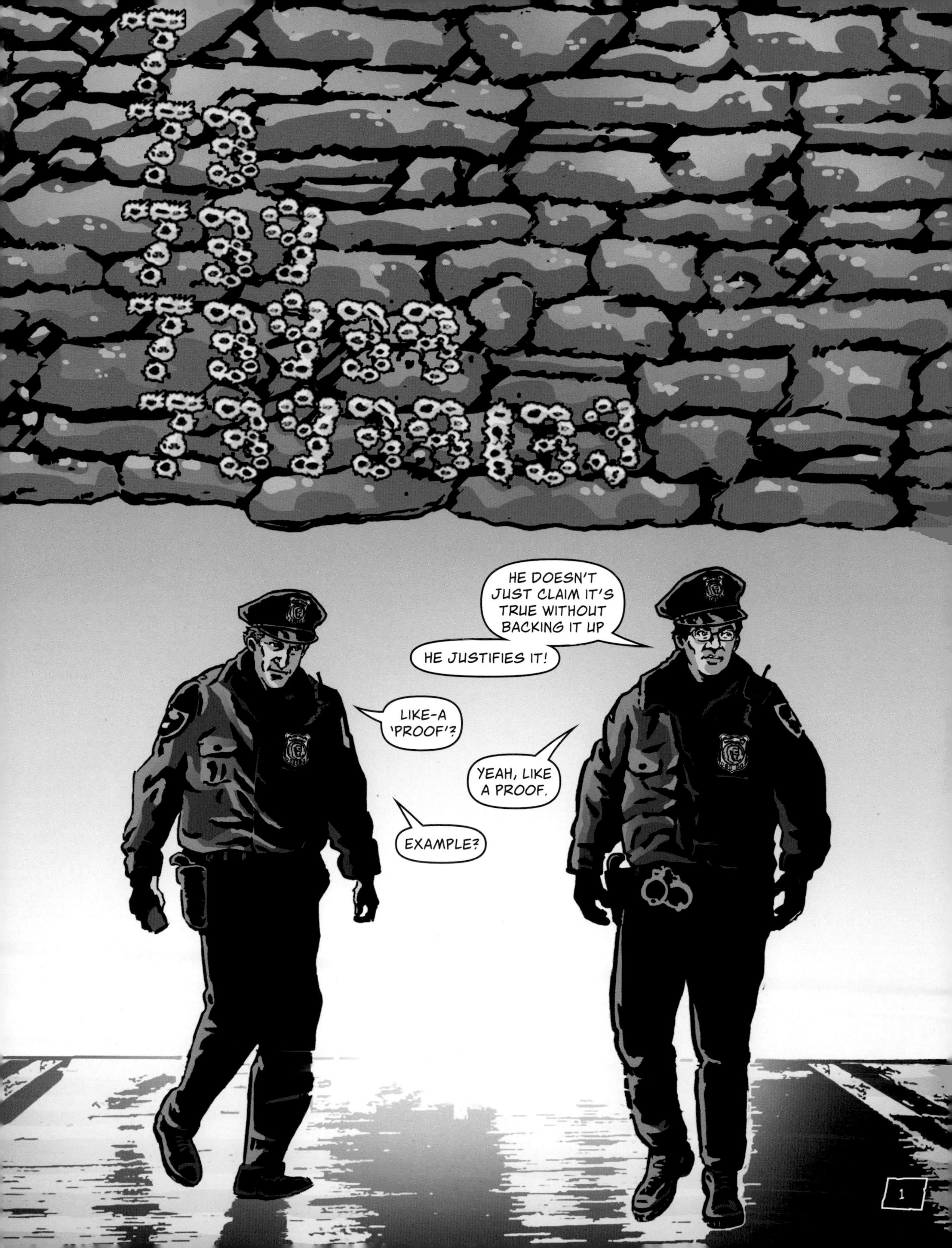
HE DOESN'T JUST CLAIM IT'S TRUE WITHOUT BACKING IT UP
HE JUSTIFIES IT!
LIKE-A 'PROOF'?
YEAH, LIKE A PROOF.
EXAMPLE?

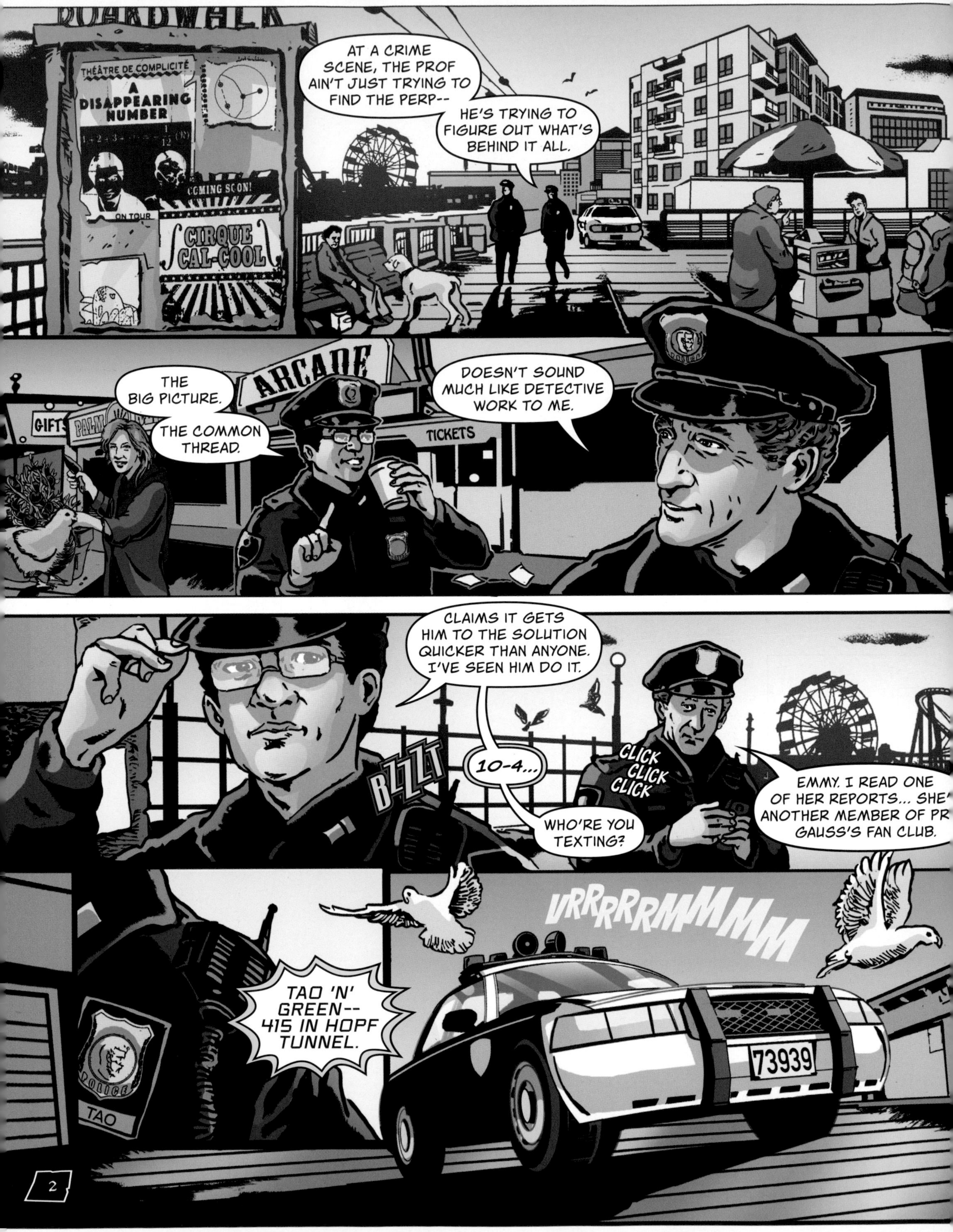
BOARDWALK
THÉÂTRE DE COMPLICITÉ
A DISAPPEARING NUMBER
ON TOUR
COMING SOON!
CIRQUE CAL-COOL
AT A CRIME SCENE, THE PROF AIN'T JUST TRYING TO FIND THE PERP--
HE'S TRYING TO FIGURE OUT WHAT'S BEHIND IT ALL.
GIFTS
ARCADE
TICKETS
THE BIG PICTURE.
THE COMMON THREAD.
DOESN'T SOUND MUCH LIKE DETECTIVE WORK TO ME.
CLAIMS IT GETS HIM TO THE SOLUTION QUICKER THAN ANYONE. I'VE SEEN HIM DO IT.
BZZZT
10-4...
CLICK CLICK CLICK
WHO'RE YOU TEXTING?
EMMY. I READ ONE OF HER REPORTS... SHE' ANOTHER MEMBER OF PR GAUSS'S FAN CLUB.
TAO 'N' GREEN-- 415 IN HOPF TUNNEL.
TAO
POLICE
VRRRRRMMMMM
73939

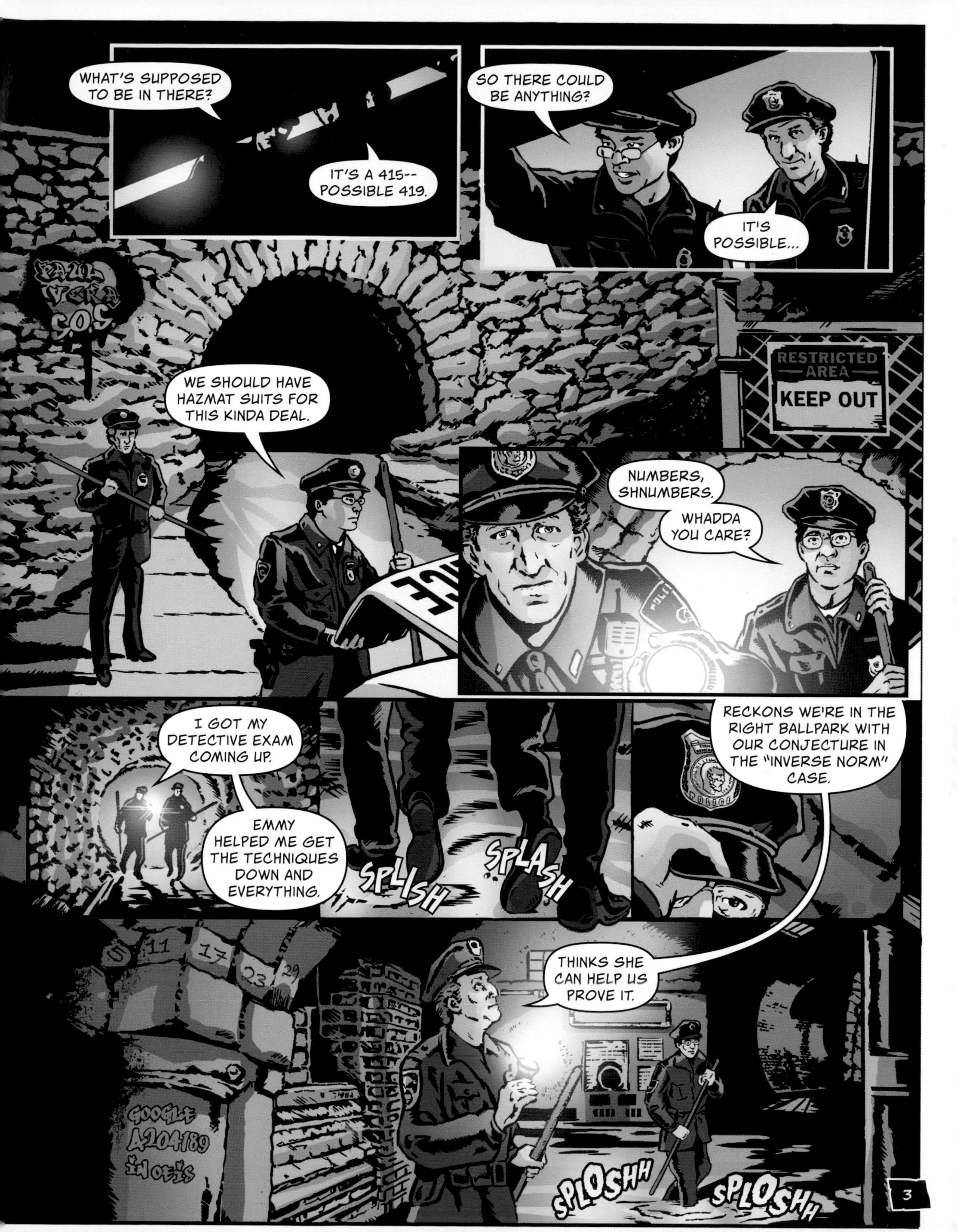

WHAT'S SUPPOSED TO BE IN THERE?
IT'S A 415-- POSSIBLE 419.
SO THERE COULD BE ANYTHING?
IT'S POSSIBLE...
RESTRICTED AREA
KEEP OUT
WE SHOULD HAVE HAZMAT SUITS FOR THIS KINDA DEAL.
NUMBERS, SHNUMBERS.
WHADDA YOU CARE?
I GOT MY DETECTIVE EXAM COMING UP.
EMMY HELPED ME GET THE TECHNIQUES DOWN AND EVERYTHING.
SPLISH
SPLASH
RECKONS WE'RE IN THE RIGHT BALLPARK WITH OUR CONJECTURE IN THE "INVERSE NORM" CASE.
THINKS SHE CAN HELP US PROVE IT.
5 11 17 23 29
GOOGLE A204189 IN OEIS
SPLOSHH
SPLOSHH

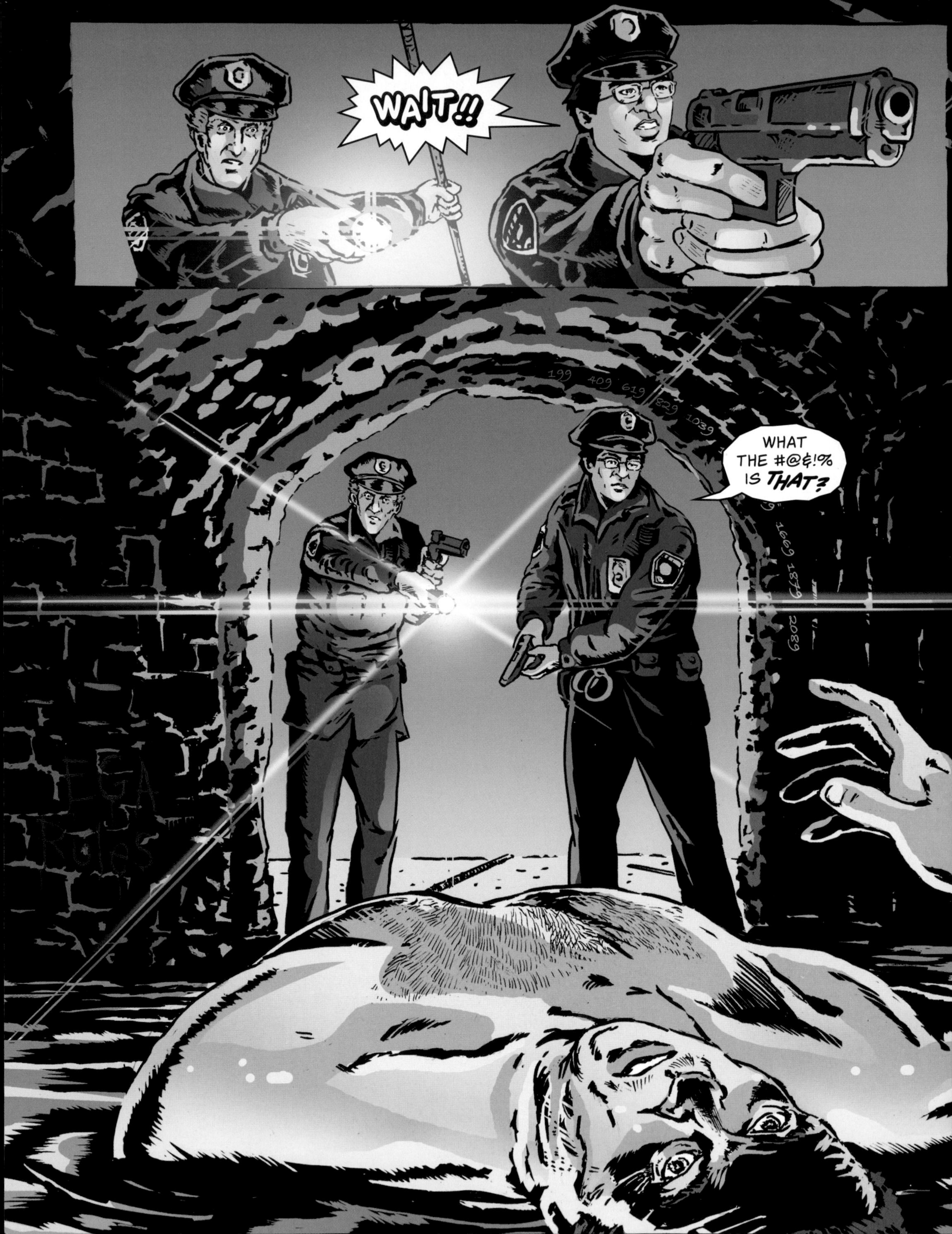
WAIT!!
WHAT THE #@&!% IS THAT?
199 409 619 829 1039
1669 1879 2089
EGA Rules

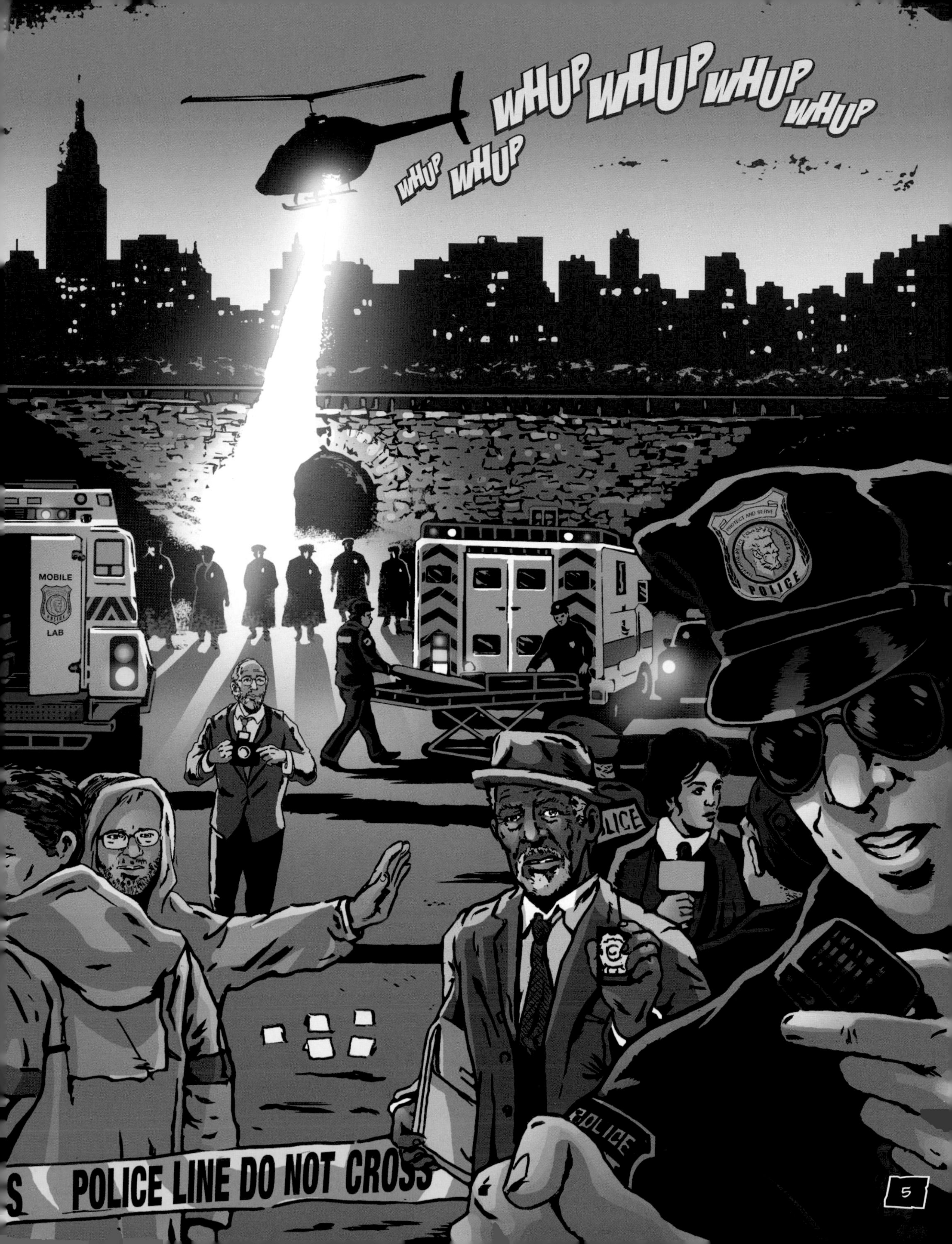
WHUP WHUP WHUP WHUP WHUP WHUP
MOBILE
LAB
POLICE
PROTECT AND SERVE
POLICE
POLICE LINE DO NOT CROSS

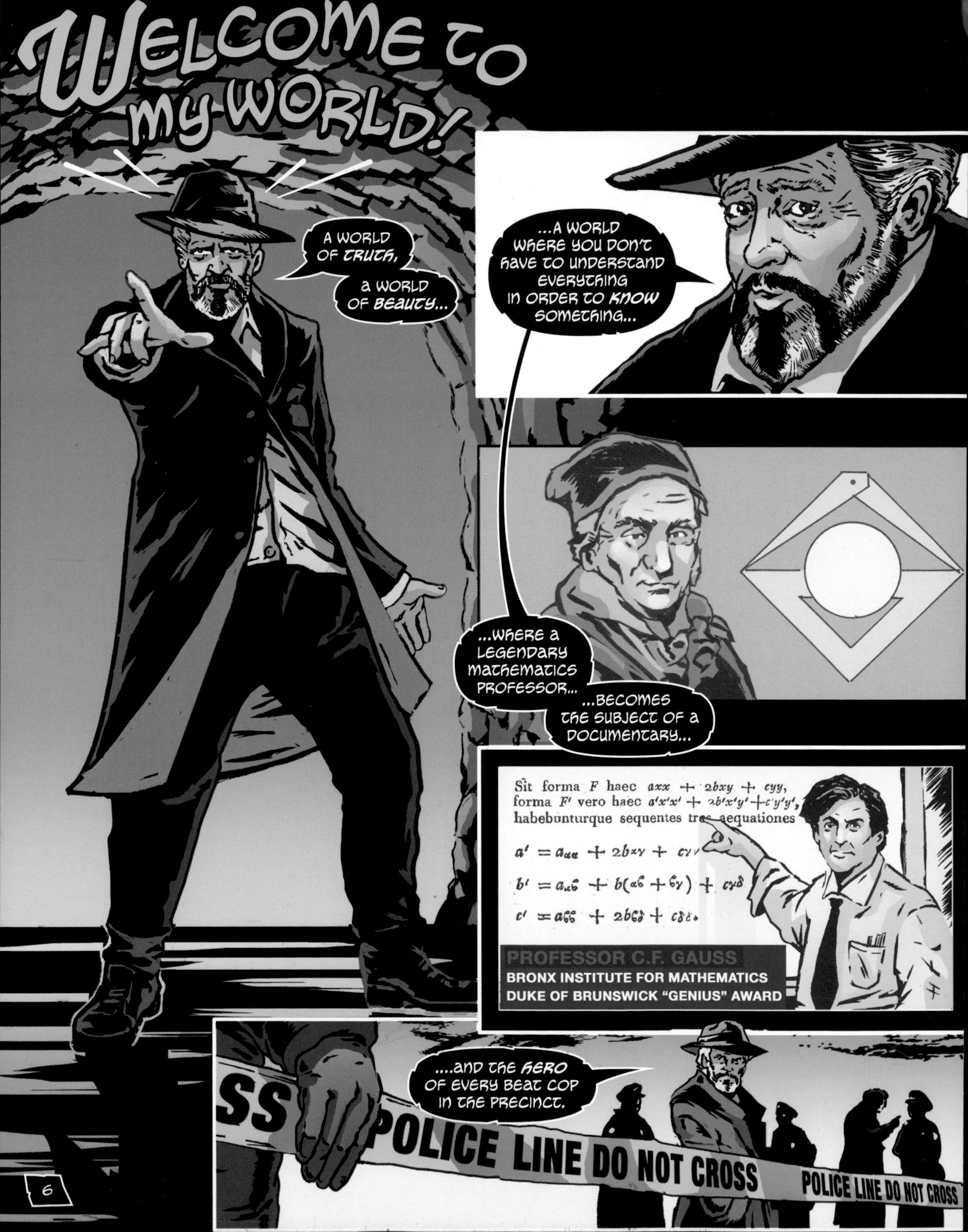
WELCOME TO MY WORLD!
A WORLD OF TRUTH,
A WORLD OF BEAUTY...
...A WORLD WHERE YOU DON'T HAVE TO UNDERSTAND EVERYTHING IN ORDER TO KNOW SOMETHING...
...WHERE A LEGENDARY MATHEMATICS PROFESSOR...
...BECOMES THE SUBJECT OF A DOCUMENTARY...
Sit forma F haec axx + 2bxy + cyy,
forma F′ vero haec a′x′x′ + 2b′x′y′ + c′y′y′,
habebunturque sequentes tres aequationes
PROFESSOR C.F. GAUSS
BRONX INSTITUTE FOR MATHEMATICS
DUKE OF BRUNSWICK "GENIUS" AWARD
...AND THE HERO OF EVERY BEAT COP IN THE PRECINCT.
POLICE LINE DO NOT CROSS
POLICE LINE DO NOT CROSS

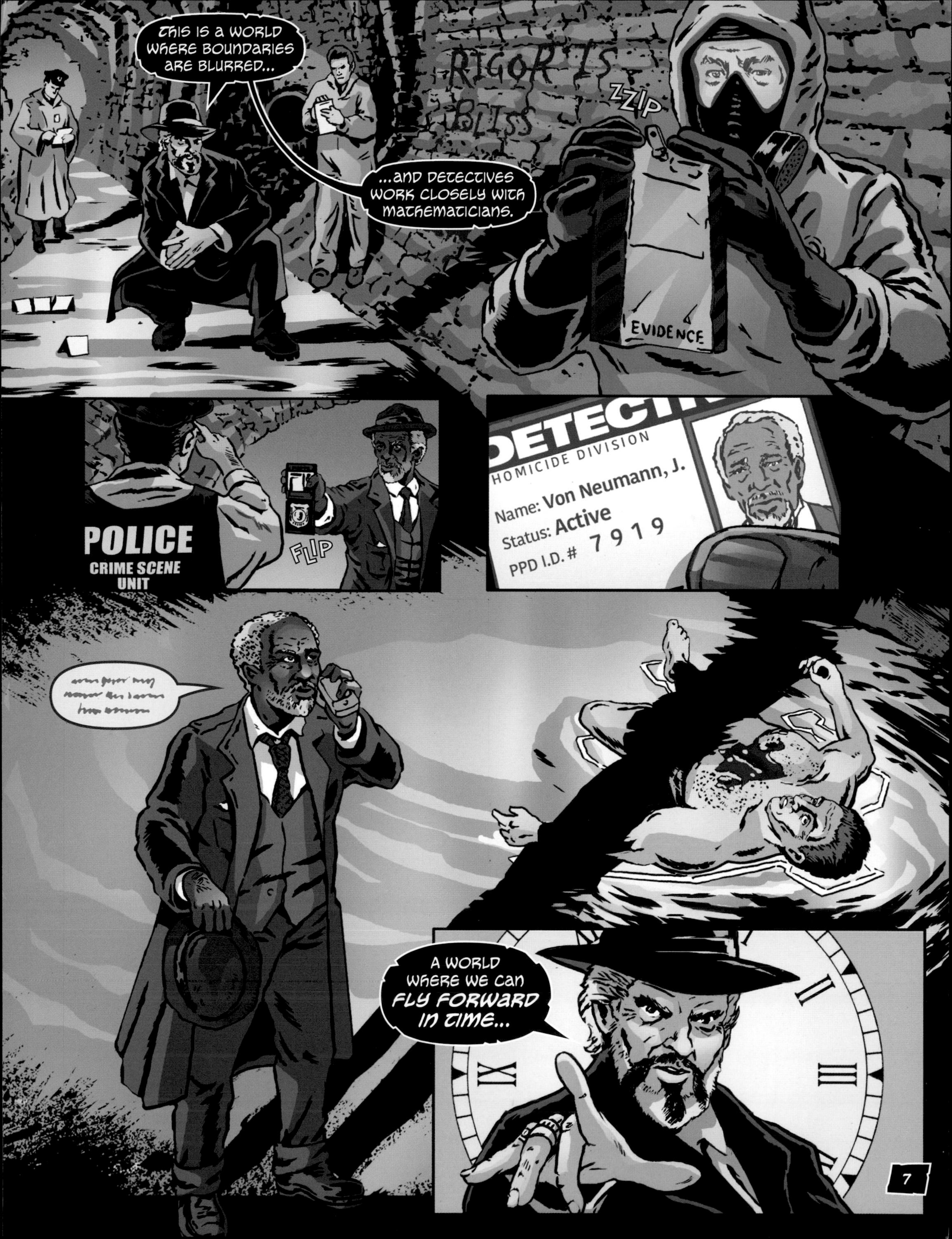
THIS IS A WORLD WHERE BOUNDARIES ARE BLURRED...
RIGOR IS BLISS
ZZIP
...AND DETECTIVES WORK CLOSELY WITH MATHEMATICIANS.
EVIDENCE
POLICE
CRIME SCENE UNIT
FLIP
HOMICIDE DIVISION
Name: Von Neumann, J.
Status: Active
PPD I.D. # 7919
A WORLD WHERE WE CAN FLY FORWARD IN TIME...

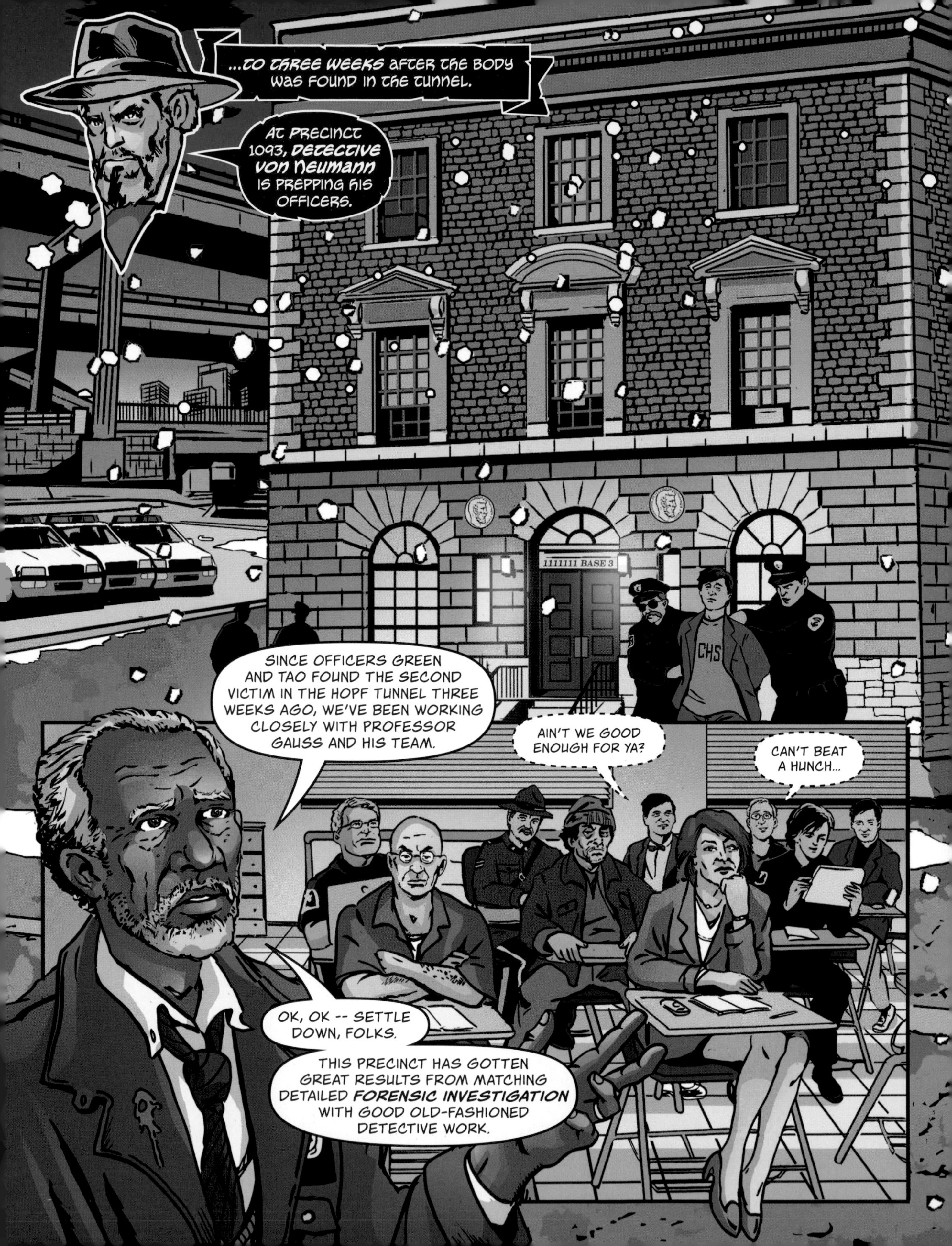
...TO THREE WEEKS AFTER THE BODY WAS FOUND IN THE TUNNEL.
AT PRECINCT 1093, DETECTIVE VON NEUMANN IS PREPPING HIS OFFICERS.
1111111 BASE 3
CHS
SINCE OFFICERS GREEN AND TAO FOUND THE SECOND VICTIM IN THE HOPF TUNNEL THREE WEEKS AGO, WE'VE BEEN WORKING CLOSELY WITH PROFESSOR GAUSS AND HIS TEAM.
AIN'T WE GOOD ENOUGH FOR YA?
CAN'T BEAT A HUNCH...
OK, OK -- SETTLE DOWN, FOLKS.
THIS PRECINCT HAS GOTTEN GREAT RESULTS FROM MATCHING DETAILED FORENSIC INVESTIGATION WITH GOOD OLD-FASHIONED DETECTIVE WORK.

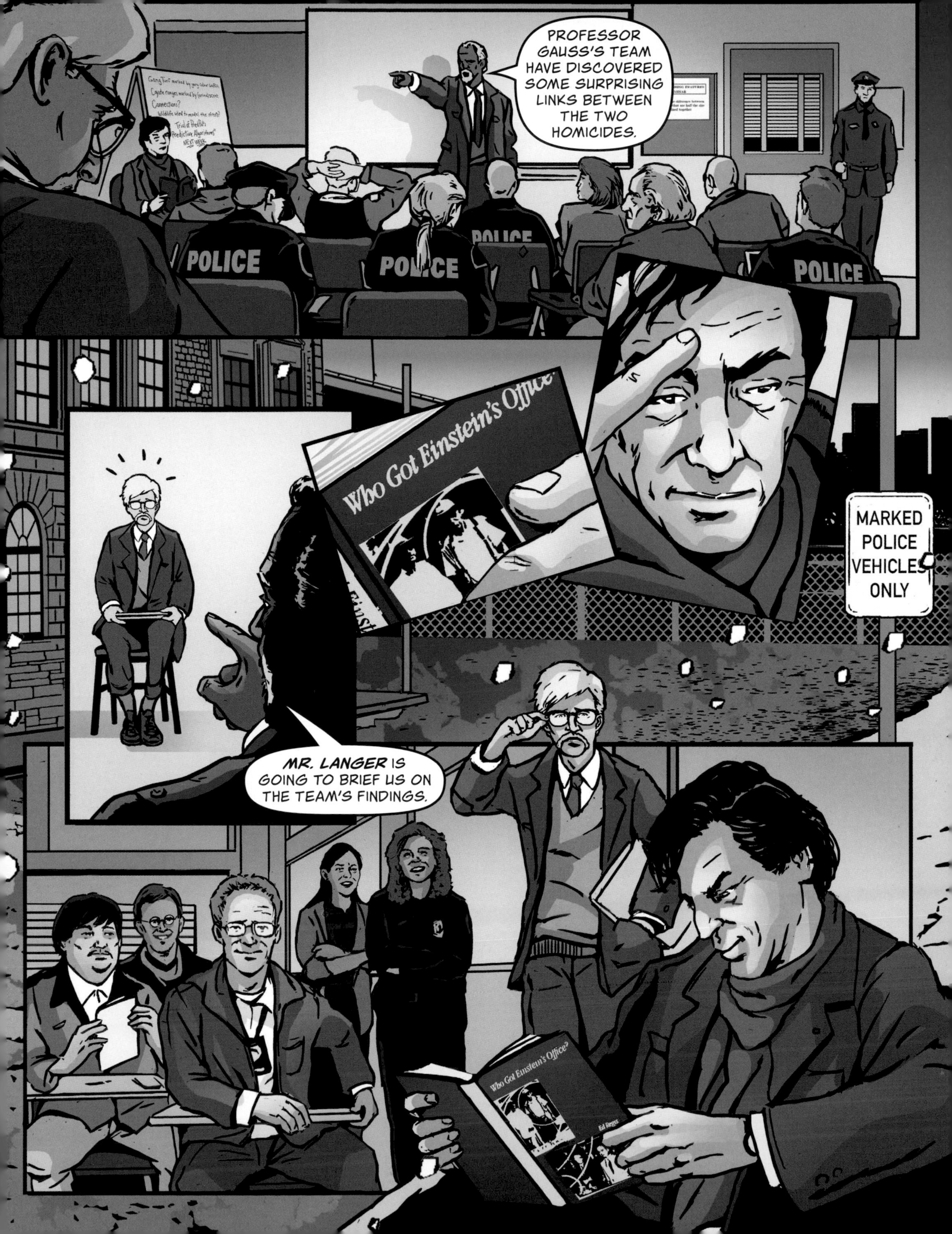

PROFESSOR GAUSS'S TEAM HAVE DISCOVERED SOME SURPRISING LINKS BETWEEN THE TWO HOMICIDES.
POLICE
POLICE
POLICE
POLICE
Who Got Einstein's Office?
MARKED POLICE VEHICLES ONLY
MR. LANGER IS GOING TO BRIEF US ON THE TEAM'S FINDINGS.
Who Got Einstein's Office?
Ed Regis

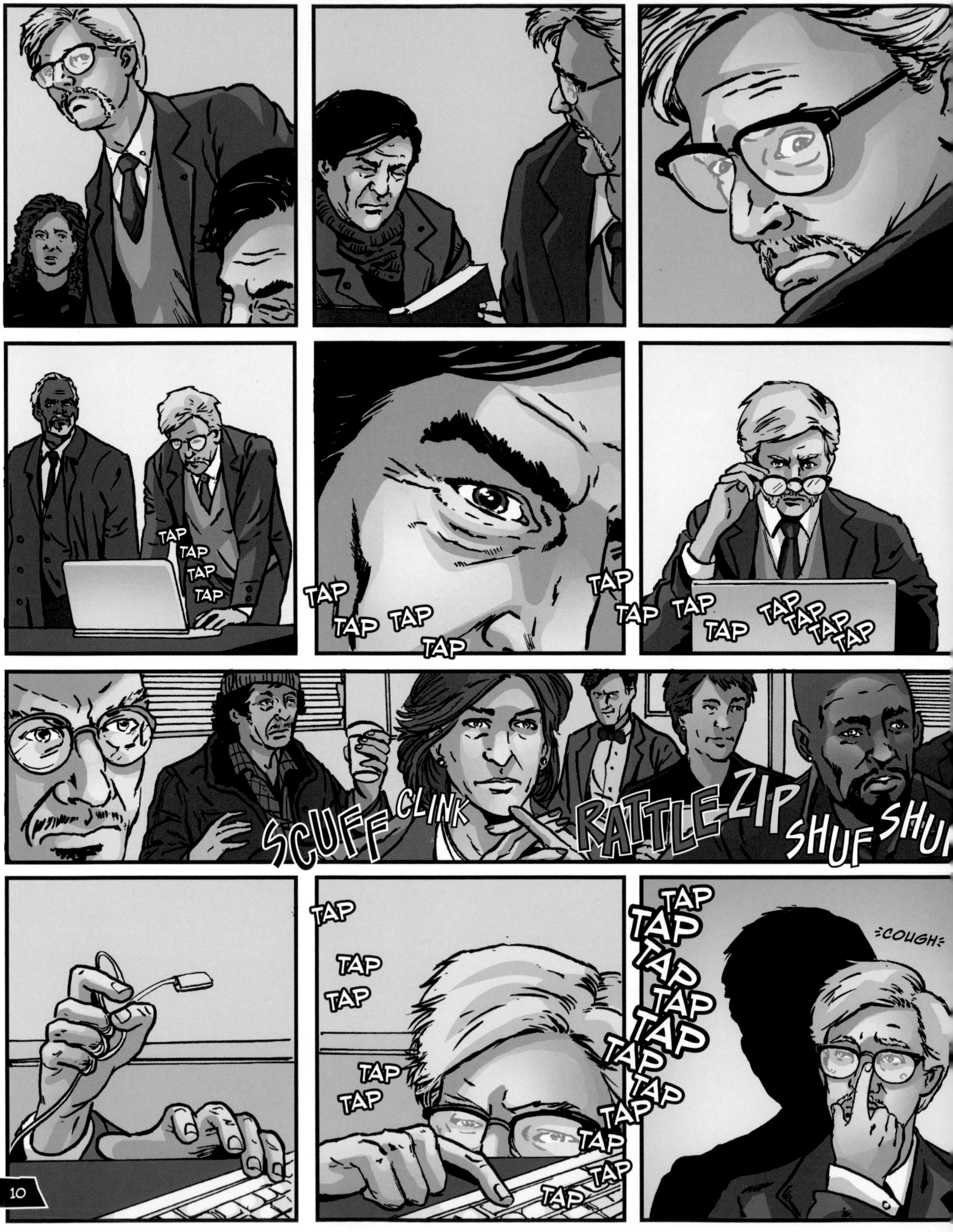

TAP
TAP
TAP
TAP
TAP
TAP TAP
TAP
TAP
TAP TAP
TAP
TAP
TAP
TAP
TAP
SCUFF
CLINK
RATTLE
ZIP
SHUF
TAP
TAP
TAP
TAP
TAP
TAP
TAP
TAP
TAP
TAP
TAP
TAP
TAP
TAP
TAP
TAP
TAP
TAP
COUGH

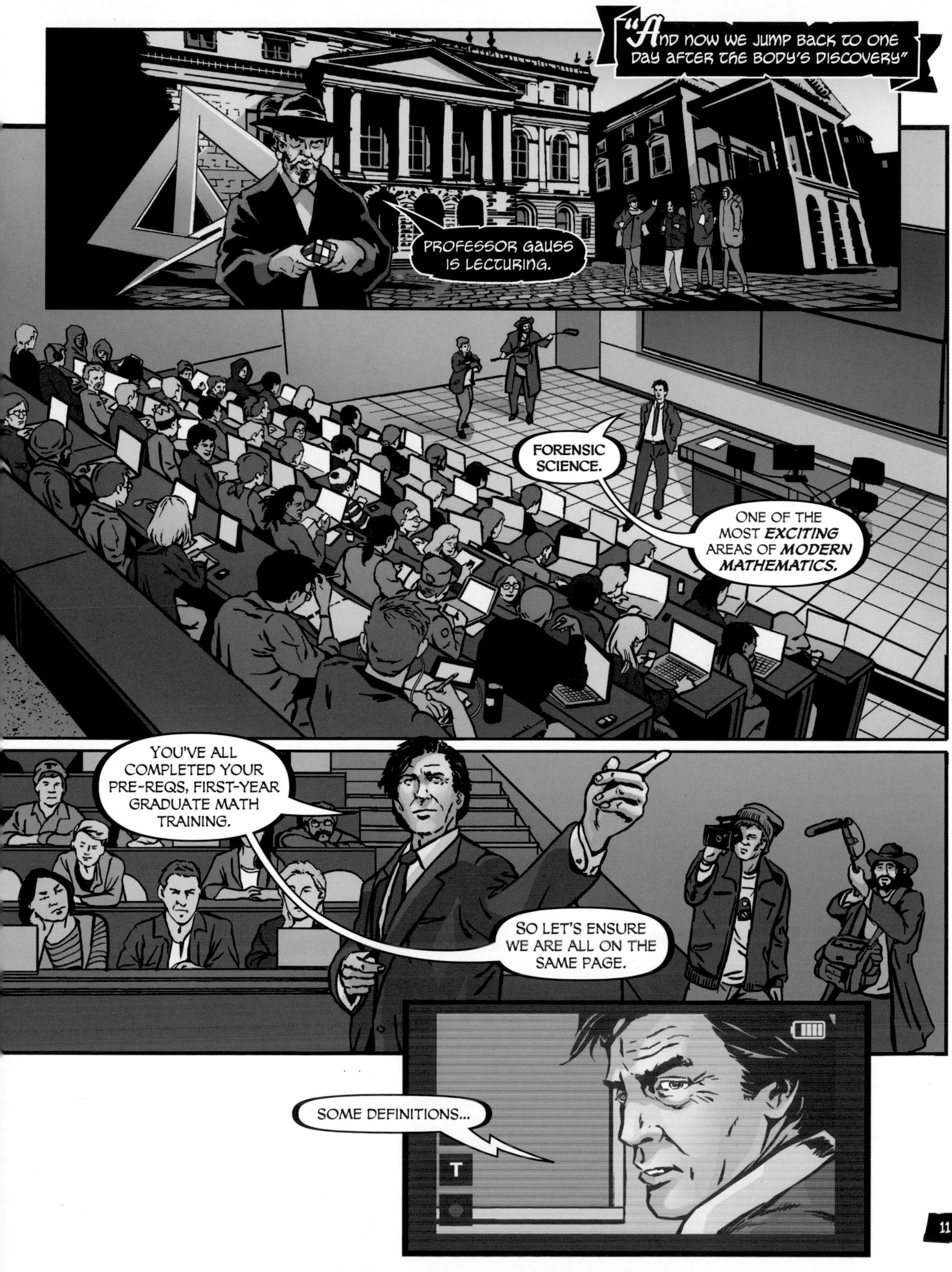
"And now we jump back to one day after the body's discovery"
PROFESSOR GAUSS IS LECTURING.
FORENSIC SCIENCE.
ONE OF THE MOST *EXCITING* AREAS OF *MODERN MATHEMATICS.*
YOU'VE ALL COMPLETED YOUR PRE-REQS, FIRST-YEAR GRADUATE MATH TRAINING.
SO LET'S ENSURE WE ARE ALL ON THE SAME PAGE.
SOME DEFINITIONS...
T

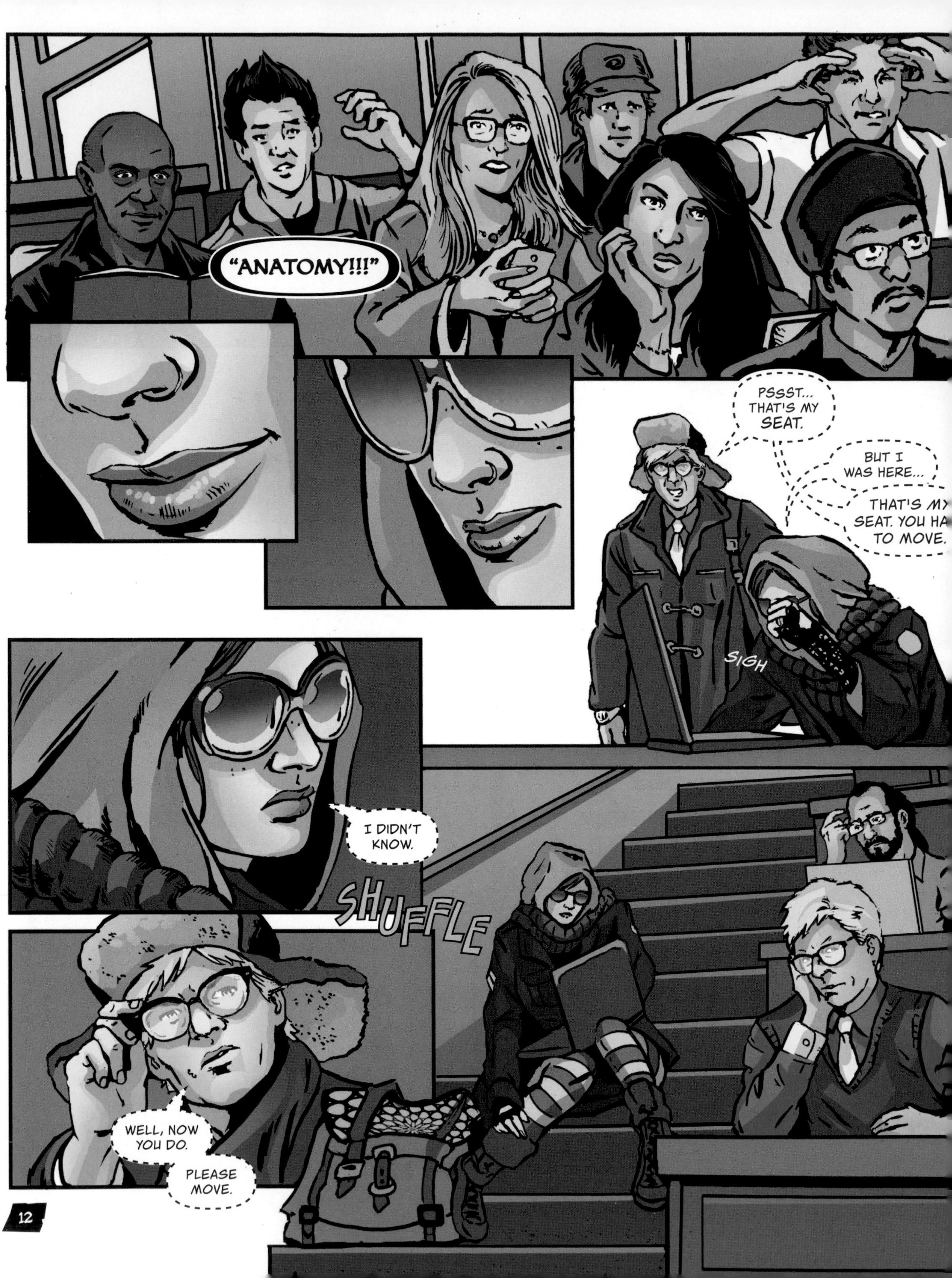
"ANATOMY!!!"
PSSST... THAT'S MY SEAT.
BUT I WAS HERE...
THAT'S MY SEAT. YOU HA TO MOVE.
SIGH
I DIDN'T KNOW.
SHUFFLE
WELL, NOW YOU DO.
PLEASE MOVE.

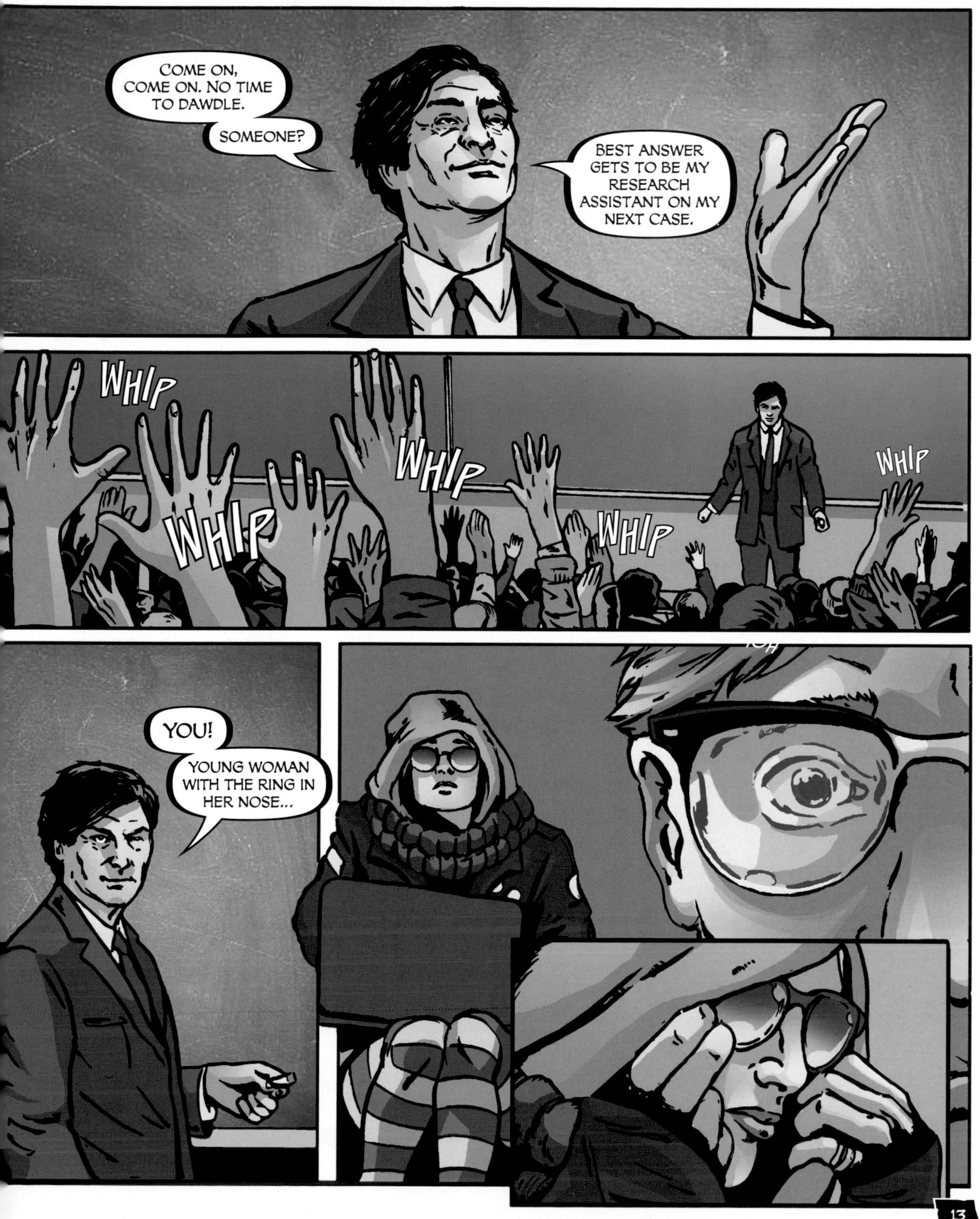
COME ON, COME ON. NO TIME TO DAWDLE.
SOMEONE?
BEST ANSWER GETS TO BE MY RESEARCH ASSISTANT ON MY NEXT CASE.
WHIP
WHIP
WHIP
WHIP
WHIP
YOU!
YOUNG WOMAN WITH THE RING IN HER NOSE...

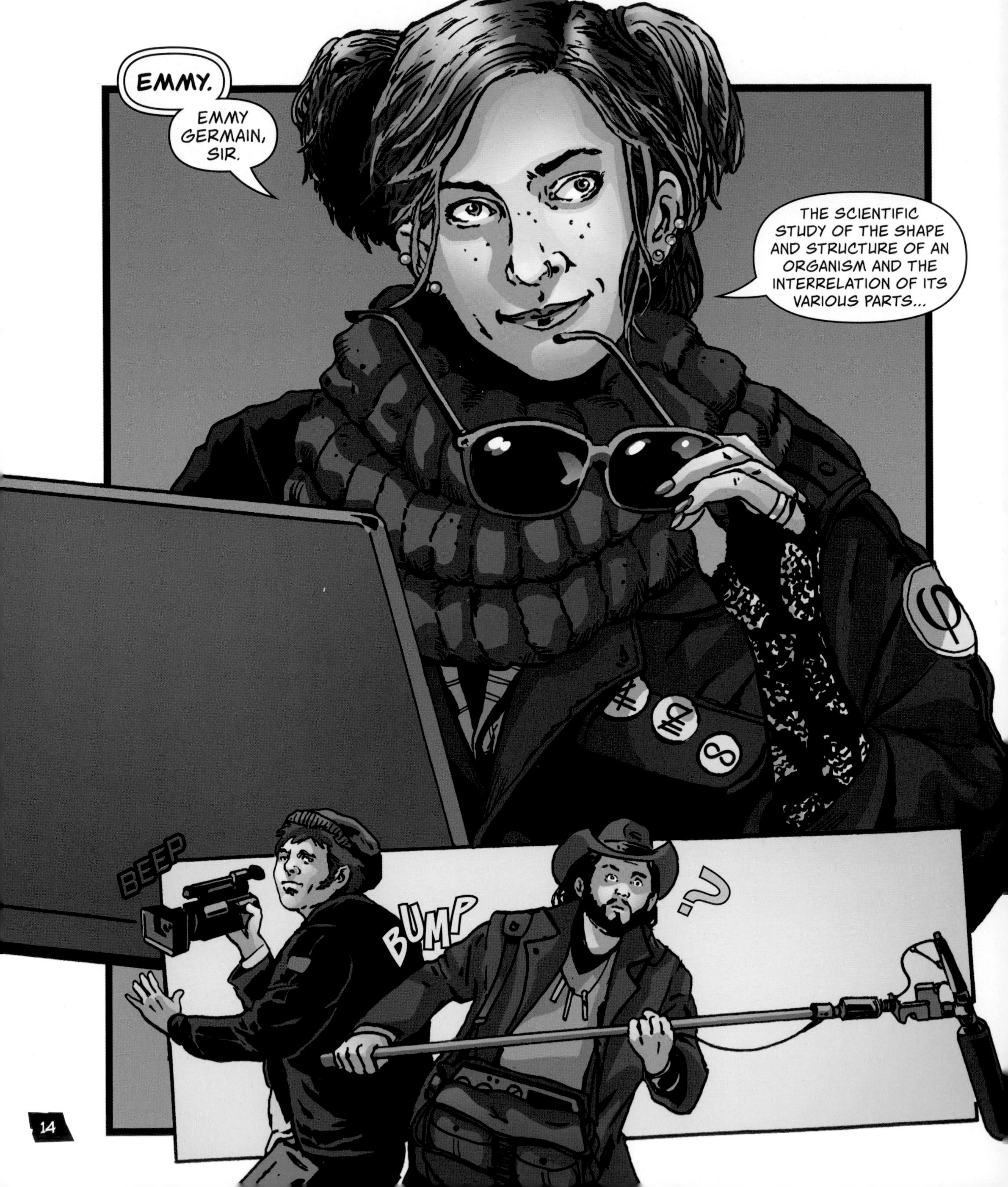
EMMY.
EMMY GERMAIN, SIR.
THE SCIENTIFIC STUDY OF THE SHAPE AND STRUCTURE OF AN ORGANISM AND THE INTERRELATION OF ITS VARIOUS PARTS...
BEEP
BUMP
?

... IN ORDER TO ASCERTAIN THEIR POSITIONS, RELATIONS, STRUCTURES, AND FUNCTIONS.

EXCELLENT. A PERFECT DEFINITION.

SEEMS LIKE YOU WILL BE MY NEW RESEARCH ASSISTANT.

CLAP

CLAP

CLAP

SO, CAN YOU CONSOLIDATE YOUR SUCCESS WITH A DEFINITION OF "FORENSIC"?

...?

ANATOMY: "The internal shape and structure... object of stru...

FOREN...

NO TIME FOR FALSE MODESTY.

DO YOU KNOW OR DON'T YOU?

RELATING TO THE **USE OF SCIENCE OR TECHNOLOGY** IN THE INVESTIGATION AND ESTABLISHMENT OF **FACTS OR EVIDENCE.**

*I* HAVE TO BE YOUR RESEARCH ASSISTANT, PROFESSOR GAUSS!!!

ANATOMY: "The internal shape and structure of the oo..."

FORE...

I DO NOT BELIEVE I ASKED YOU, MR....?

PLEASE BE SEATED.

LANGER

"LANGER"

HMMM...

WELL, MR. LANGER IS CORRECT.

ANATOMY: "The internal shape and structure of the object of study."

FORENSIC: "Deductive scientific investigation of the ...nce"

AND NOW THAT WE'RE CLEAR ON WHAT WE'RE DOING, WE CAN BEGIN.

SO WHO'S GOING TO BE YOUR ASSISTANT?

I DO NOT TYPICALLY TAKE ON MORE THAN ONE ASSISTANT AT A TIME...

SO IT'S MS. GERMAIN.

HE SAID "MS. GERMAIN"...

THAT'S *ME!*

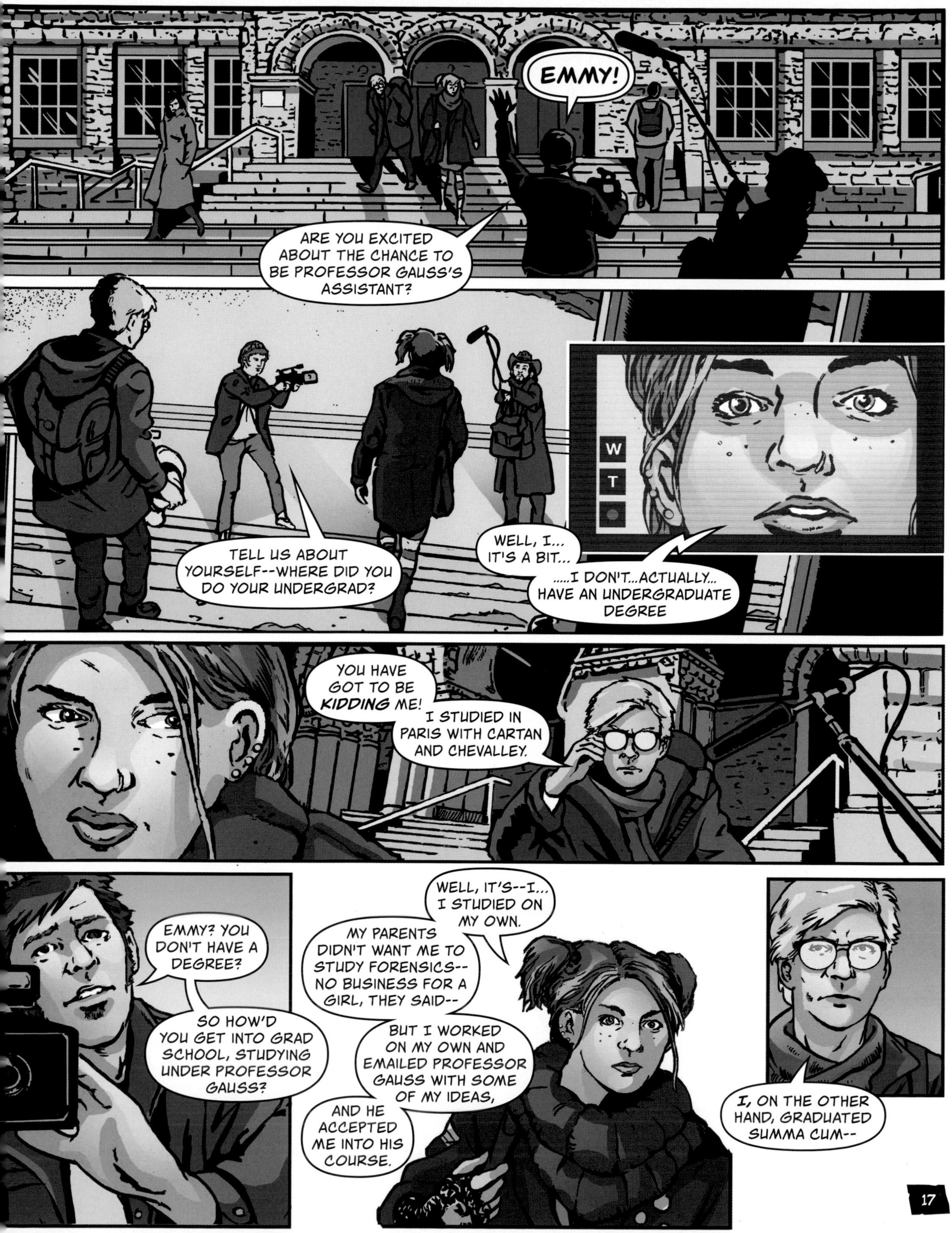
EMMY!
ARE YOU EXCITED ABOUT THE CHANCE TO BE PROFESSOR GAUSS'S ASSISTANT?
TELL US ABOUT YOURSELF--WHERE DID YOU DO YOUR UNDERGRAD?
W
T
WELL, I... IT'S A BIT...
.....I DON'T...ACTUALLY... HAVE AN UNDERGRADUATE DEGREE
YOU HAVE GOT TO BE KIDDING ME!
I STUDIED IN PARIS WITH CARTAN AND CHEVALLEY.
EMMY? YOU DON'T HAVE A DEGREE?
SO HOW'D YOU GET INTO GRAD SCHOOL, STUDYING UNDER PROFESSOR GAUSS?
WELL, IT'S--I... I STUDIED ON MY OWN.
MY PARENTS DIDN'T WANT ME TO STUDY FORENSICS--NO BUSINESS FOR A GIRL, THEY SAID--
BUT I WORKED ON MY OWN AND EMAILED PROFESSOR GAUSS WITH SOME OF MY IDEAS,
AND HE ACCEPTED ME INTO HIS COURSE.
I, ON THE OTHER HAND, GRADUATED SUMMA CUM--

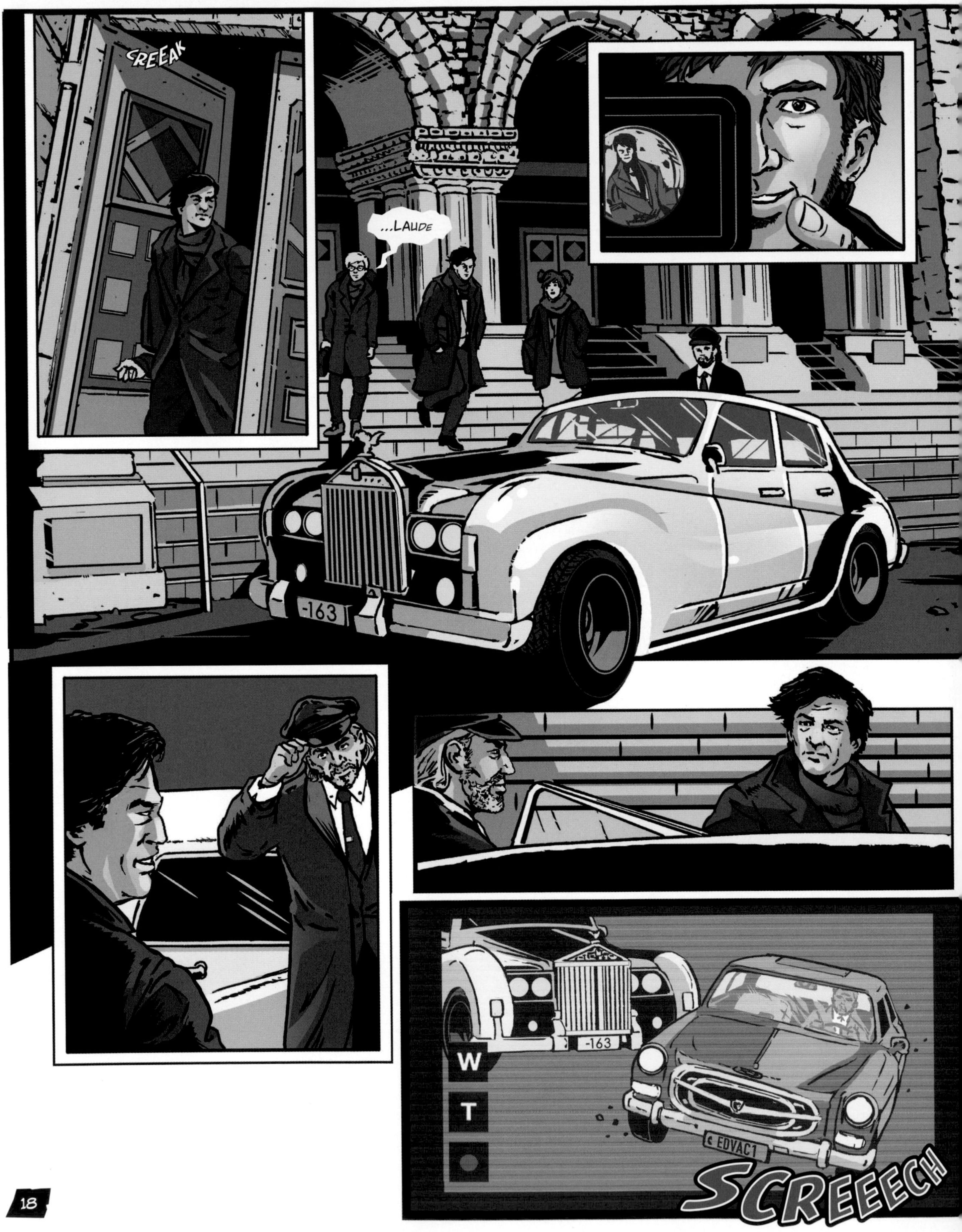
CREEAK
...LAUDE
-163
-163
EDVAC1
W
T
SCREEECH

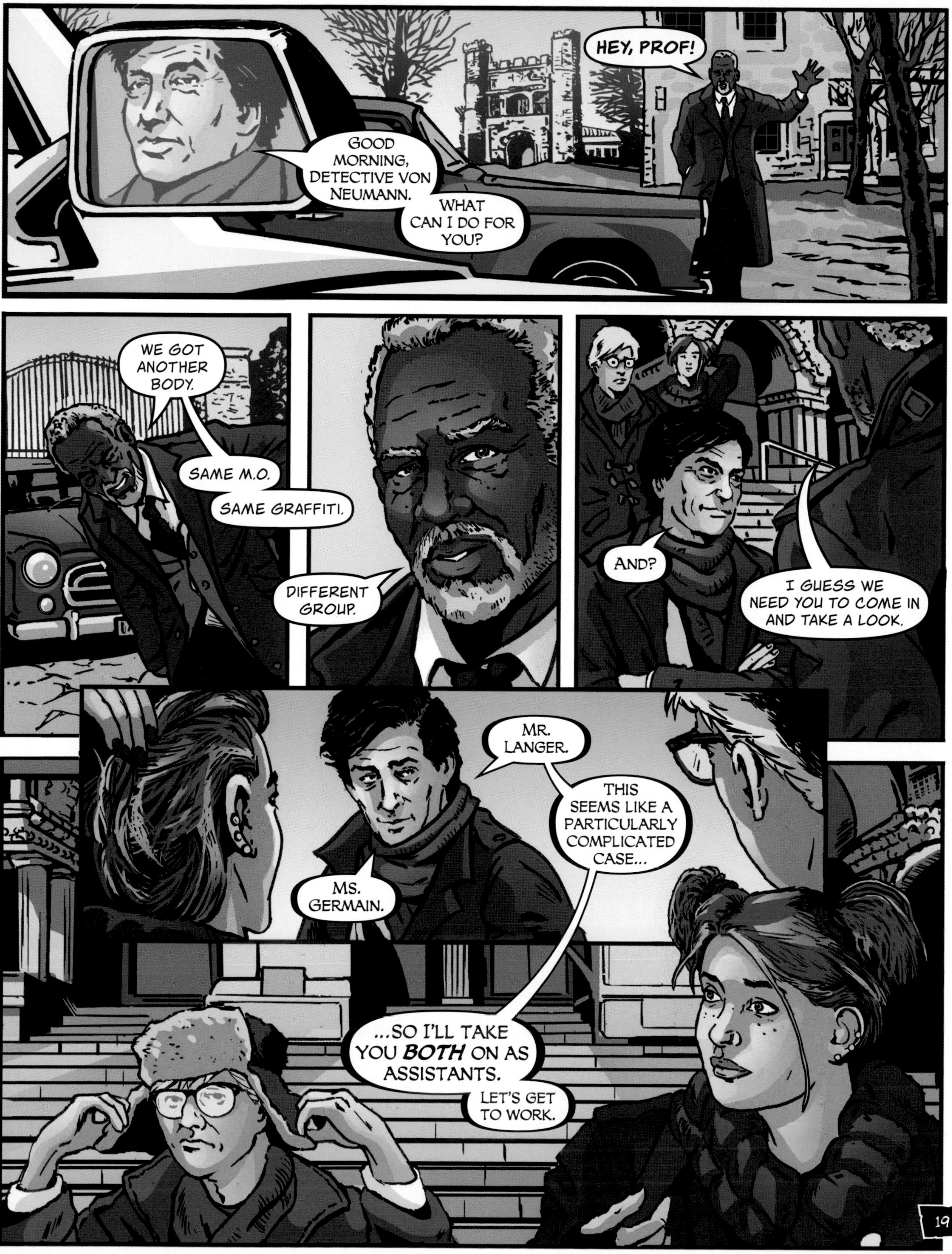
HEY, PROF!
GOOD MORNING, DETECTIVE VON NEUMANN.
WHAT CAN I DO FOR YOU?
WE GOT ANOTHER BODY.
SAME M.O.
SAME GRAFFITI.
DIFFERENT GROUP.
AND?
I GUESS WE NEED YOU TO COME IN AND TAKE A LOOK.
MR. LANGER.
MS. GERMAIN.
THIS SEEMS LIKE A PARTICULARLY COMPLICATED CASE...
...SO I'LL TAKE YOU *BOTH* ON AS ASSISTANTS.
LET'S GET TO WORK.

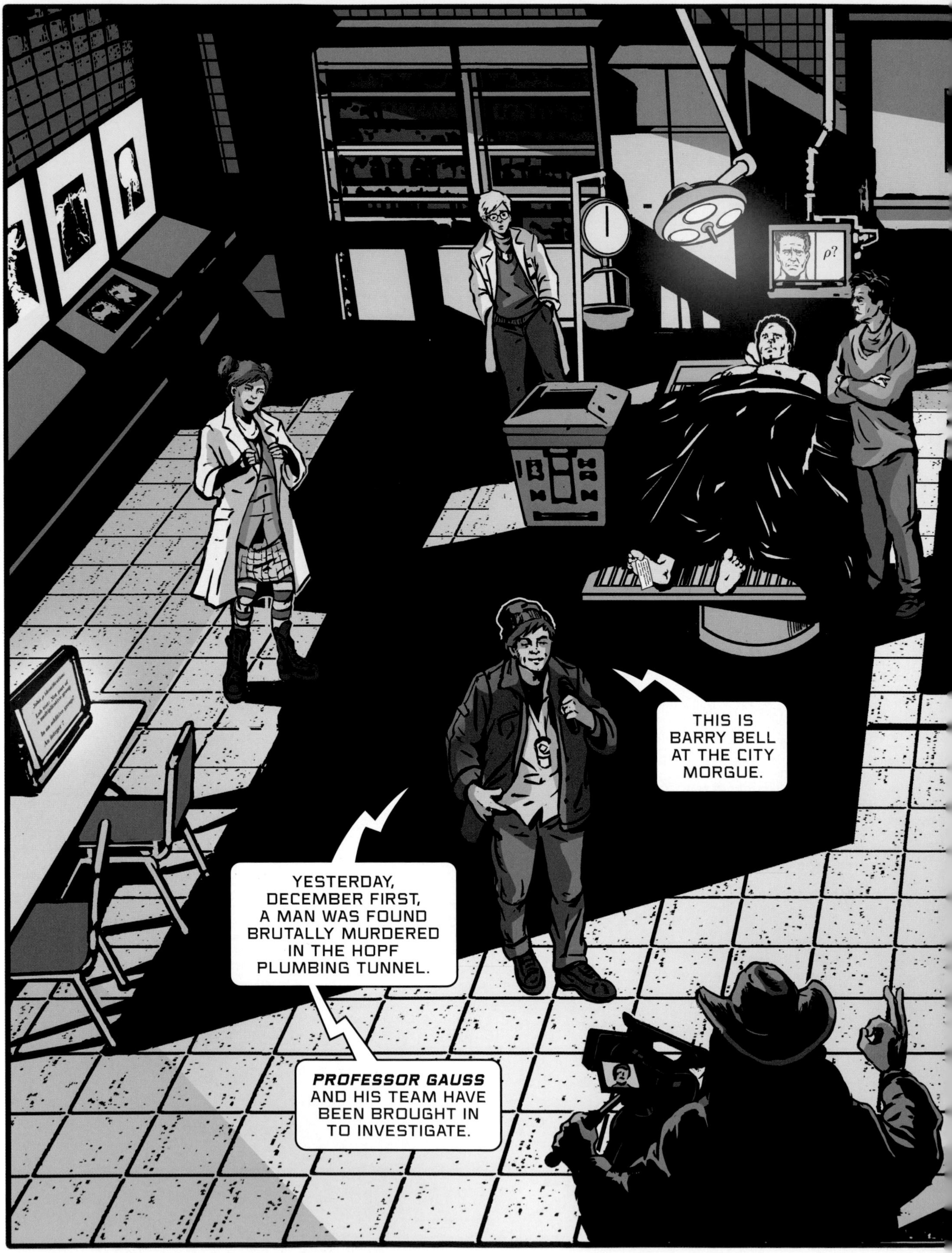

THIS IS BARRY BELL AT THE CITY MORGUE.
YESTERDAY, DECEMBER FIRST, A MAN WAS FOUND BRUTALLY MURDERED IN THE HOPF PLUMBING TUNNEL.
PROFESSOR GAUSS AND HIS TEAM HAVE BEEN BROUGHT IN TO INVESTIGATE.

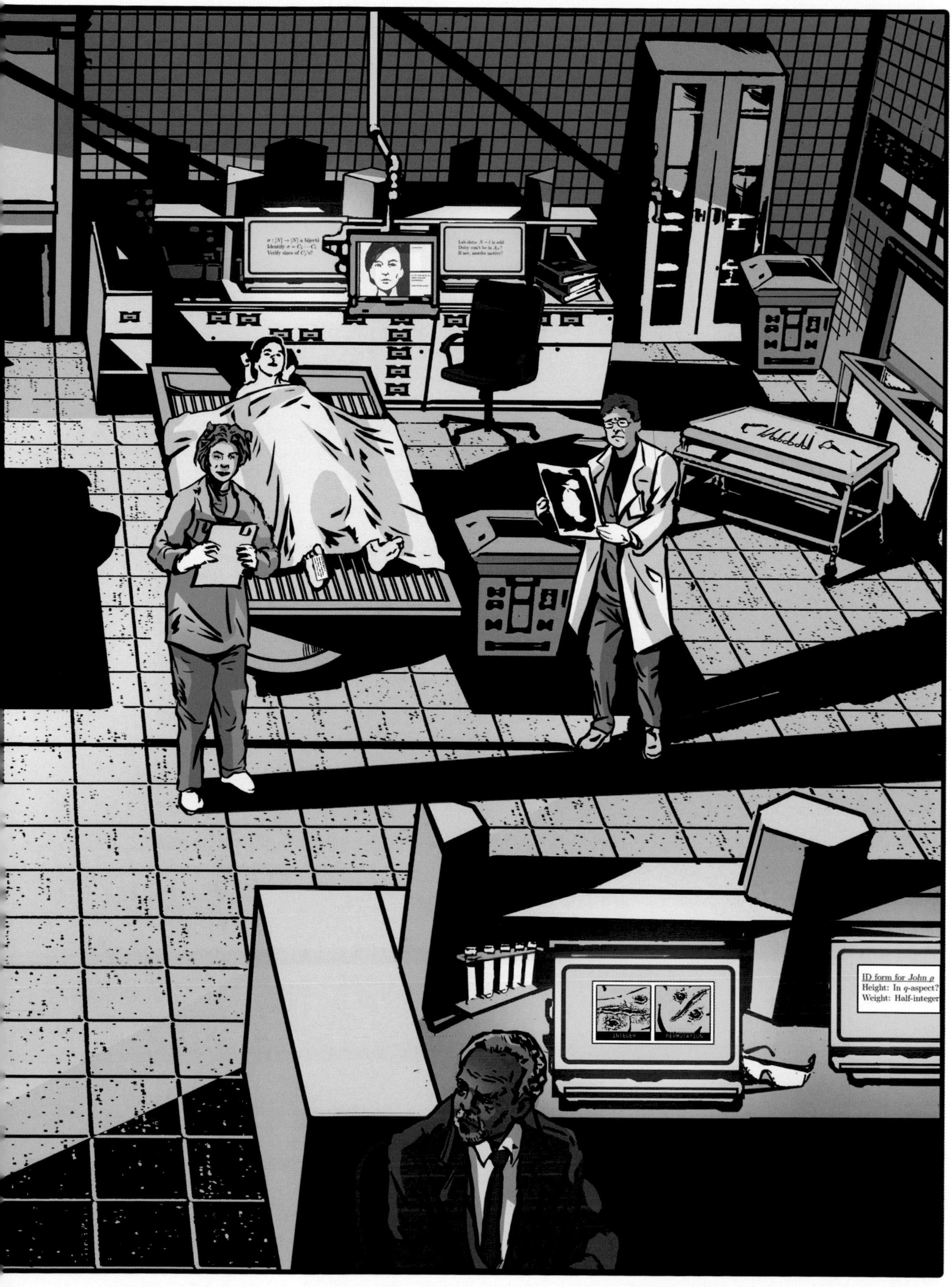
$\sigma : [N] \to [N]$ a bijecti
Identify $\sigma = C_1 \cdots C_t$
Verify sizes of $C_j$'s?
Lab data: $N - t$ is odd
Daisy can't be in $A_N$?
If not, murder motive?
INTEGER
PERMUTATION
ID form for *John ρ*
Height: In *q*-aspect?
Weight: Half-integer

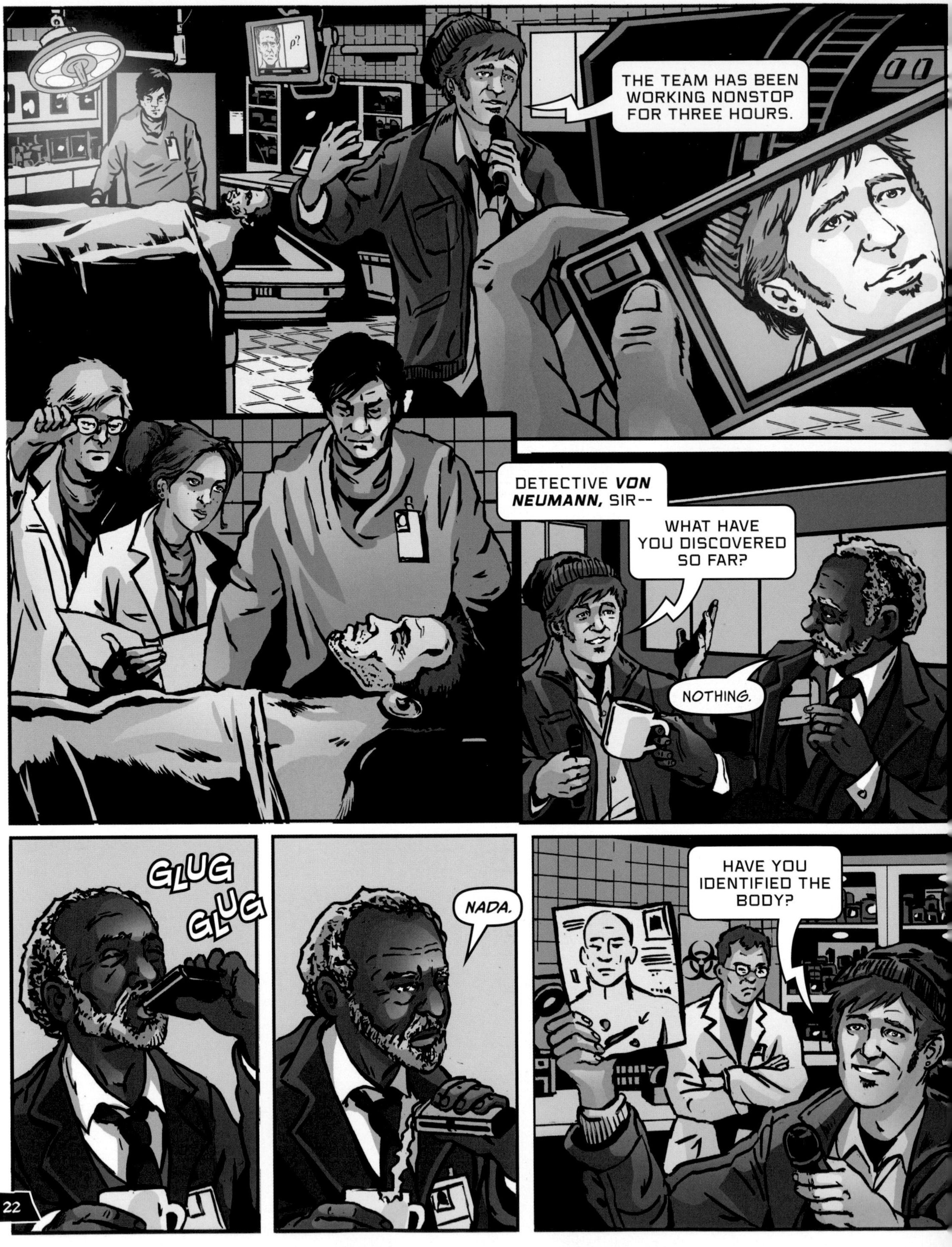
THE TEAM HAS BEEN WORKING NONSTOP FOR THREE HOURS.
DETECTIVE VON NEUMANN, SIR--
WHAT HAVE YOU DISCOVERED SO FAR?
NOTHING.
GLUG GLUG
NADA.
HAVE YOU IDENTIFIED THE BODY?

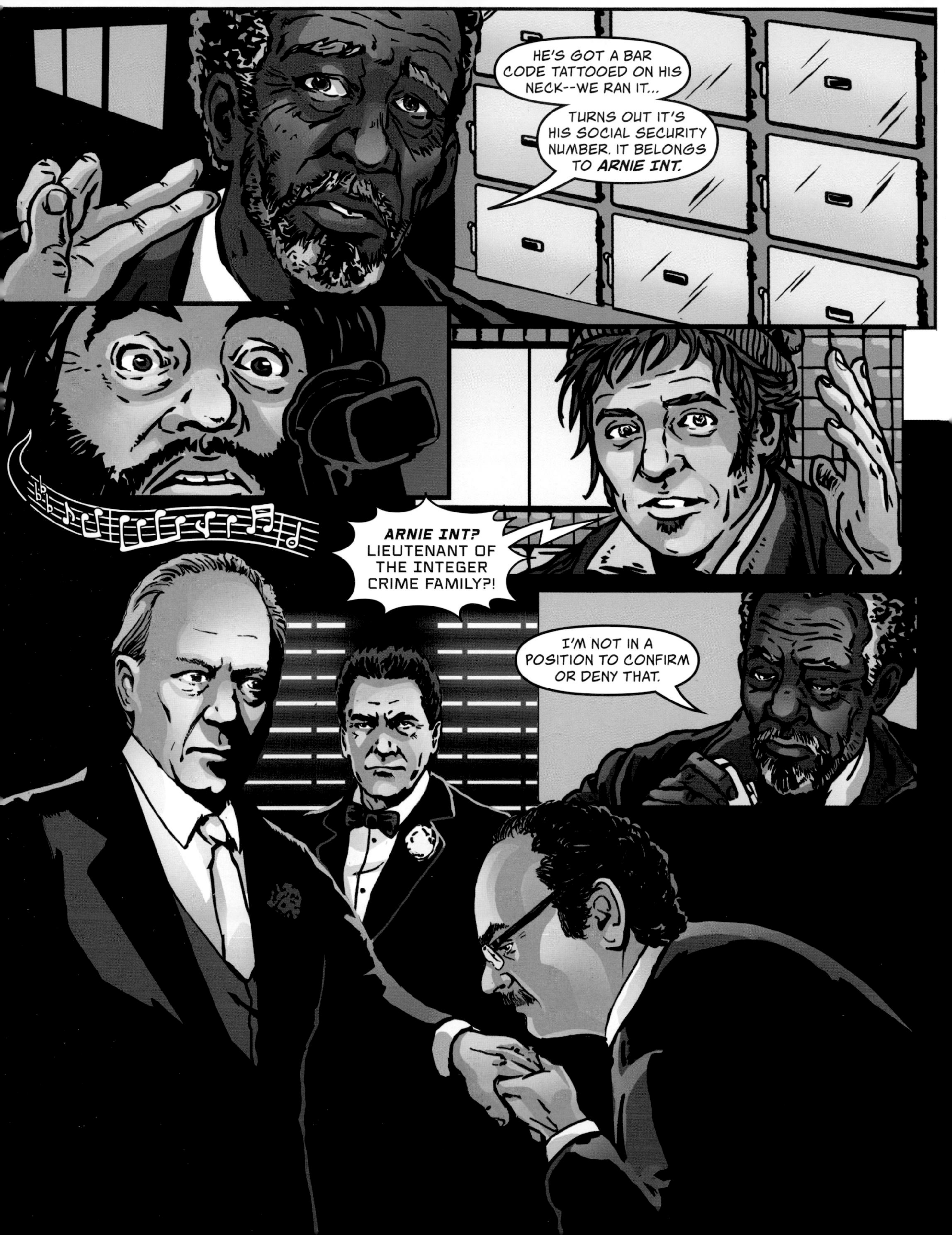
HE'S GOT A BAR CODE TATTOOED ON HIS NECK--WE RAN IT...
TURNS OUT IT'S HIS SOCIAL SECURITY NUMBER. IT BELONGS TO ARNIE INT.
ARNIE INT? LIEUTENANT OF THE INTEGER CRIME FAMILY?!
I'M NOT IN A POSITION TO CONFIRM OR DENY THAT.

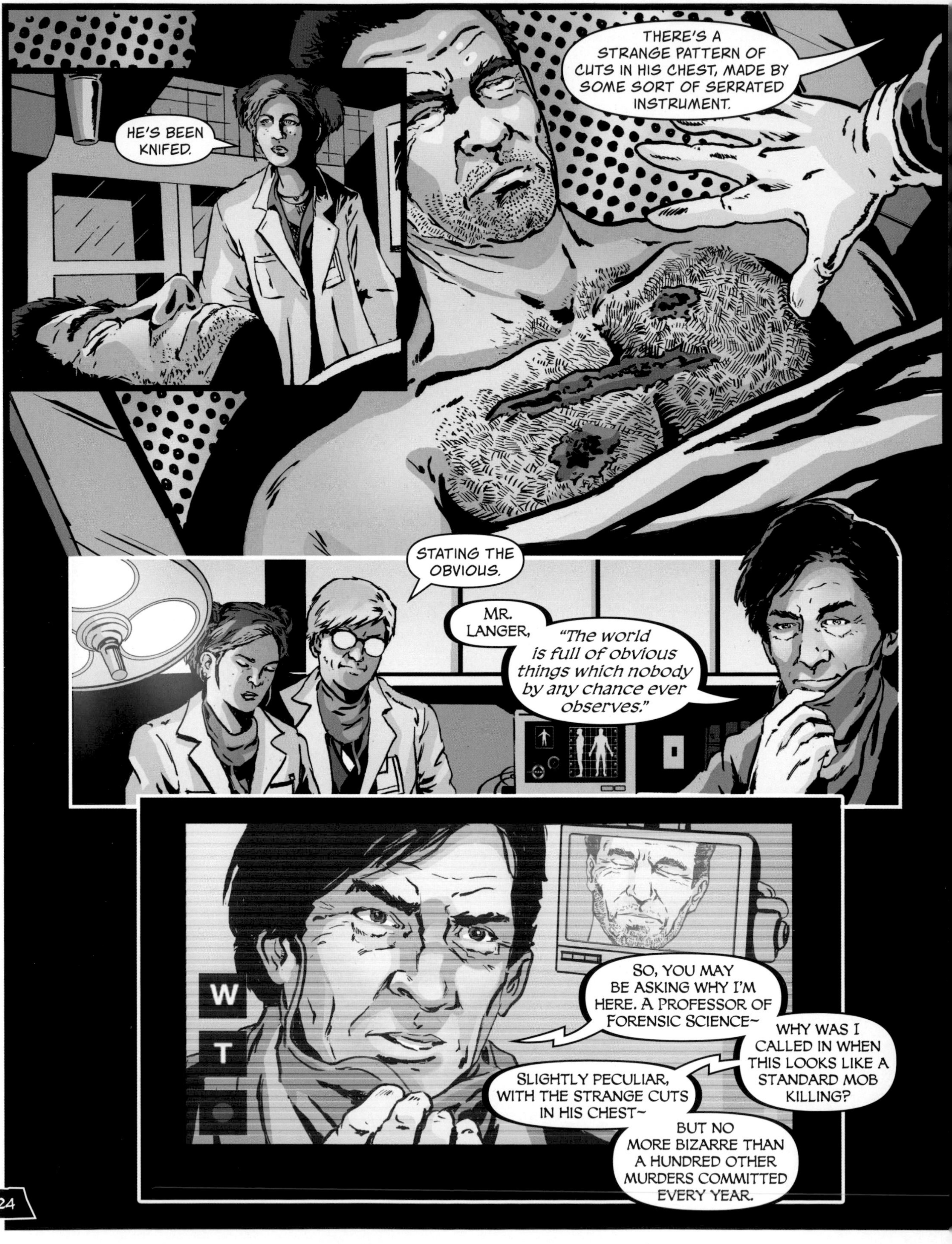
HE'S BEEN KNIFED.
THERE'S A STRANGE PATTERN OF CUTS IN HIS CHEST, MADE BY SOME SORT OF SERRATED INSTRUMENT.
STATING THE OBVIOUS.
MR. LANGER,
*"The world is full of obvious things which nobody by any chance ever observes."*
W
T
SO, YOU MAY BE ASKING WHY I'M HERE. A PROFESSOR OF FORENSIC SCIENCE~
WHY WAS I CALLED IN WHEN THIS LOOKS LIKE A STANDARD MOB KILLING?
SLIGHTLY PECULIAR, WITH THE STRANGE CUTS IN HIS CHEST~
BUT NO MORE BIZARRE THAN A HUNDRED OTHER MURDERS COMMITTED EVERY YEAR.

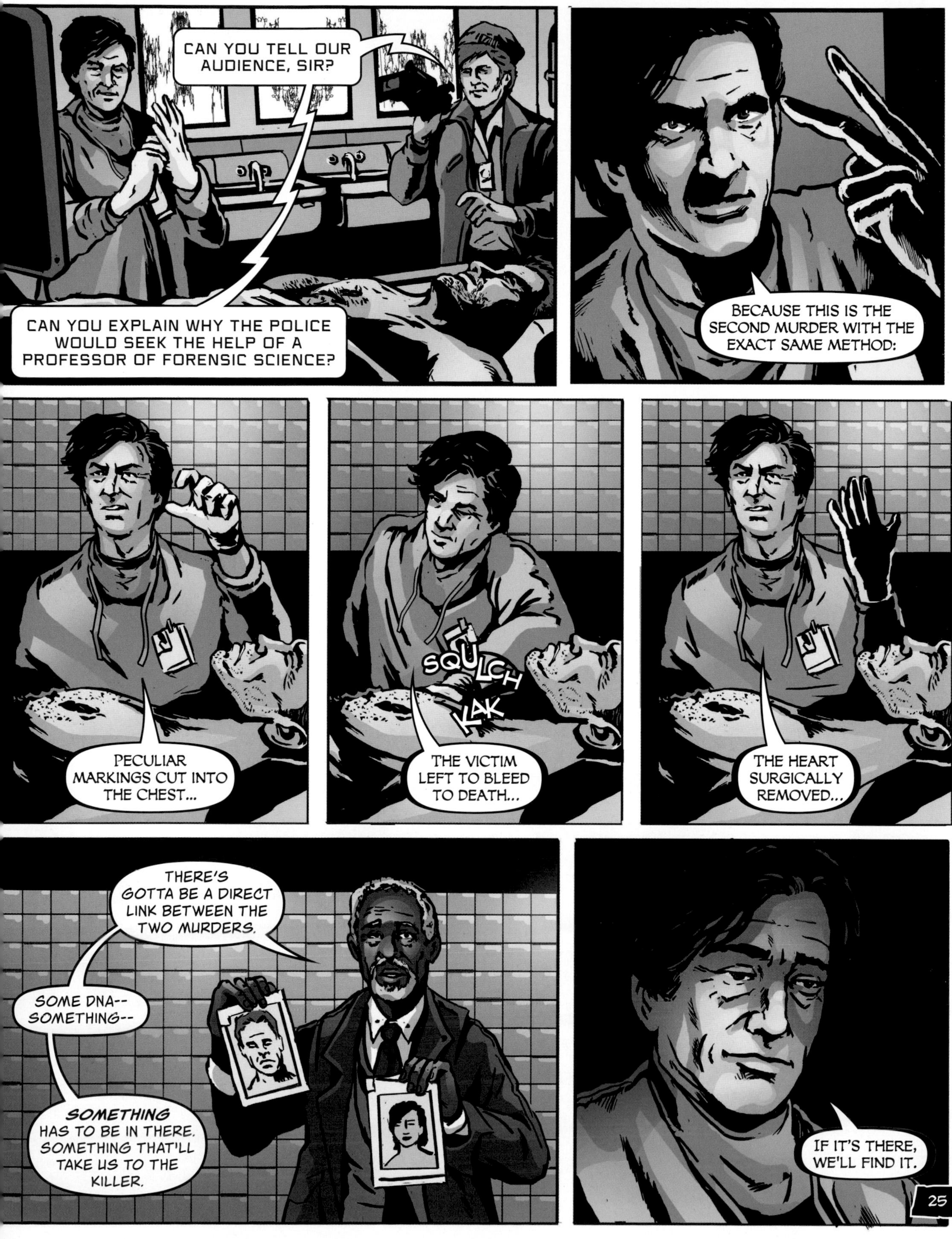
CAN YOU EXPLAIN WHY THE POLICE WOULD SEEK THE HELP OF A PROFESSOR OF FORENSIC SCIENCE?
CAN YOU TELL OUR AUDIENCE, SIR?
BECAUSE THIS IS THE SECOND MURDER WITH THE EXACT SAME METHOD:
PECULIAR MARKINGS CUT INTO THE CHEST...
SQULCH
KAK
THE VICTIM LEFT TO BLEED TO DEATH...
THE HEART SURGICALLY REMOVED...
THERE'S GOTTA BE A DIRECT LINK BETWEEN THE TWO MURDERS.
SOME DNA-- SOMETHING--
*SOMETHING* HAS TO BE IN THERE. SOMETHING THAT'LL TAKE US TO THE KILLER.
IF IT'S THERE, WE'LL FIND IT.

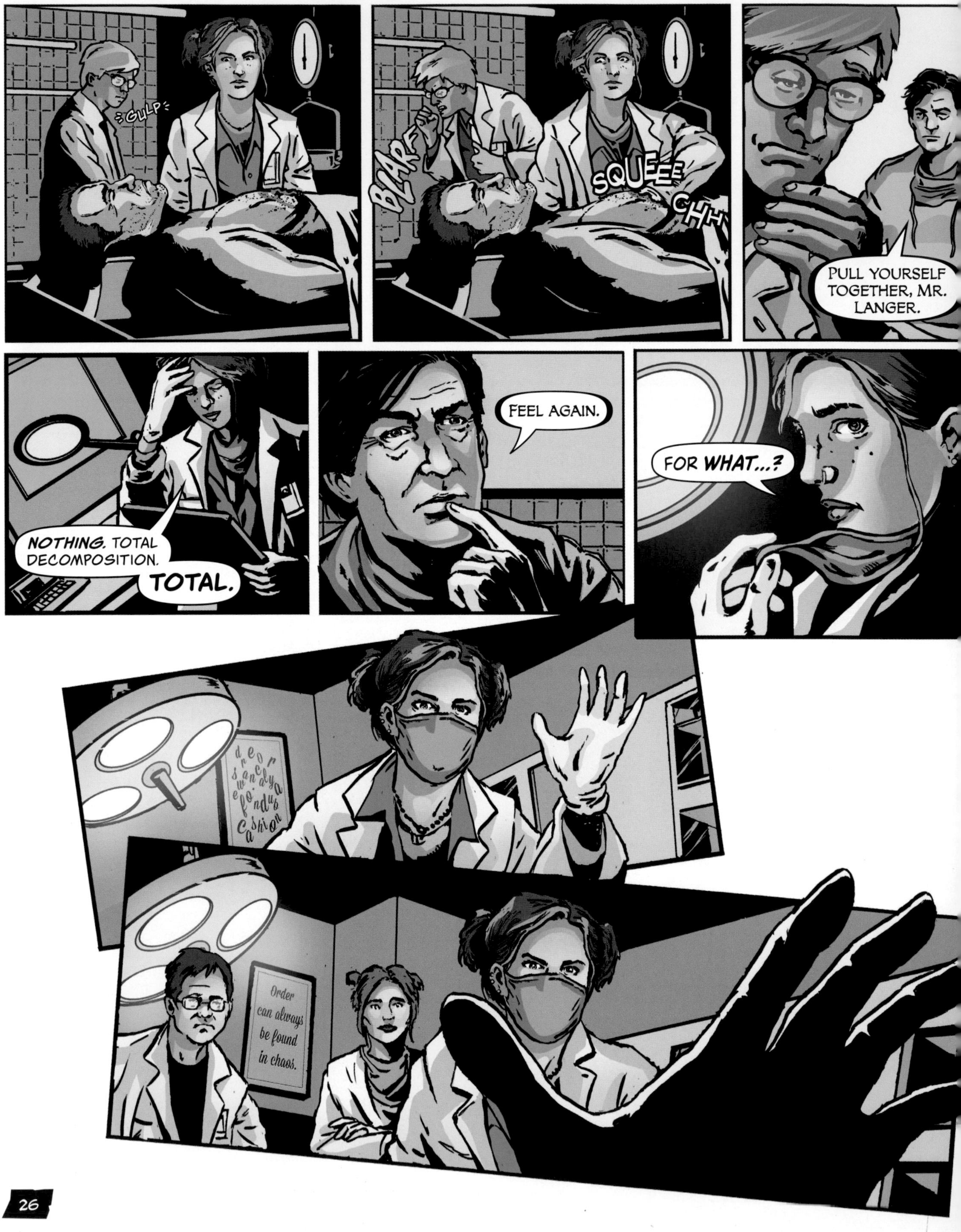
GULP
BLAR
SQUEEE
CHH
PULL YOURSELF TOGETHER, MR. LANGER.
NOTHING. TOTAL DECOMPOSITION.
TOTAL.
FEEL AGAIN.
FOR WHAT...?
Order can always be found in chaos.

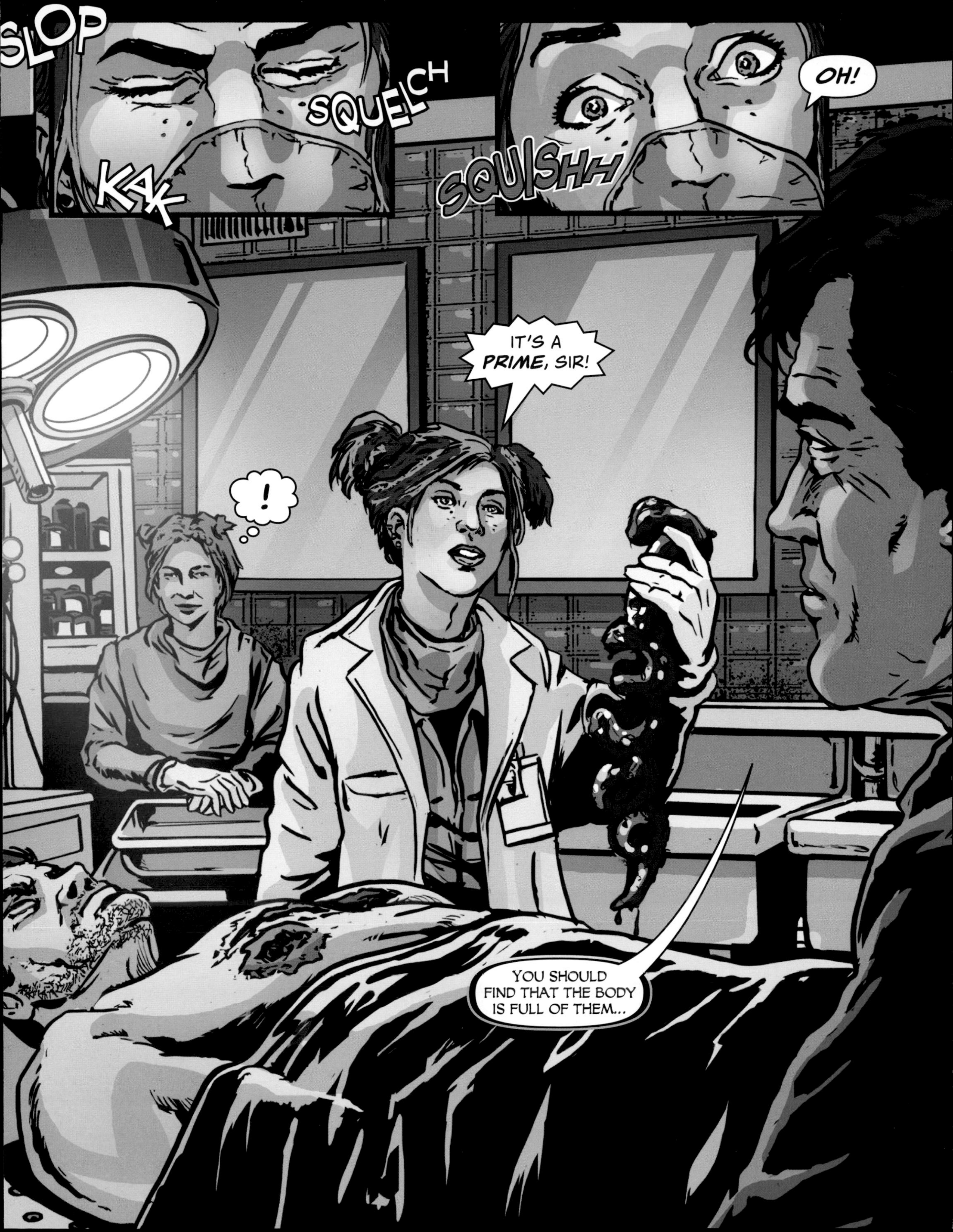
SLOP
SQUELCH
KAK
SQUISHH
OH!
IT'S A PRIME, SIR!
!
YOU SHOULD FIND THAT THE BODY IS FULL OF THEM...

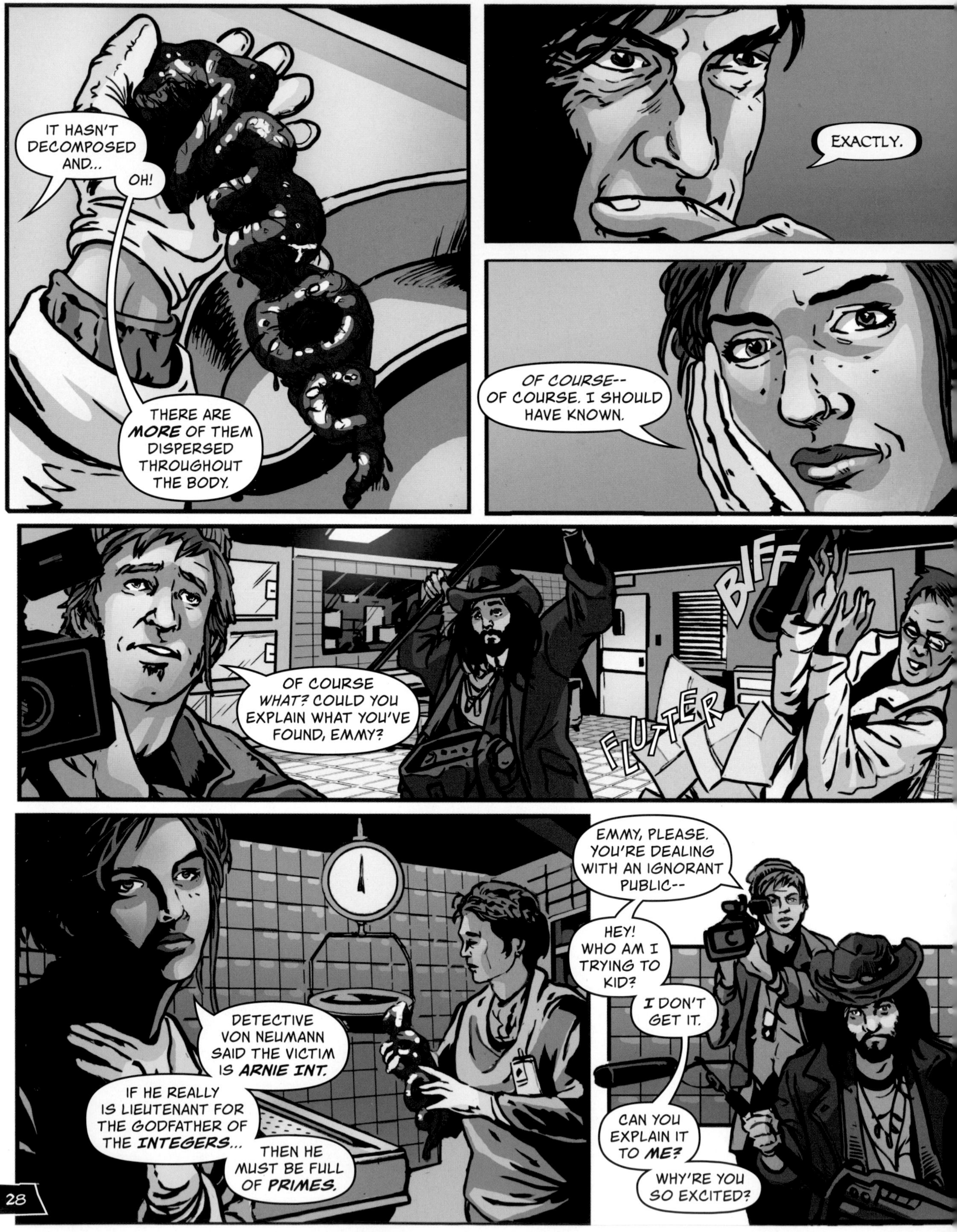
IT HASN'T DECOMPOSED AND...
OH!
THERE ARE MORE OF THEM DISPERSED THROUGHOUT THE BODY.
EXACTLY.
OF COURSE-- OF COURSE. I SHOULD HAVE KNOWN.
OF COURSE WHAT? COULD YOU EXPLAIN WHAT YOU'VE FOUND, EMMY?
BIFF
FLUTTER
DETECTIVE VON NEUMANN SAID THE VICTIM IS ARNIE INT.
IF HE REALLY IS LIEUTENANT FOR THE GODFATHER OF THE INTEGERS...
THEN HE MUST BE FULL OF PRIMES.
EMMY, PLEASE. YOU'RE DEALING WITH AN IGNORANT PUBLIC--
HEY! WHO AM I TRYING TO KID?
I DON'T GET IT.
CAN YOU EXPLAIN IT TO ME?
WHY'RE YOU SO EXCITED?

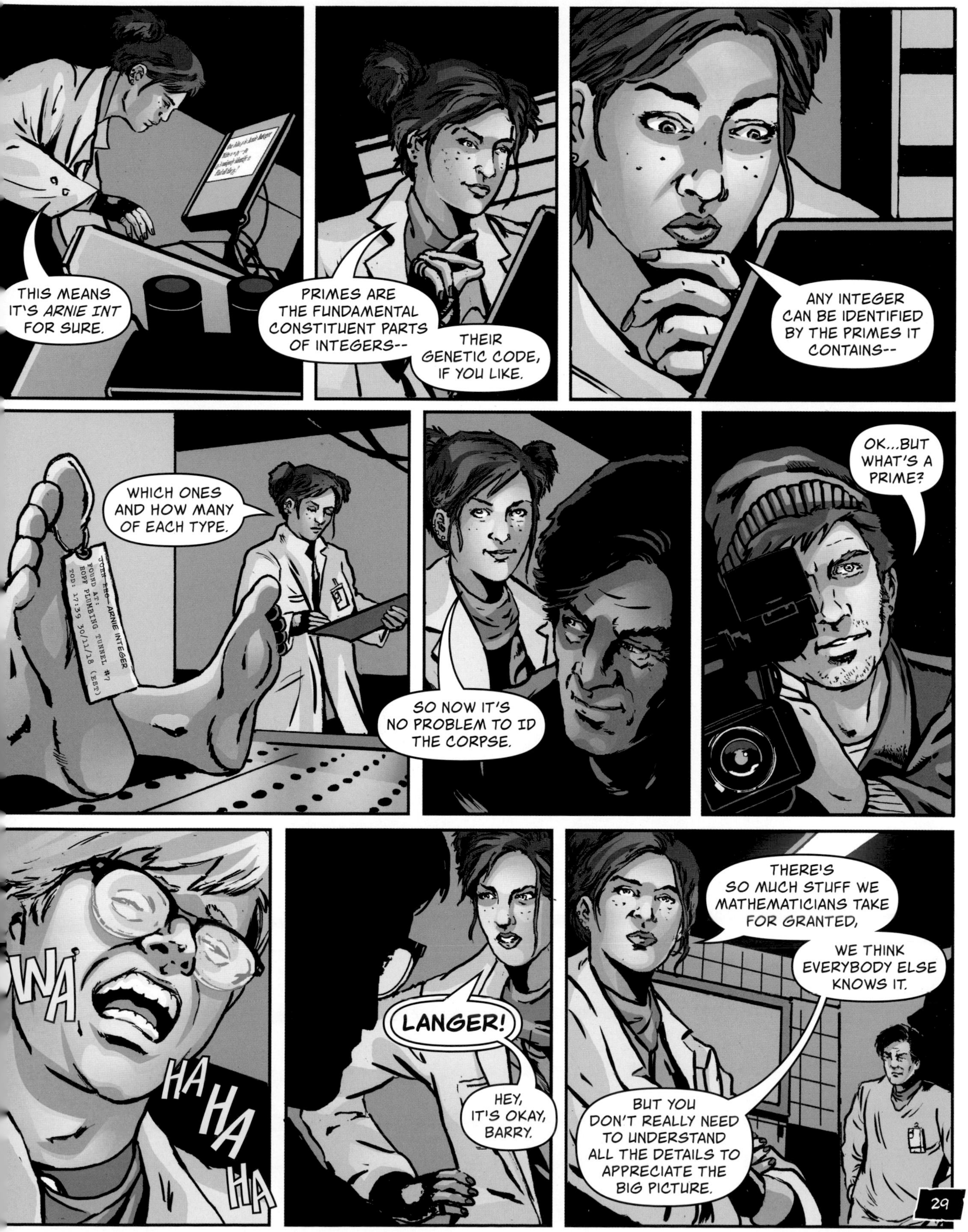
THIS MEANS IT'S *ARNIE INT* FOR SURE.
PRIMES ARE THE FUNDAMENTAL CONSTITUENT PARTS OF INTEGERS--
THEIR GENETIC CODE, IF YOU LIKE.
ANY INTEGER CAN BE IDENTIFIED BY THE PRIMES IT CONTAINS--
WHICH ONES AND HOW MANY OF EACH TYPE.
JOHN DOE ARNIE INTEGER
FOUND AT: HOPF PLUMBING TUNNEL #7
TOD: 17:39 30/11/18 (EST)
SO NOW IT'S NO PROBLEM TO ID THE CORPSE.
OK...BUT WHAT'S A PRIME?
WA HA HA HA HA
LANGER!
HEY, IT'S OKAY, BARRY.
THERE'S SO MUCH STUFF WE MATHEMATICIANS TAKE FOR GRANTED,
WE THINK EVERYBODY ELSE KNOWS IT.
BUT YOU DON'T REALLY NEED TO UNDERSTAND ALL THE DETAILS TO APPRECIATE THE BIG PICTURE.

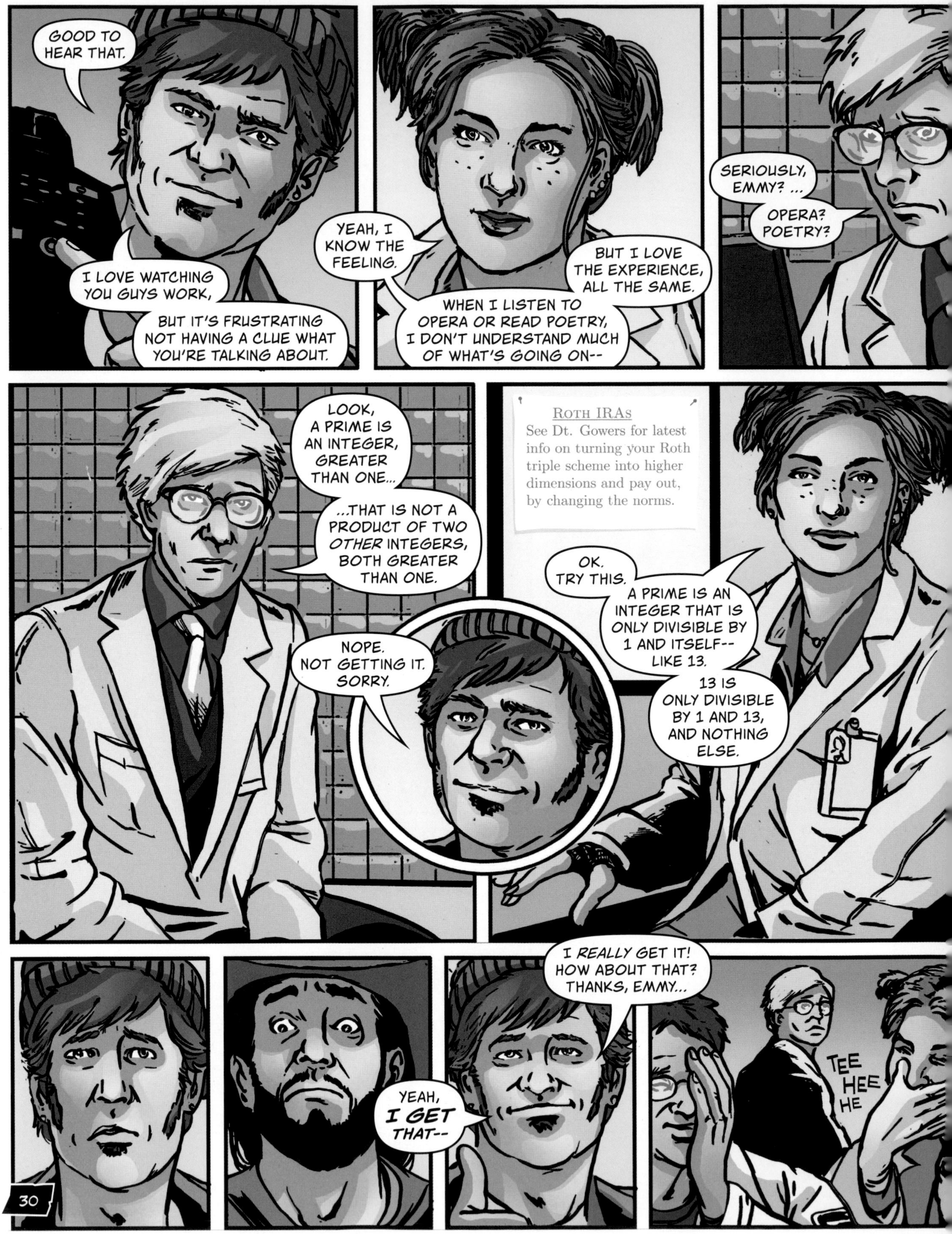
GOOD TO HEAR THAT.
I LOVE WATCHING YOU GUYS WORK,
BUT IT'S FRUSTRATING NOT HAVING A CLUE WHAT YOU'RE TALKING ABOUT.
YEAH, I KNOW THE FEELING.
WHEN I LISTEN TO OPERA OR READ POETRY, I DON'T UNDERSTAND MUCH OF WHAT'S GOING ON--
BUT I LOVE THE EXPERIENCE, ALL THE SAME.
SERIOUSLY, EMMY? ...
OPERA? POETRY?
LOOK, A PRIME IS AN INTEGER, GREATER THAN ONE...
...THAT IS NOT A PRODUCT OF TWO *OTHER* INTEGERS, BOTH GREATER THAN ONE.
ROTH IRAs
See Dt. Gowers for latest info on turning your Roth triple scheme into higher dimensions and pay out, by changing the norms.
NOPE. NOT GETTING IT. SORRY.
OK. TRY THIS.
A PRIME IS AN INTEGER THAT IS ONLY DIVISIBLE BY 1 AND ITSELF-- LIKE 13.
13 IS ONLY DIVISIBLE BY 1 AND 13, AND NOTHING ELSE.
YEAH, **I GET THAT--**
I *REALLY* GET IT! HOW ABOUT THAT? THANKS, EMMY...
TEE HEE HE

HONESTLY! SHE'S NOT TELLING YOU ANYTHING *DEEP*.

EVEN THE ANCIENT GREEK PHILOSOPHERS KNEW "THE FUNDAMENTAL THEOREM OF ARITHMETIC."

IT'S THE *QUALITY* OF THE IDEAS THAT IS IMPORTANT, MR. LANGER,

NOT HOW LONG THEY HAVE BEEN IN CIRCULATION.

BRING ME THE PAPERWORK ON THE PREVIOUS VICTIM...

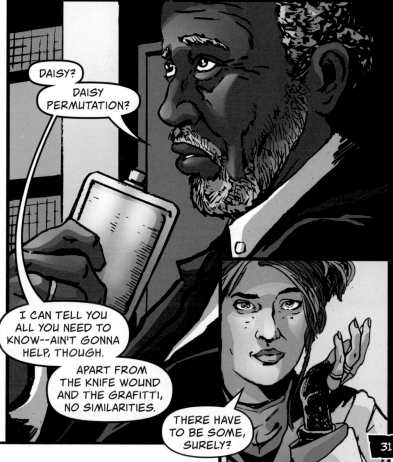

DAISY? DAISY PERMUTATION?

I CAN TELL YOU ALL YOU NEED TO KNOW--AIN'T GONNA HELP, THOUGH.

APART FROM THE KNIFE WOUND AND THE GRAFITTI, NO SIMILARITIES.

THERE HAVE TO BE SOME, SURELY?

31

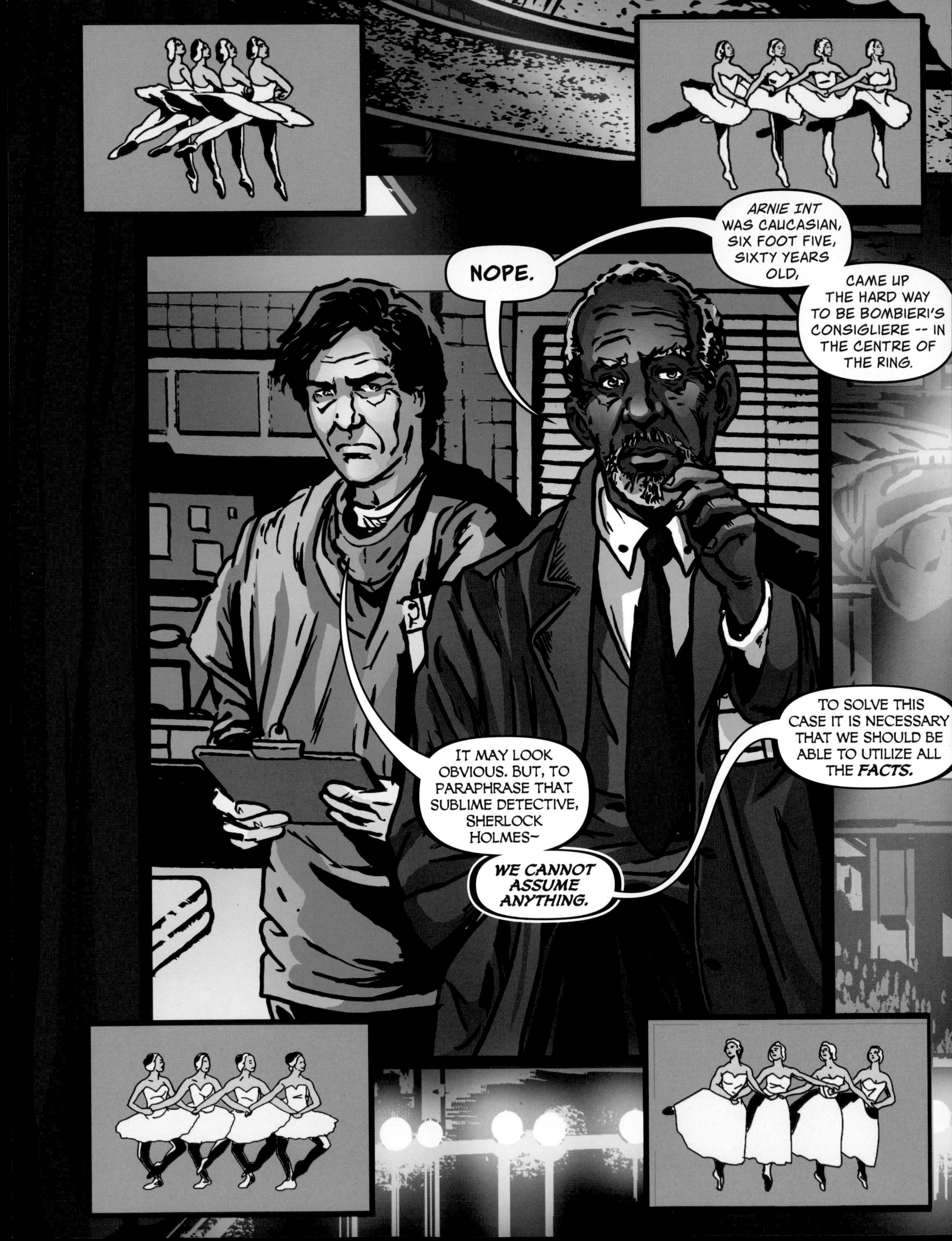
NOPE.
ARNIE INT WAS CAUCASIAN, SIX FOOT FIVE, SIXTY YEARS OLD,
CAME UP THE HARD WAY TO BE BOMBIERI'S CONSIGLIERE -- IN THE CENTRE OF THE RING.
IT MAY LOOK OBVIOUS. BUT, TO PARAPHRASE THAT SUBLIME DETECTIVE, SHERLOCK HOLMES~
WE CANNOT ASSUME ANYTHING.
TO SOLVE THIS CASE IT IS NECESSARY THAT WE SHOULD BE ABLE TO UTILIZE ALL THE FACTS.

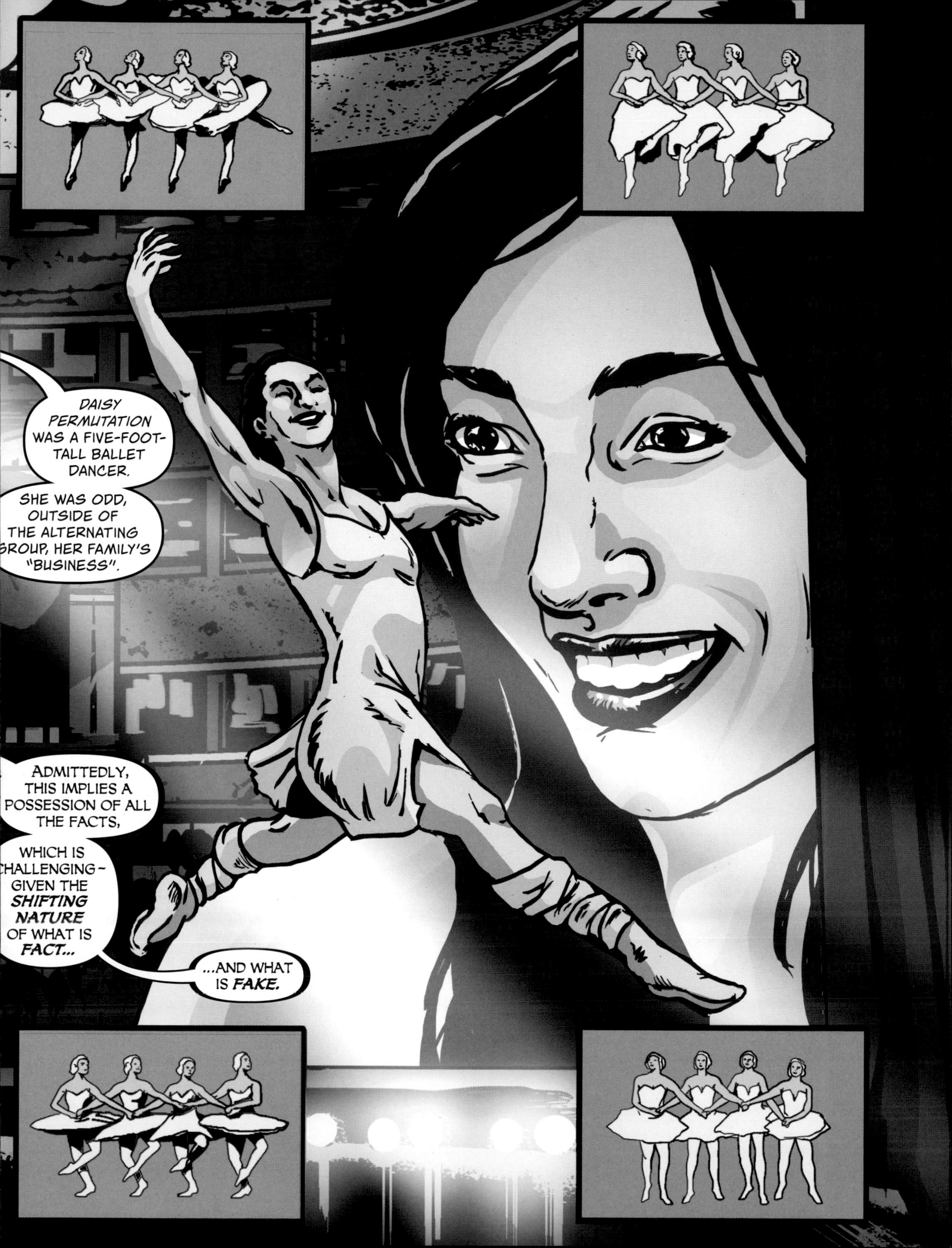

DAISY PERMUTATION WAS A FIVE-FOOT-TALL BALLET DANCER.
SHE WAS ODD, OUTSIDE OF THE ALTERNATING GROUP, HER FAMILY'S "BUSINESS".
ADMITTEDLY, THIS IMPLIES A POSSESSION OF ALL THE FACTS,
WHICH IS CHALLENGING– GIVEN THE SHIFTING NATURE OF WHAT IS FACT...
...AND WHAT IS FAKE.

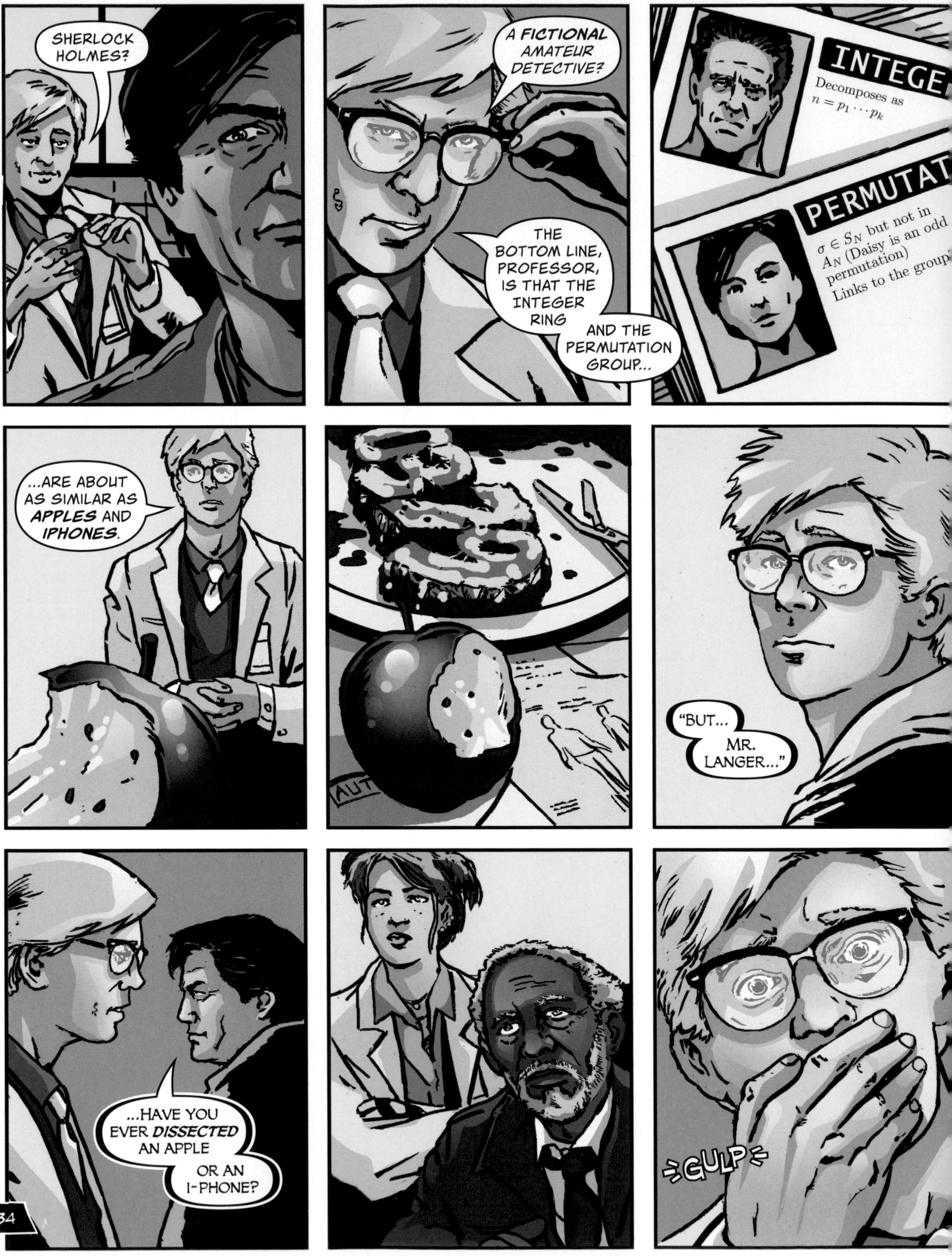

SHERLOCK HOLMES?
A FICTIONAL AMATEUR DETECTIVE?
THE BOTTOM LINE, PROFESSOR, IS THAT THE INTEGER RING
AND THE PERMUTATION GROUP...
INTEGE
Decomposes as $n = p_1 \cdots p_k$
PERMUTAT
$\sigma \in S_N$ but not in $A_N$ (Daisy is an odd permutation)
Links to the group
...ARE ABOUT AS SIMILAR AS APPLES AND IPHONES.
AUT
"BUT...
MR. LANGER..."
...HAVE YOU EVER DISSECTED AN APPLE
OR AN I-PHONE?
GULP

"Later that evening..."
Something's eating von Neumann
DETECTIVE VON NEUMANN
CLICK
CLACK
CLICK
CLICK
CLICK-ITY
CLACK
CLACK
THUNK
Dear Joe,

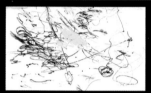

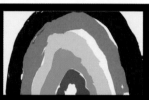

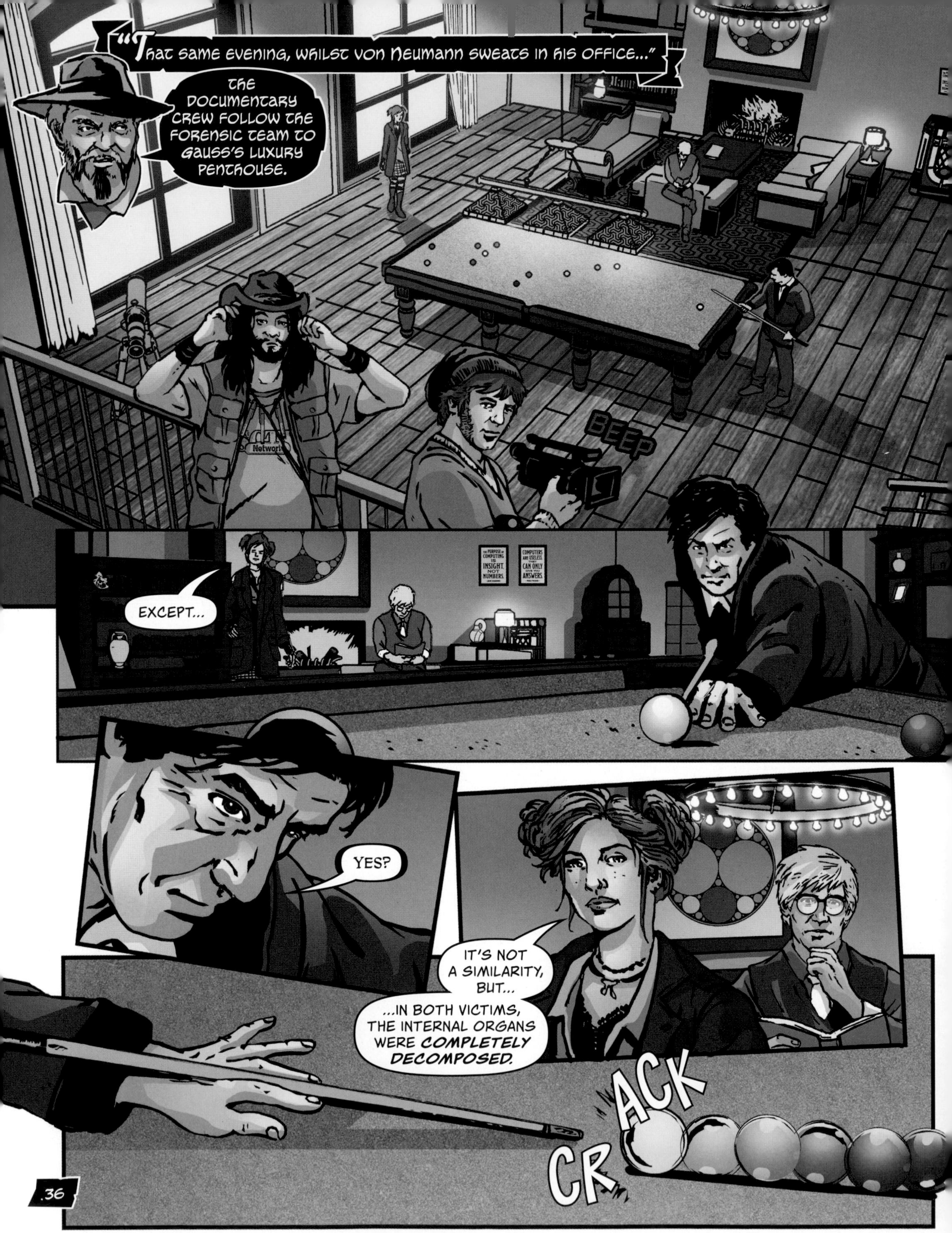
"That same evening, whilst von Neumann sweats in his office..."
The documentary crew follow the forensic team to Gauss's luxury penthouse.
Network
BEEP
THE PURPOSE OF COMPUTING IS INSIGHT, NOT NUMBERS.
COMPUTERS ARE USELESS. THEY CAN ONLY GIVE YOU ANSWERS.
EXCEPT...
YES?
IT'S NOT A SIMILARITY, BUT...
...IN BOTH VICTIMS, THE INTERNAL ORGANS WERE ***COMPLETELY DECOMPOSED.***
CRACK

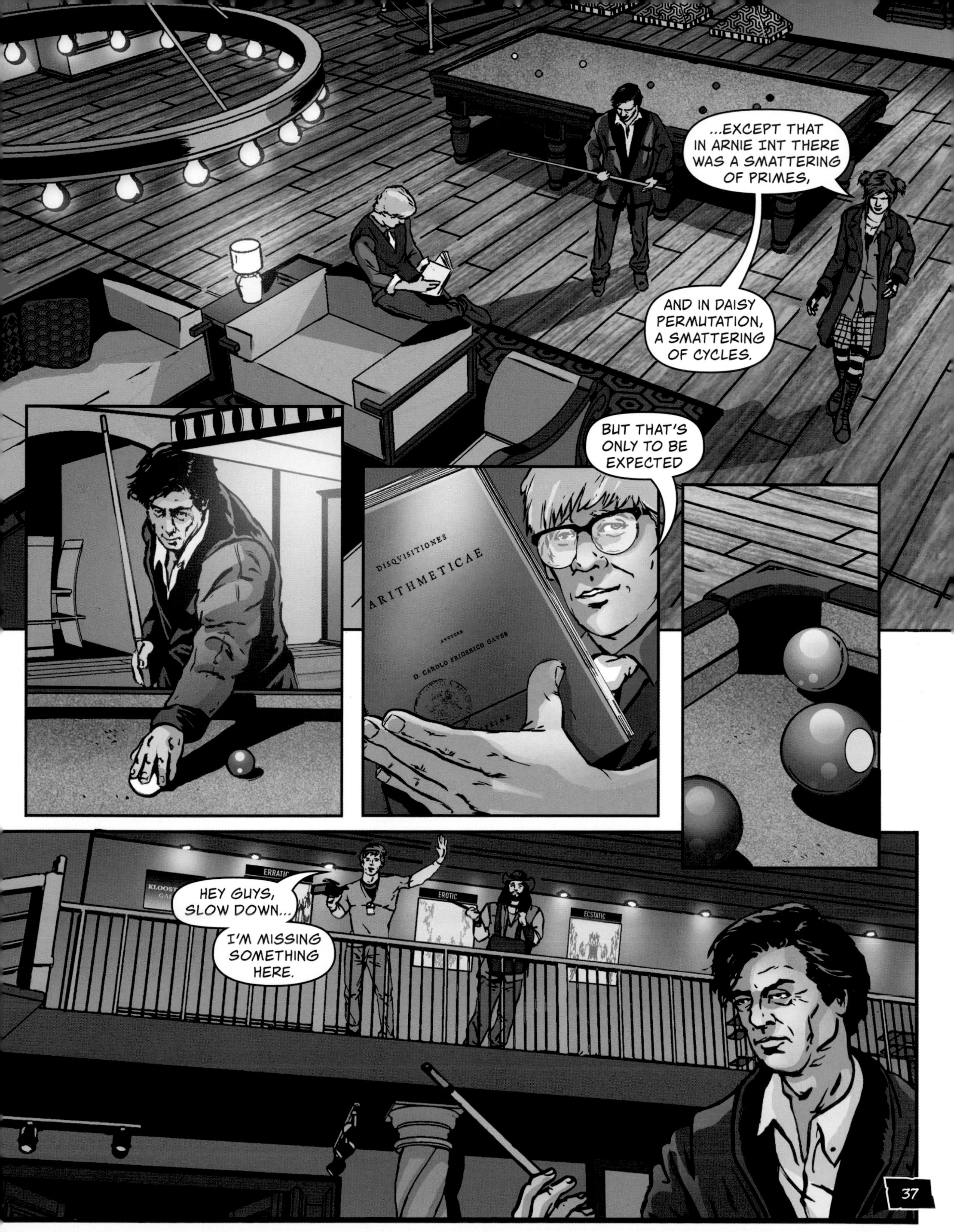
...EXCEPT THAT IN ARNIE INT THERE WAS A SMATTERING OF PRIMES,
AND IN DAISY PERMUTATION, A SMATTERING OF CYCLES.
BUT THAT'S ONLY TO BE EXPECTED
DISQUISITIONES ARITHMETICAE
AVCTORE D. CAROLO FRIDERICO GAVSS
HEY GUYS, SLOW DOWN...
I'M MISSING SOMETHING HERE.
ERRATIC
EROTIC
ECSTATIC

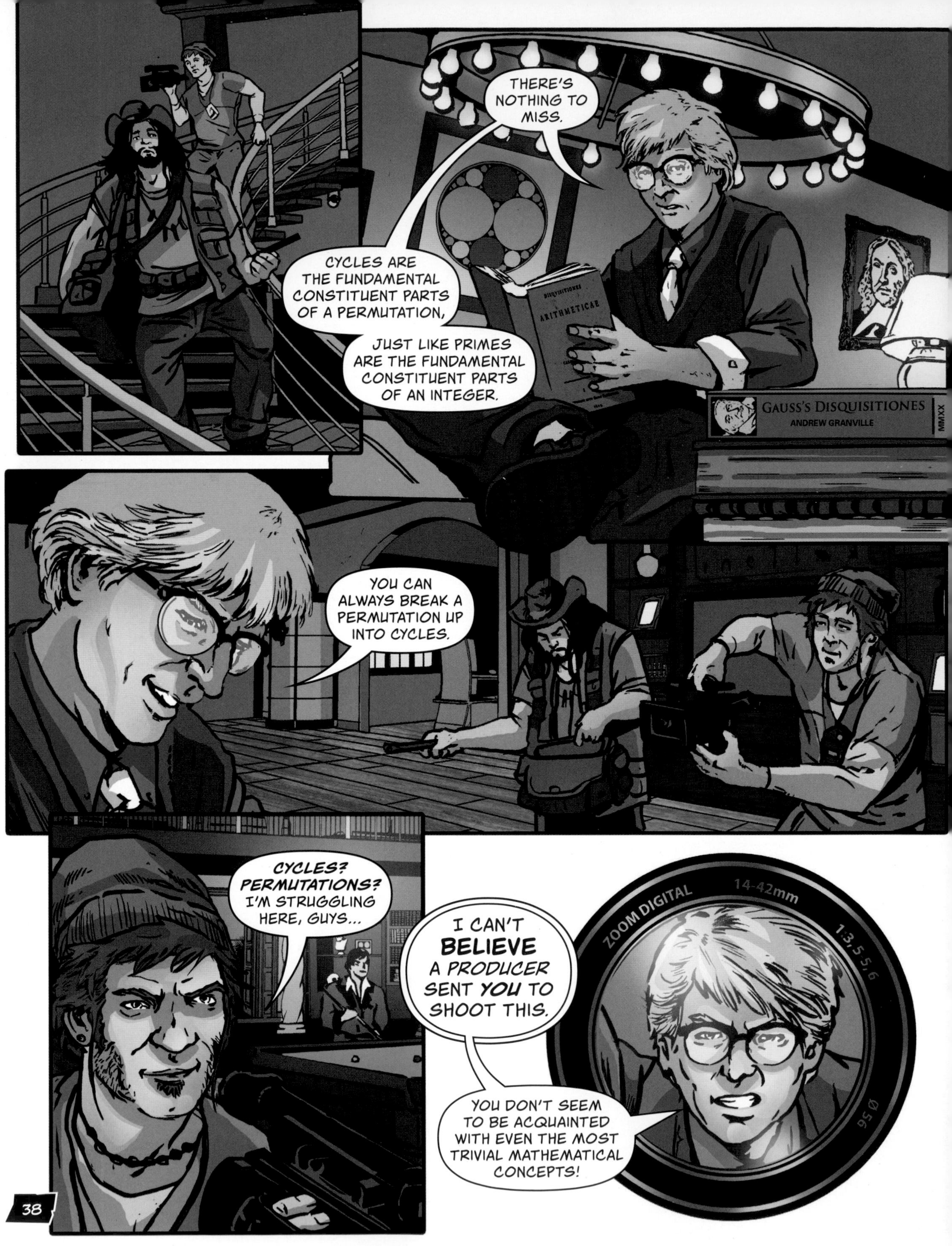
THERE'S NOTHING TO MISS.
CYCLES ARE THE FUNDAMENTAL CONSTITUENT PARTS OF A PERMUTATION,
JUST LIKE PRIMES ARE THE FUNDAMENTAL CONSTITUENT PARTS OF AN INTEGER.
GAUSS'S DISQUISITIONES
ANDREW GRANVILLE
MMXX
YOU CAN ALWAYS BREAK A PERMUTATION UP INTO CYCLES.
*CYCLES? PERMUTATIONS?* I'M STRUGGLING HERE, GUYS...
I CAN'T **BELIEVE** A *PRODUCER* SENT ***YOU*** TO SHOOT THIS.
YOU DON'T SEEM TO BE ACQUAINTED WITH EVEN THE MOST TRIVIAL MATHEMATICAL CONCEPTS!
ZOOM DIGITAL
14-42mm
1:3,5-5,6
Ø56

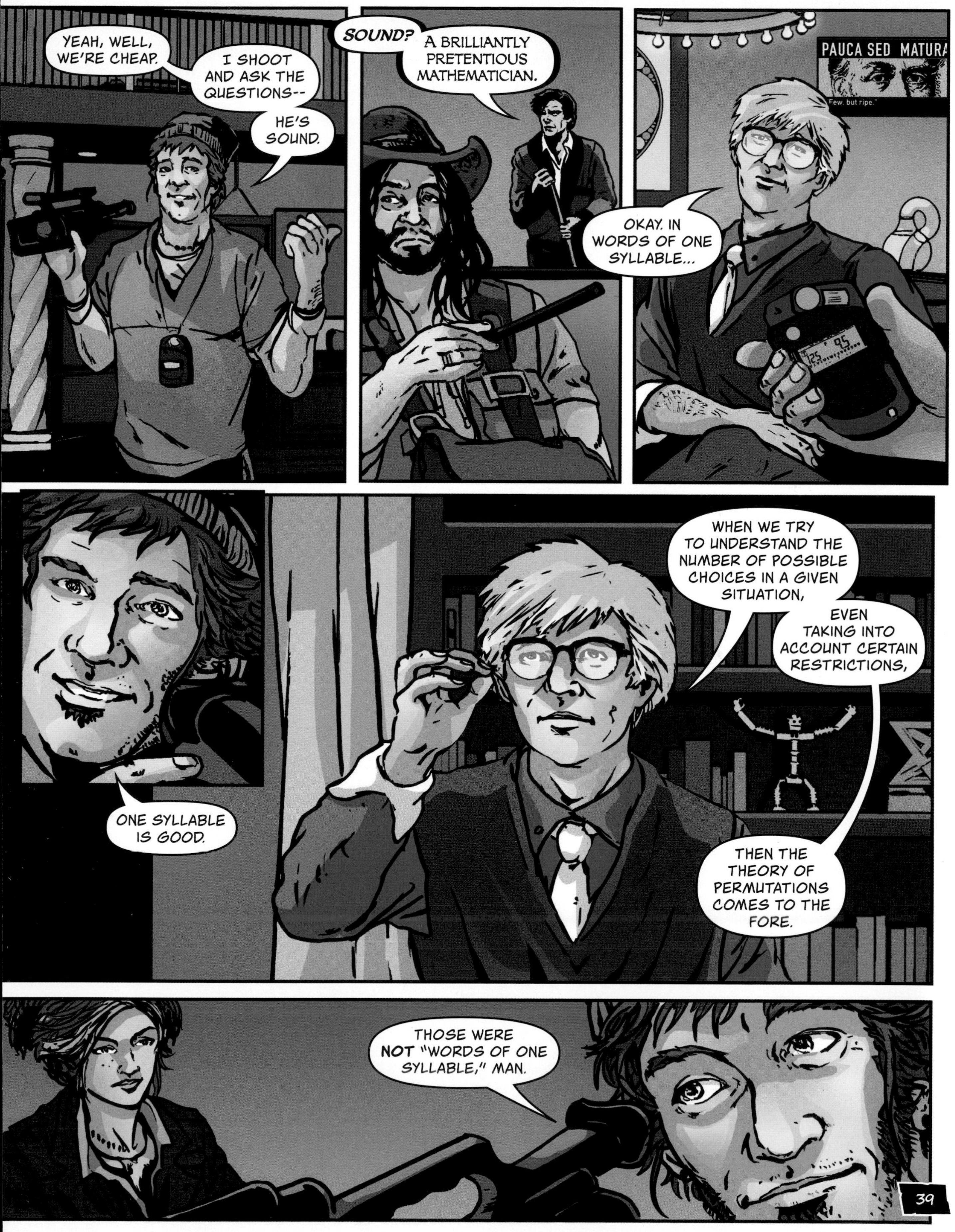
YEAH, WELL, WE'RE CHEAP.
I SHOOT AND ASK THE QUESTIONS--
HE'S SOUND.
SOUND?
A BRILLIANTLY PRETENTIOUS MATHEMATICIAN.
PAUCA SED MATURA
"Few, but ripe."
OKAY. IN WORDS OF ONE SYLLABLE...
WHEN WE TRY TO UNDERSTAND THE NUMBER OF POSSIBLE CHOICES IN A GIVEN SITUATION,
EVEN TAKING INTO ACCOUNT CERTAIN RESTRICTIONS,
THEN THE THEORY OF PERMUTATIONS COMES TO THE FORE.
ONE SYLLABLE IS GOOD.
THOSE WERE NOT "WORDS OF ONE SYLLABLE," MAN.

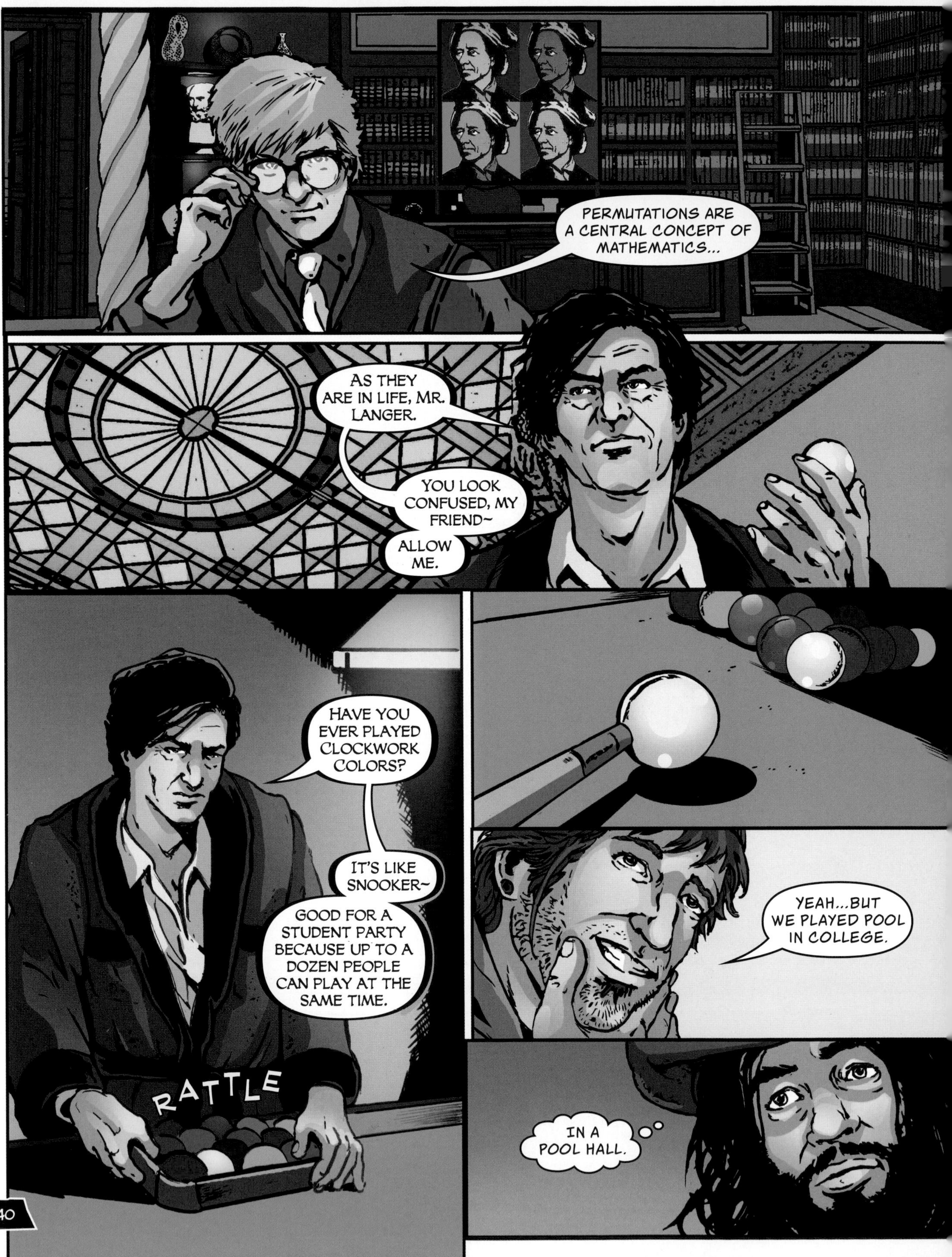
PERMUTATIONS ARE A CENTRAL CONCEPT OF MATHEMATICS...
AS THEY ARE IN LIFE, MR. LANGER.
YOU LOOK CONFUSED, MY FRIEND~
ALLOW ME.
HAVE YOU EVER PLAYED CLOCKWORK COLORS?
IT'S LIKE SNOOKER~
GOOD FOR A STUDENT PARTY BECAUSE UP TO A DOZEN PEOPLE CAN PLAY AT THE SAME TIME.
RATTLE
YEAH...BUT WE PLAYED POOL IN COLLEGE.
IN A POOL HALL.

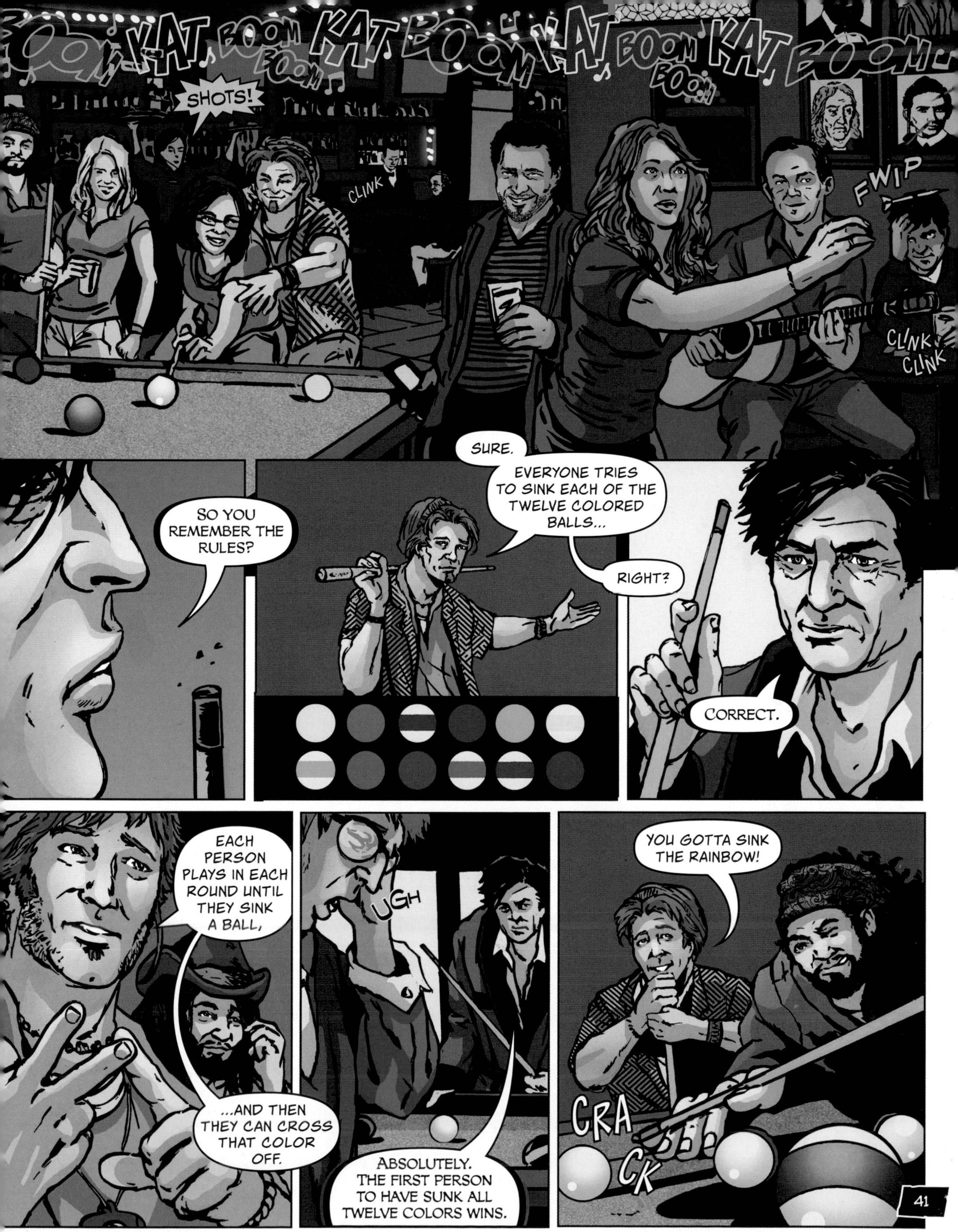
KAT BOOM KAT BOOM KAT BOOM KAT BOOM
SHOTS!
CLINK
FWIP
CLINK CLINK
SO YOU REMEMBER THE RULES?
SURE.
EVERYONE TRIES TO SINK EACH OF THE TWELVE COLORED BALLS...
RIGHT?
CORRECT.
EACH PERSON PLAYS IN EACH ROUND UNTIL THEY SINK A BALL,
...AND THEN THEY CAN CROSS THAT COLOR OFF.
UGH
ABSOLUTELY. THE FIRST PERSON TO HAVE SUNK ALL TWELVE COLORS WINS.
YOU GOTTA SINK THE RAINBOW!
CRACK

WE ASK OURSELVES~

IN HOW MANY WAYS CAN THE COLORS BE SUNK BY THE DIFFERENT PLAYERS?

EXACTLY. AND WHILST WE MATHEMATICIANS AWAIT OUR TURN,

JEEZ, DON'T YOU GUYS *EVER* TAKE A BREAK?

CR ACK

SO~WHAT WOULD YOU ESTIMATE?

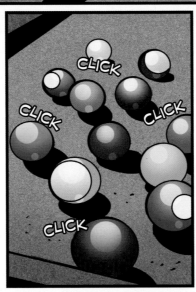

CLICK

CLICK

CLICK

CLICK

YOU GOTTA BE *KIDDING* ME? YOU WANT ME TO *GUESS*?

I HAVE **NO CLUE**...

SAY... A COUPLE A HUNDRED, MAYBE?

HEH HEH HEH HEH

IT'S A LITTLE MORE THAN YOU MIGHT IMAGINE, MR. BELL.

EMMY~CAN YOU EXPLAIN?

WELL, THE FIRST PERSON COULD HAVE SUNK ANY OF THE TWELVE POSSIBLE COLORS--

SAY, RED.

THE SECOND PERSON SELECTS FROM THE REMAINING ELEVEN BALLS--

ANY COLOR OTHER THAN RED, THE COLOR THE FIRST PERSON SUNK--

SO THERE ARE ELEVEN POSSIBILITIES.

THE NEXT PERSON PICKS FROM TEN COLORS, THE NEXT FROM NINE, AND SO ON...

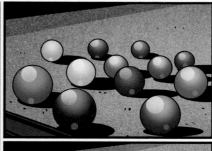

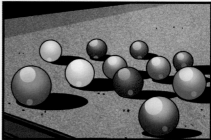

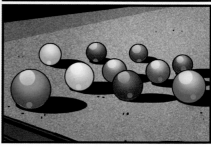

DOES THAT MAKE SENSE?

I THINK SO.

WE CALL THIS A "PERMUTATION"--

AND THE TOTAL NUMBER OF PERMUTATIONS OF TWELVE COLORS IS 12 TIMES 11 TIMES 10, AND SO ON,

ALL THE WAY DOWN TO TIMES 1.

WE CALL THIS PRODUCT "TWELVE FACTORIAL" AND... WAIT FOR IT...

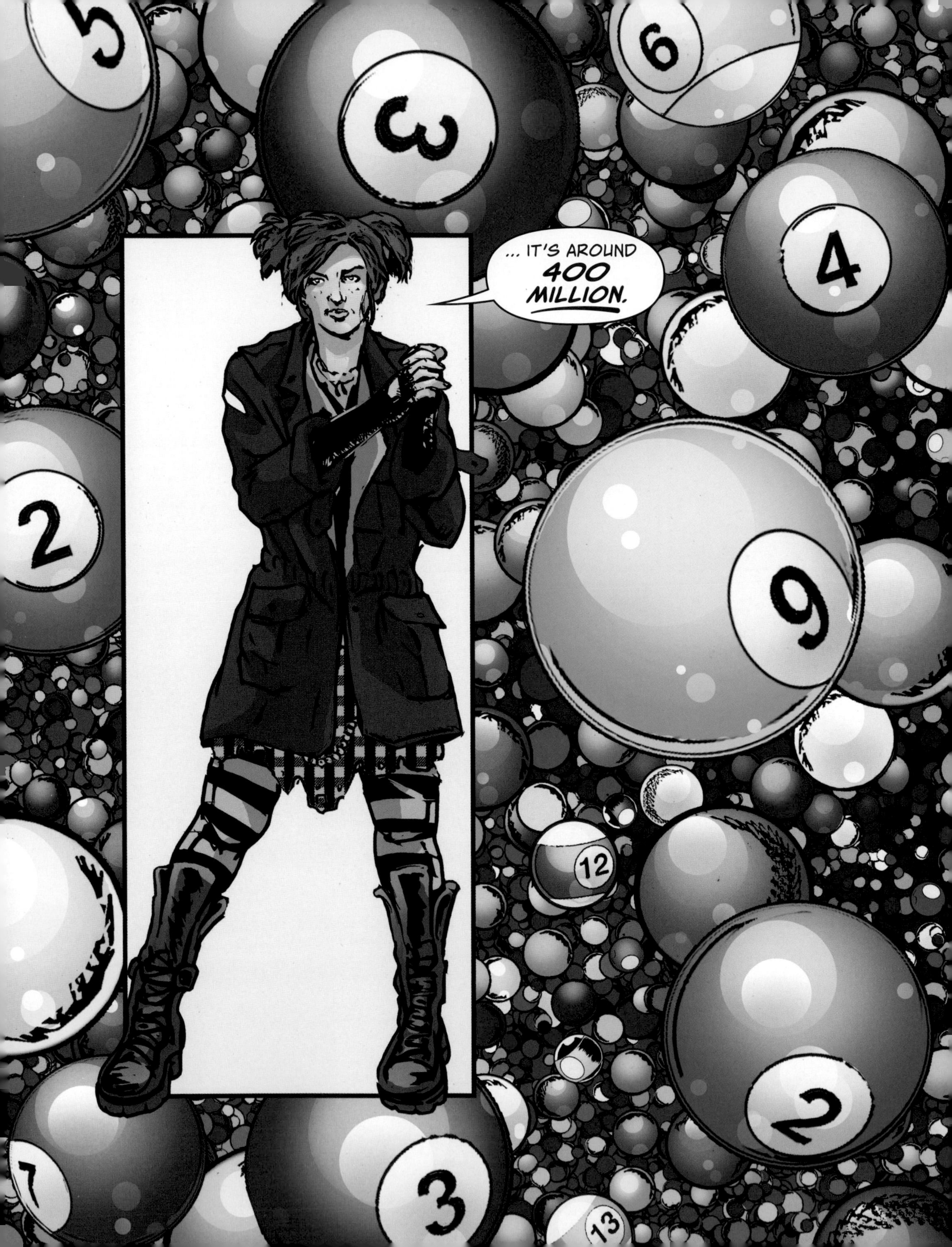
... IT'S AROUND 400 MILLION.

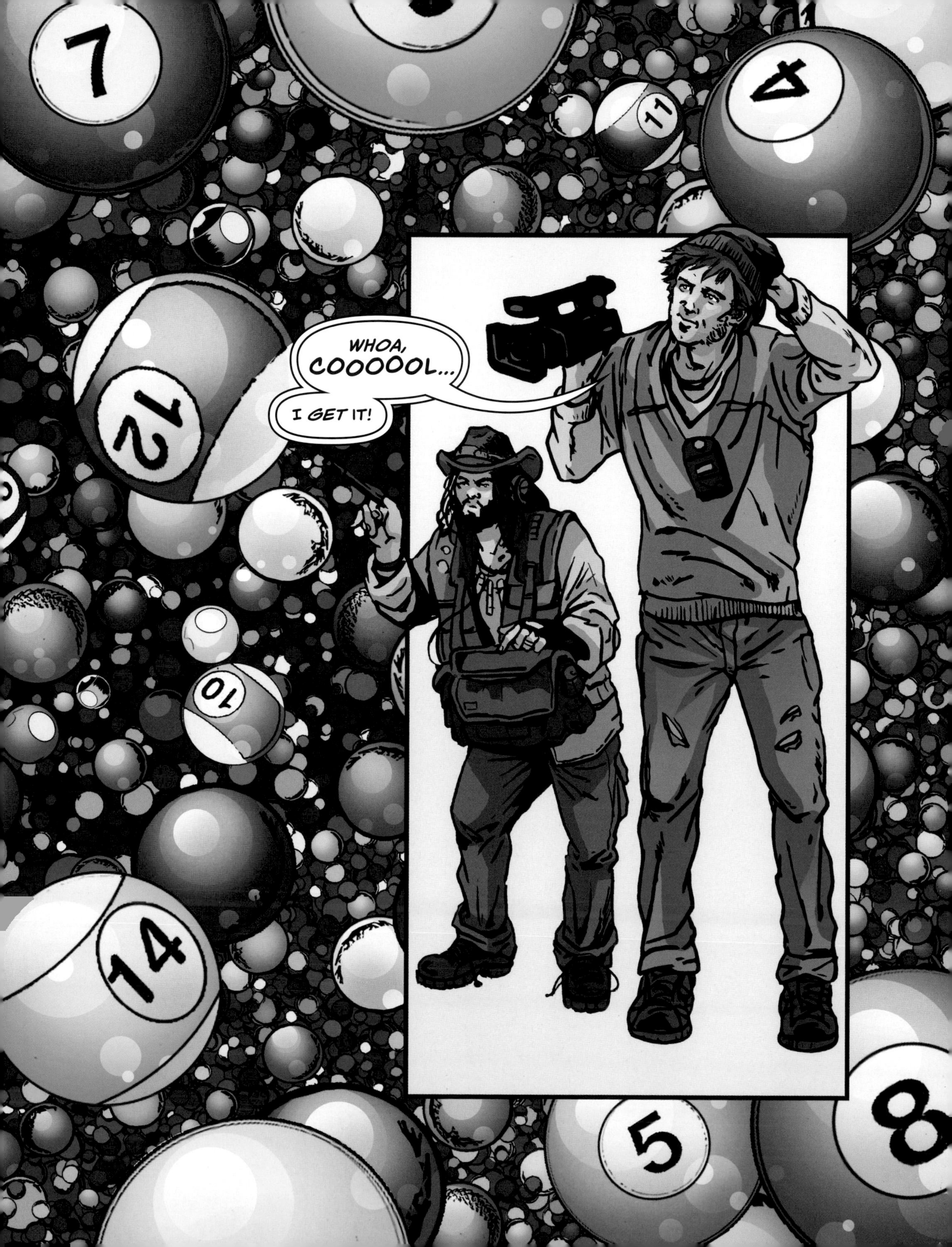
WHOA, COOOOOL...
I GET IT!

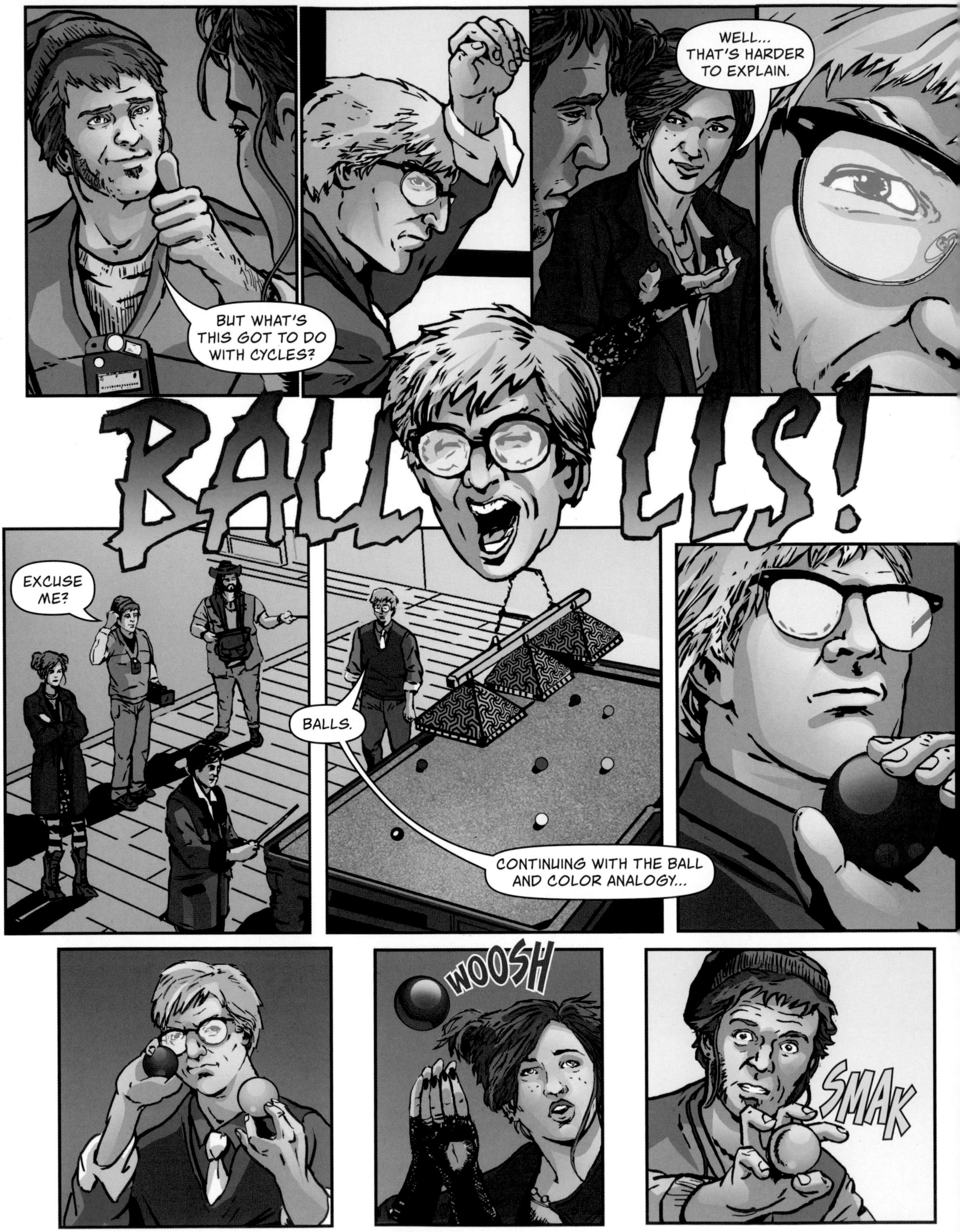
BUT WHAT'S THIS GOT TO DO WITH CYCLES?
WELL... THAT'S HARDER TO EXPLAIN.
BALLS!
EXCUSE ME?
BALLS.
CONTINUING WITH THE BALL AND COLOR ANALOGY...
WOOSH
SMAK

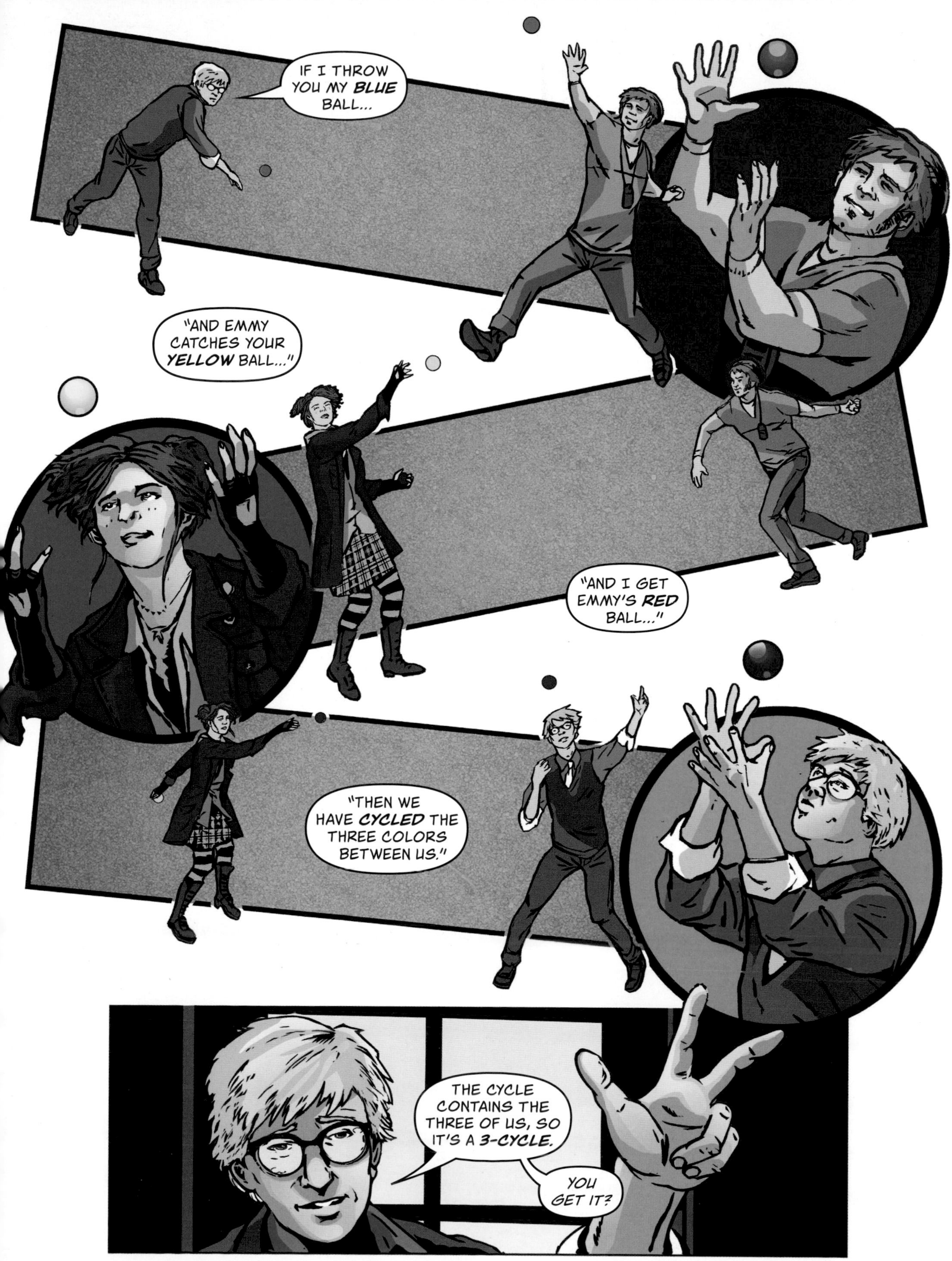
IF I THROW YOU MY BLUE BALL...
"AND EMMY CATCHES YOUR YELLOW BALL..."
"AND I GET EMMY'S RED BALL..."
"THEN WE HAVE CYCLED THE THREE COLORS BETWEEN US."
THE CYCLE CONTAINS THE THREE OF US, SO IT'S A 3-CYCLE.
YOU GET IT?

"RIGHT! EMMY PASSES..."

TO ME...

"ME TO YOU, MR. LANGER..."

"AND THEN YOU..."

"BACK TO EMMY AGAIN."

MOREOVER, YOU CAN BREAK DOWN ANY PERMUTATION INTO CYCLES IN EXACTLY THE SAME WAY.

THUS, **CYCLES** ARE THE FUNDAMENTAL CONSTITUENT PARTS OF A **PERMUTATION**,

JUST LIKE **PRIMES** ARE THE *FUNDAMENTAL CONSTITUENT PARTS OF AN* **INTEGER**.

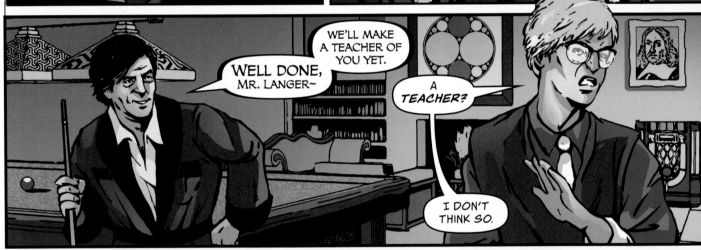

WELL DONE, MR. LANGER~

WE'LL MAKE A TEACHER OF YOU YET.

A TEACHER?

I DON'T THINK SO.

SO WE'RE AGREED THAT IT MAKES SENSE TO CALIBRATE THE PRIMES WITH THE CYCLES...

BUT HOW?

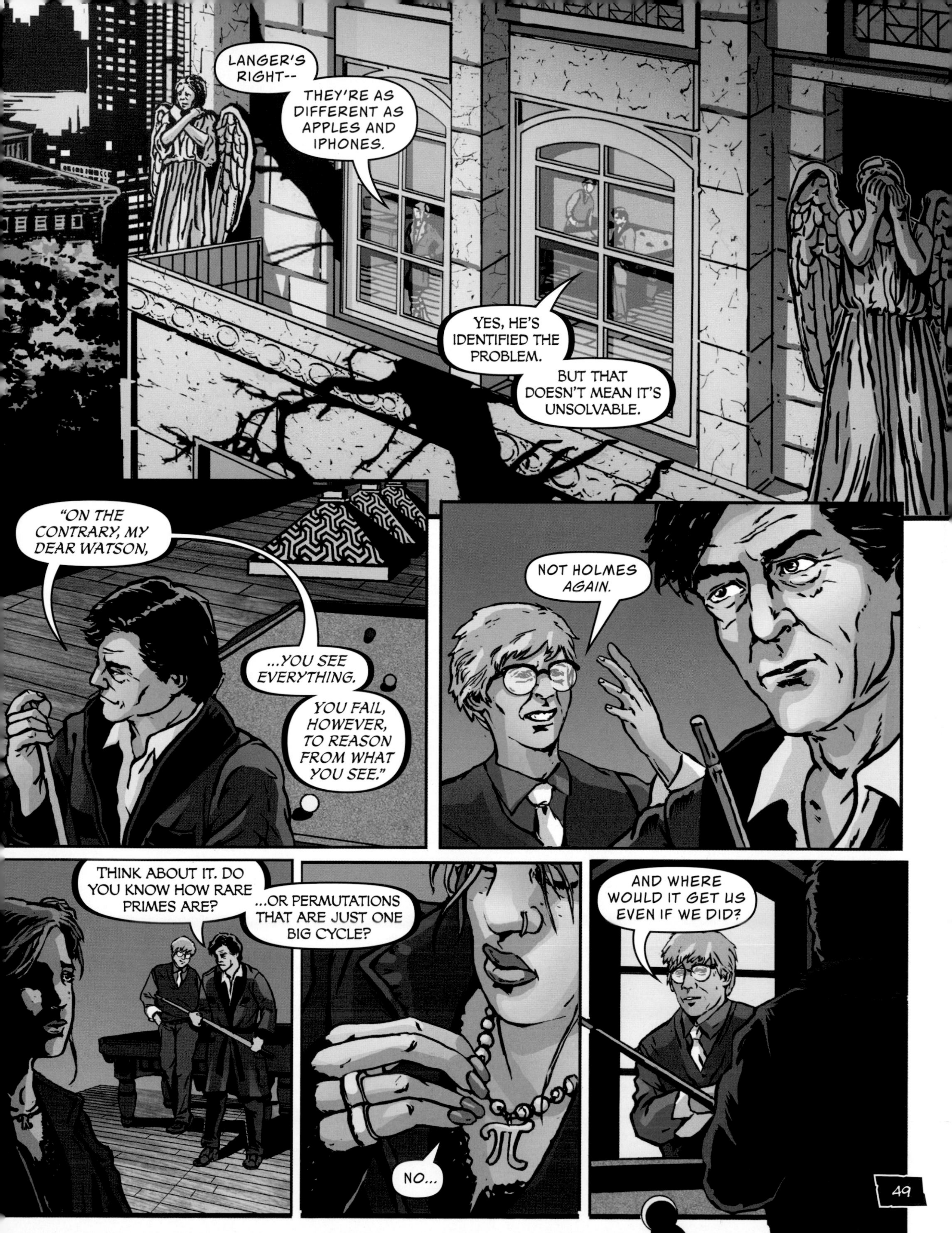

LANGER'S RIGHT--
THEY'RE AS DIFFERENT AS APPLES AND IPHONES.
YES, HE'S IDENTIFIED THE PROBLEM.
BUT THAT DOESN'T MEAN IT'S UNSOLVABLE.
"ON THE CONTRARY, MY DEAR WATSON,
...YOU SEE EVERYTHING.
YOU FAIL, HOWEVER, TO REASON FROM WHAT YOU SEE."
NOT HOLMES AGAIN.
THINK ABOUT IT. DO YOU KNOW HOW RARE PRIMES ARE?
...OR PERMUTATIONS THAT ARE JUST ONE BIG CYCLE?
NO...
AND WHERE WOULD IT GET US EVEN IF WE DID?

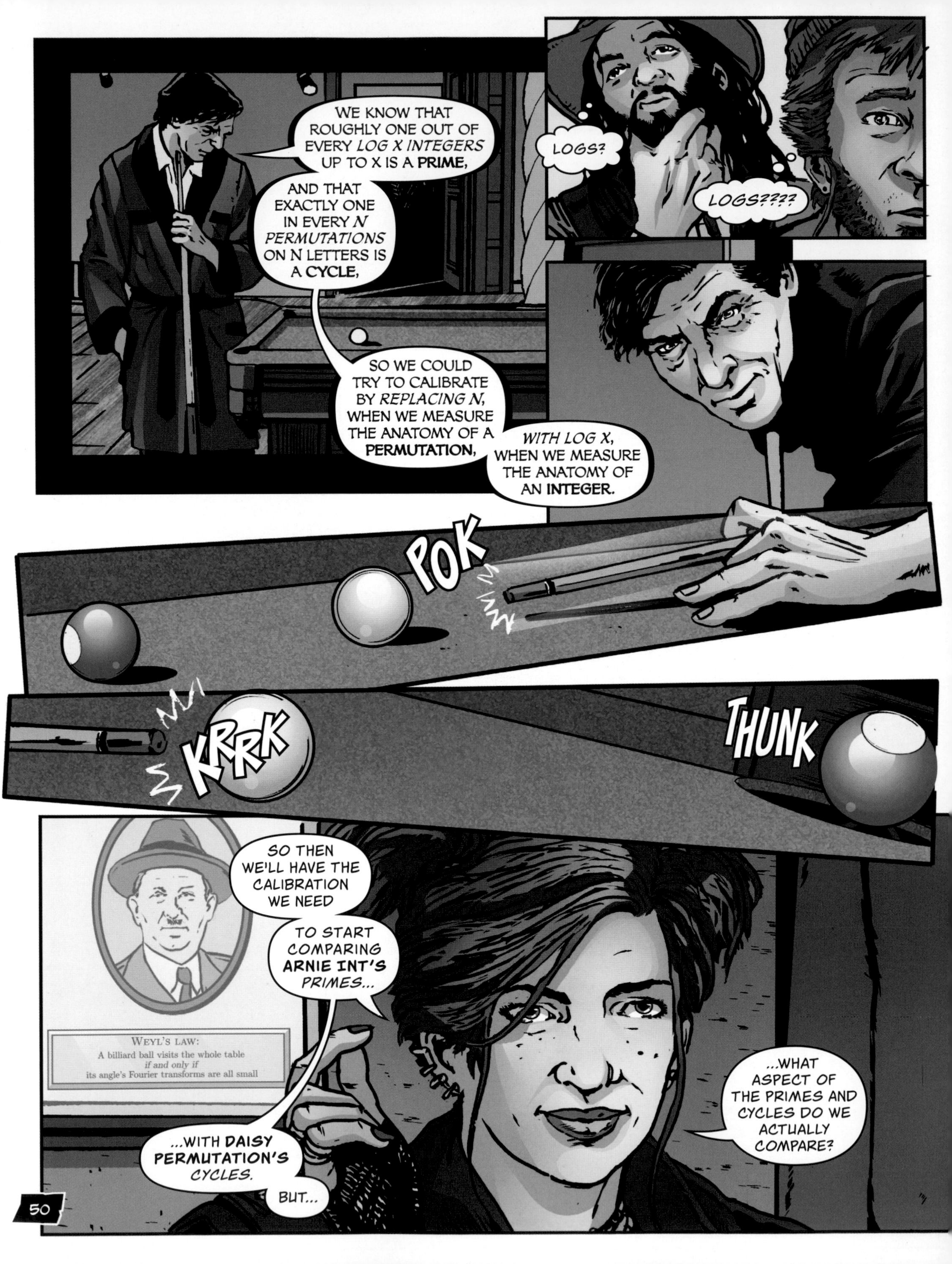
WE KNOW THAT ROUGHLY ONE OUT OF EVERY *LOG X INTEGERS* UP TO X IS A **PRIME**,
AND THAT EXACTLY ONE IN EVERY *N PERMUTATIONS* ON N LETTERS IS A **CYCLE**,
LOGS?
LOGS????
SO WE COULD TRY TO CALIBRATE BY *REPLACING N*, WHEN WE MEASURE THE ANATOMY OF A **PERMUTATION**,
*WITH LOG X*, WHEN WE MEASURE THE ANATOMY OF AN **INTEGER**.
POK
KRRK
THUNK
SO THEN WE'LL HAVE THE CALIBRATION WE NEED
TO START COMPARING **ARNIE INT'S** *PRIMES*...
WEYL'S LAW:
A billiard ball visits the whole table
*if and only if*
its angle's Fourier transforms are all small
...WITH **DAISY PERMUTATION'S** *CYCLES*.
BUT...
...WHAT ASPECT OF THE PRIMES AND CYCLES DO WE ACTUALLY COMPARE?

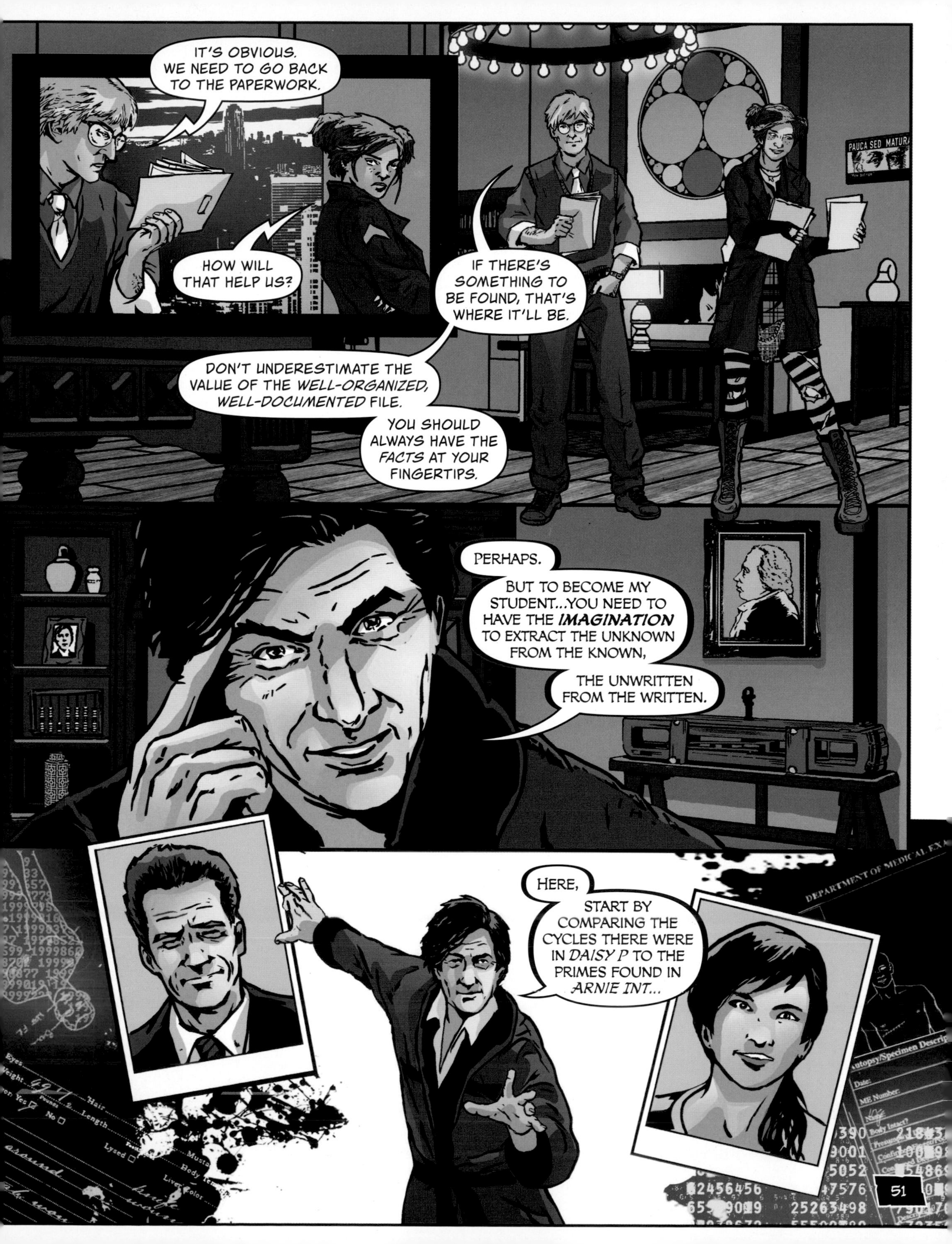
IT'S OBVIOUS. WE NEED TO GO BACK TO THE PAPERWORK.
HOW WILL THAT HELP US?
IF THERE'S SOMETHING TO BE FOUND, THAT'S WHERE IT'LL BE.
DON'T UNDERESTIMATE THE VALUE OF THE *WELL-ORGANIZED, WELL-DOCUMENTED* FILE.
YOU SHOULD ALWAYS HAVE THE *FACTS* AT YOUR FINGERTIPS.
PAUCA SED MATURA
PERHAPS.
BUT TO BECOME MY STUDENT...YOU NEED TO HAVE THE ***IMAGINATION*** TO EXTRACT THE UNKNOWN FROM THE KNOWN,
THE UNWRITTEN FROM THE WRITTEN.
HERE,
START BY COMPARING THE CYCLES THERE WERE IN *DAISY P* TO THE PRIMES FOUND IN *ARNIE INT*...
DEPARTMENT OF MEDICAL EX
Autopsy/Specimen Descrip

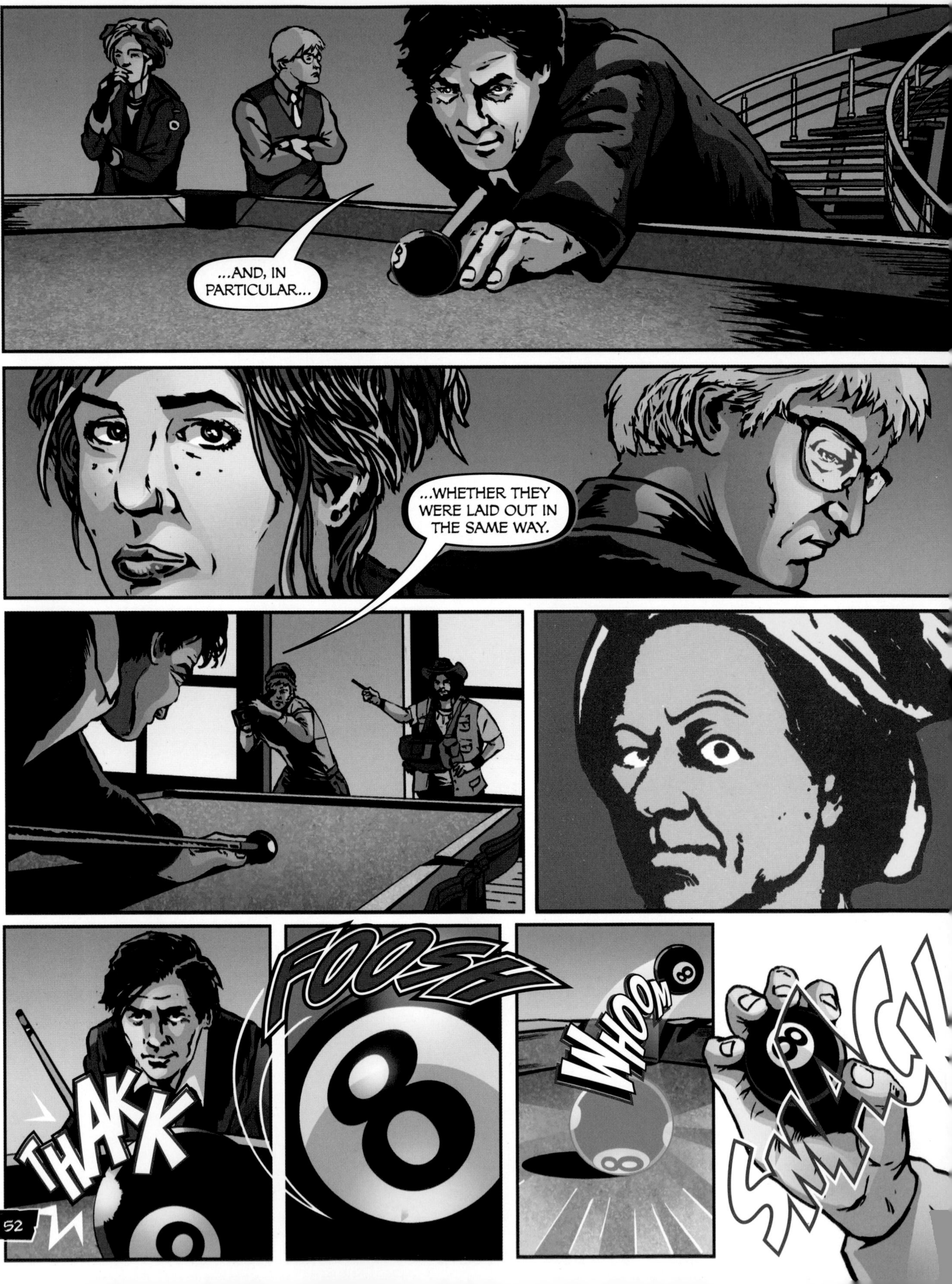

...AND, IN PARTICULAR...
...WHETHER THEY WERE LAID OUT IN THE SAME WAY.
THAKK
FOOSH
WHOOM

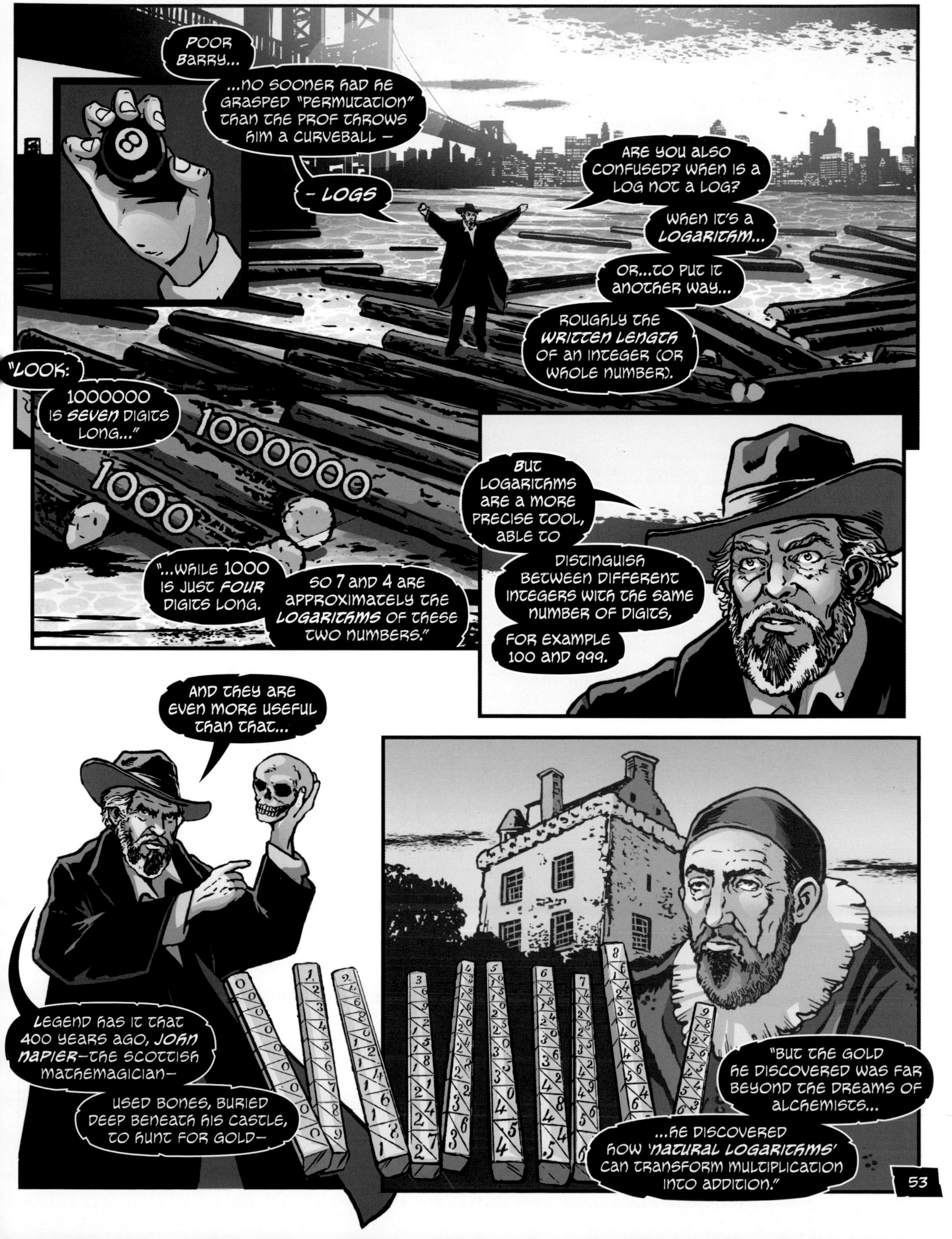
POOR BARRY...
...NO SOONER HAD HE GRASPED "PERMUTATION" THAN THE PROF THROWS HIM A CURVEBALL —
— LOGS
ARE YOU ALSO CONFUSED? WHEN IS A LOG NOT A LOG?
WHEN IT'S A LOGARITHM...
OR...TO PUT IT ANOTHER WAY...
ROUGHLY THE WRITTEN LENGTH OF AN INTEGER (OR WHOLE NUMBER).
"LOOK:
1000000 IS SEVEN DIGITS LONG..."
1000000
1000
"...WHILE 1000 IS JUST FOUR DIGITS LONG.
SO 7 AND 4 ARE APPROXIMATELY THE LOGARITHMS OF THESE TWO NUMBERS."
BUT LOGARITHMS ARE A MORE PRECISE TOOL, ABLE TO
DISTINGUISH BETWEEN DIFFERENT INTEGERS WITH THE SAME NUMBER OF DIGITS,
FOR EXAMPLE 100 AND 999.
AND THEY ARE EVEN MORE USEFUL THAN THAT...
LEGEND HAS IT THAT 400 YEARS AGO, JOHN NAPIER—THE SCOTTISH MATHEMAGICIAN—
USED BONES, BURIED DEEP BENEATH HIS CASTLE, TO HUNT FOR GOLD—
"BUT THE GOLD HE DISCOVERED WAS FAR BEYOND THE DREAMS OF ALCHEMISTS...
...HE DISCOVERED HOW 'NATURAL LOGARITHMS' CAN TRANSFORM MULTIPLICATION INTO ADDITION."

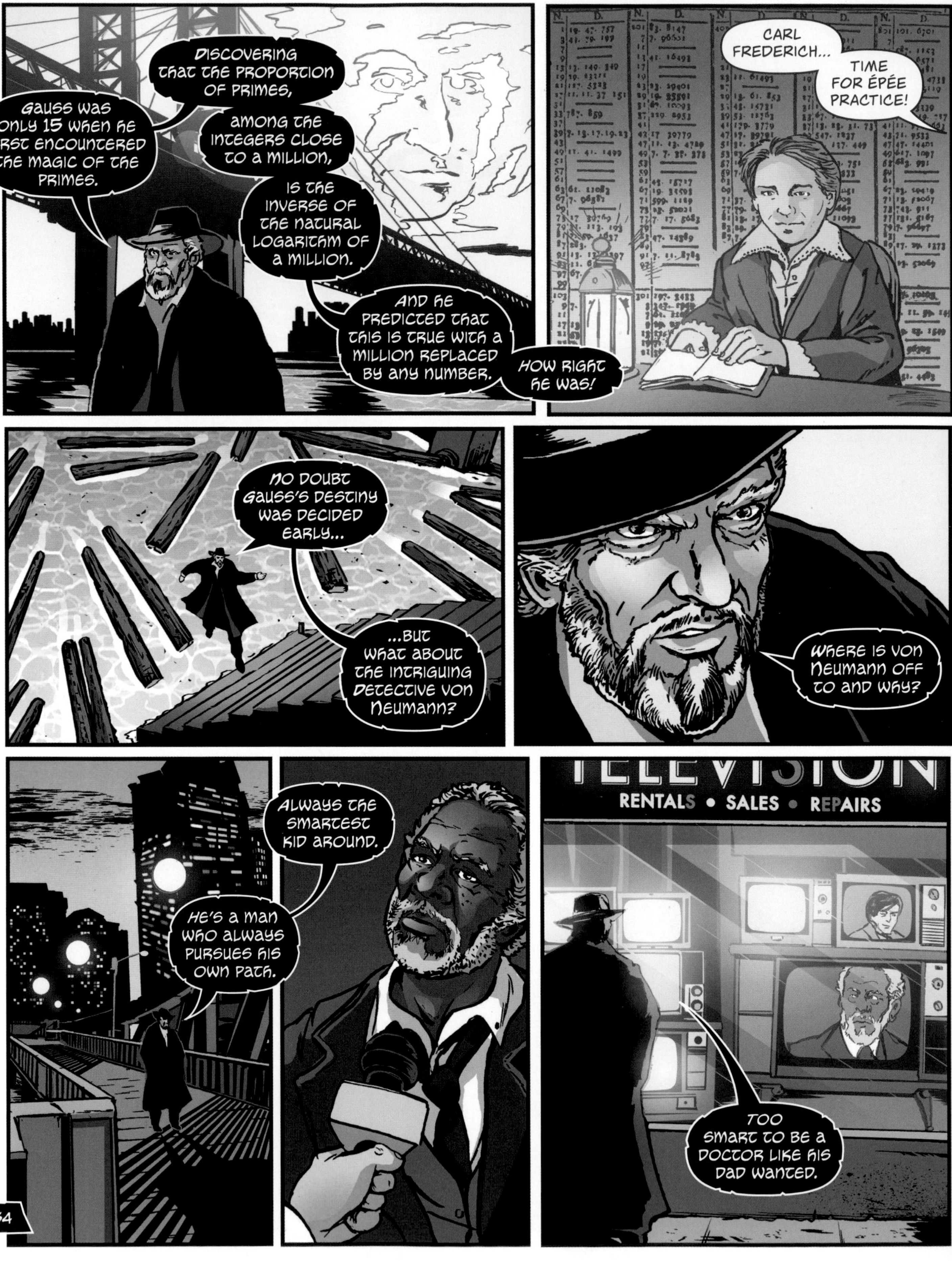
Gauss was only 15 when he first encountered the magic of the primes.
Discovering that the proportion of primes,
among the integers close to a million,
is the inverse of the natural logarithm of a million.
And he predicted that this is true with a million replaced by any number.
How right he was!
CARL FREDERICH...
TIME FOR ÉPÉE PRACTICE!
No doubt Gauss's destiny was decided early...
...But what about the intriguing detective von Neumann?
Where is von Neumann off to and why?
Always the smartest kid around.
He's a man who always pursues his own path.
TELEVISION
RENTALS • SALES • REPAIRS
Too smart to be a doctor like his dad wanted.

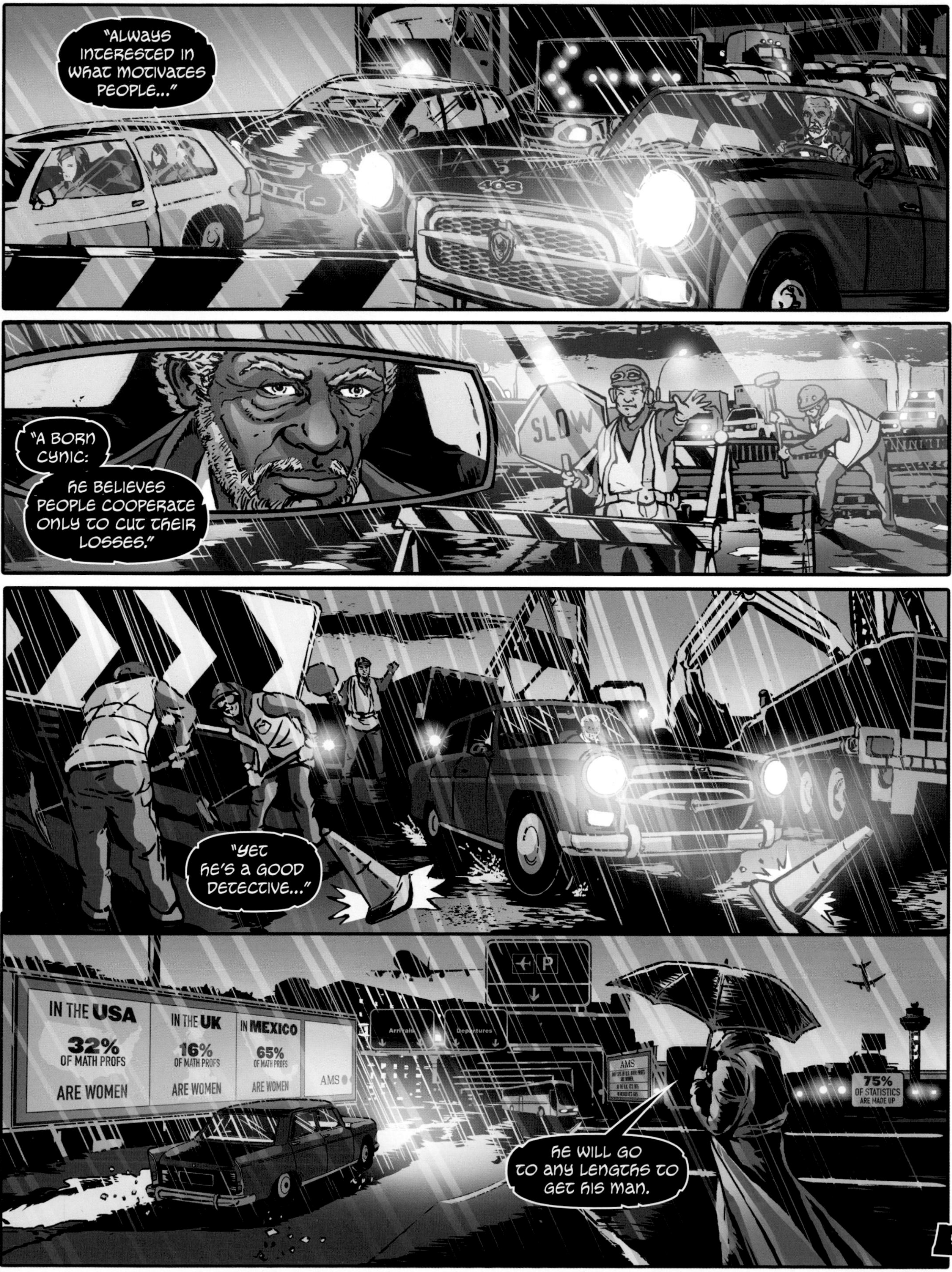
"ALWAYS INTERESTED IN WHAT MOTIVATES PEOPLE..."
403
"A BORN CYNIC:
HE BELIEVES PEOPLE COOPERATE ONLY TO CUT THEIR LOSSES."
SLOW
"YET HE'S A GOOD DETECTIVE..."
IN THE USA 32% OF MATH PROFS ARE WOMEN
IN THE UK 16% OF MATH PROFS ARE WOMEN
IN MEXICO 65% OF MATH PROFS ARE WOMEN
AMS
P
Arrivals
Departures
AMS
75% OF STATISTICS ARE MADE UP
HE WILL GO TO ANY LENGTHS TO GET HIS MAN.

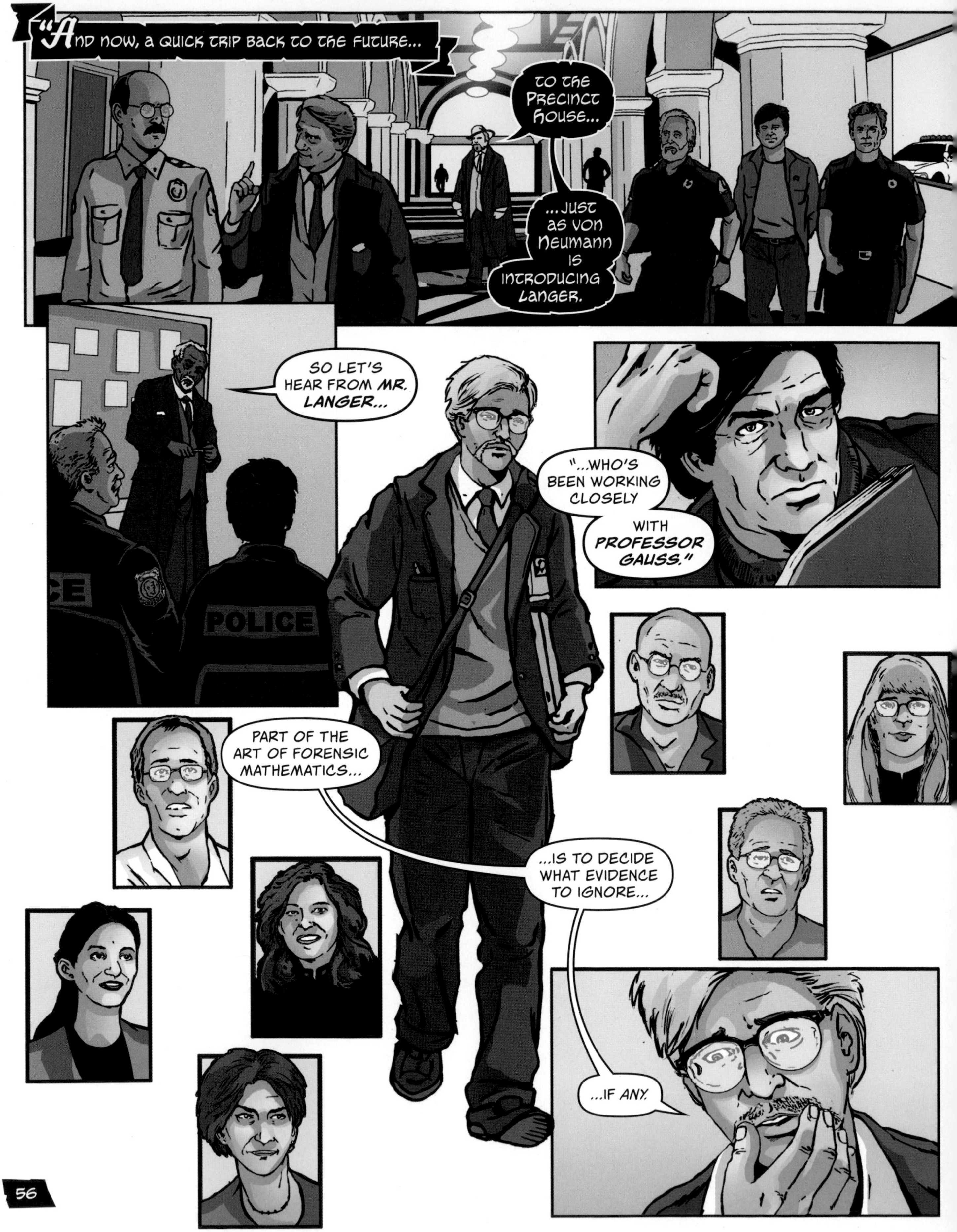
"And now, a quick trip back to the future...
to the Precinct House...
...just as von Neumann is introducing Langer.
SO LET'S HEAR FROM *MR. LANGER...*
"...WHO'S BEEN WORKING CLOSELY
WITH ***PROFESSOR GAUSS."***
POLICE
PART OF THE ART OF FORENSIC MATHEMATICS...
...IS TO DECIDE WHAT EVIDENCE TO IGNORE...
...IF *ANY.*

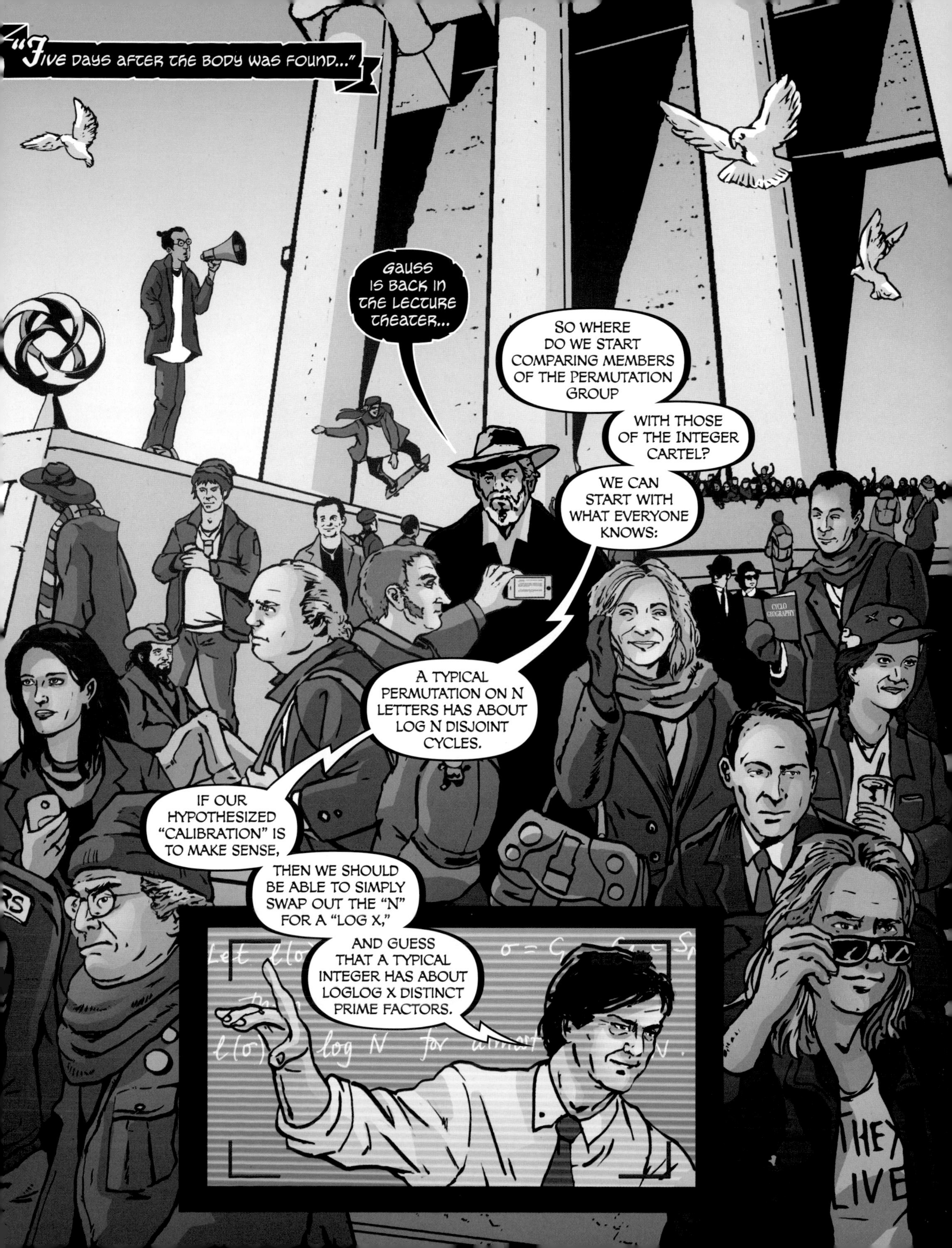
"Five days after the body was found..."
Gauss is back in the lecture theater...
So where do we start comparing members of the Permutation Group
with those of the Integer Cartel?
We can start with what everyone knows:
A typical permutation on N letters has about log N disjoint cycles.
If our hypothesized "calibration" is to make sense,
then we should be able to simply swap out the "N" for a "log X,"
and guess that a typical integer has about loglog X distinct prime factors.
CYCLO GEOGRAPHY
THEY LIVE

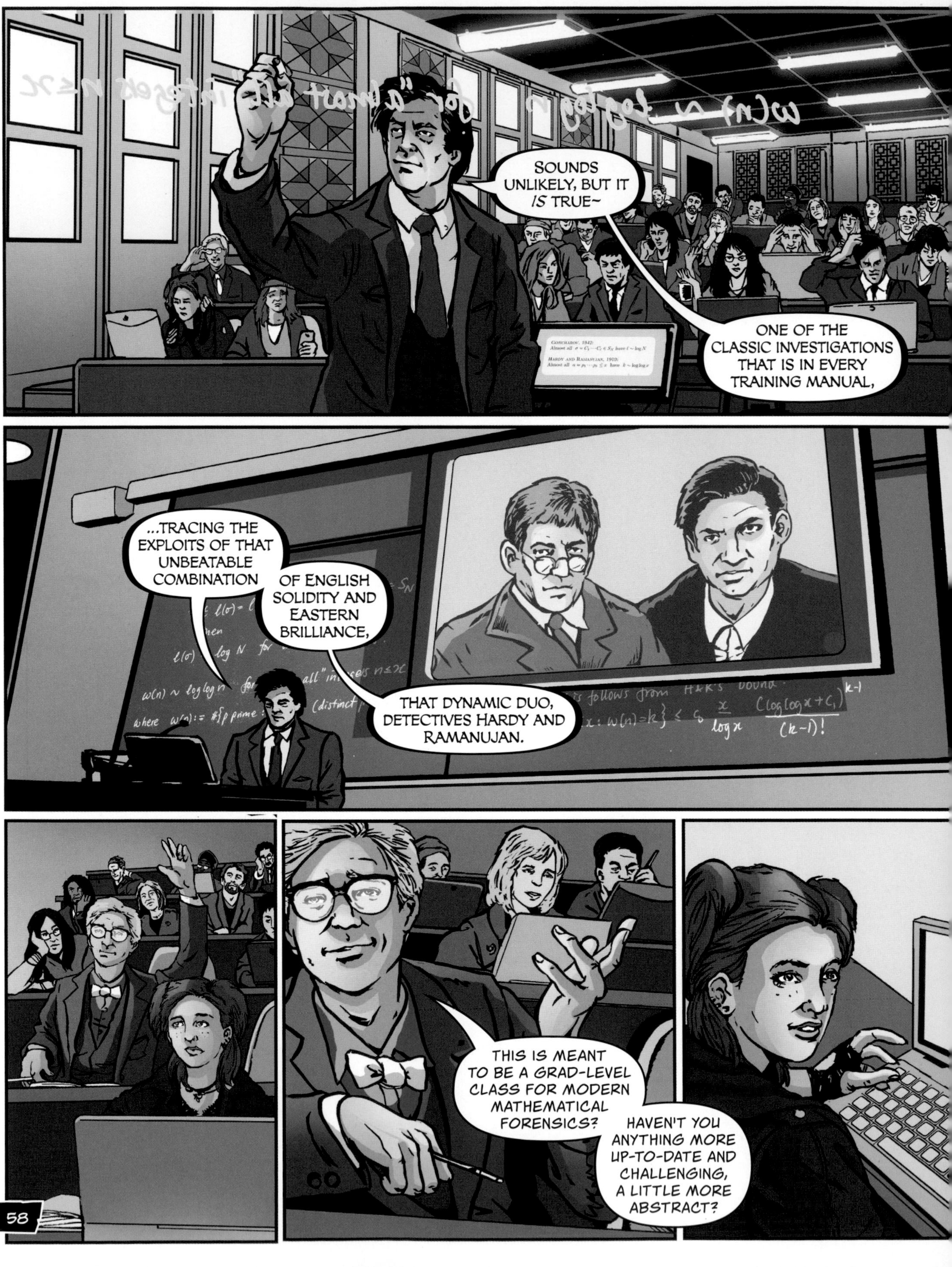
SOUNDS UNLIKELY, BUT IT *IS* TRUE~
ONE OF THE CLASSIC INVESTIGATIONS THAT IS IN EVERY TRAINING MANUAL,
...TRACING THE EXPLOITS OF THAT UNBEATABLE COMBINATION
OF ENGLISH SOLIDITY AND EASTERN BRILLIANCE,
THAT DYNAMIC DUO, DETECTIVES HARDY AND RAMANUJAN.
THIS IS MEANT TO BE A GRAD-LEVEL CLASS FOR MODERN MATHEMATICAL FORENSICS?
HAVEN'T YOU ANYTHING MORE UP-TO-DATE AND CHALLENGING, A LITTLE MORE ABSTRACT?

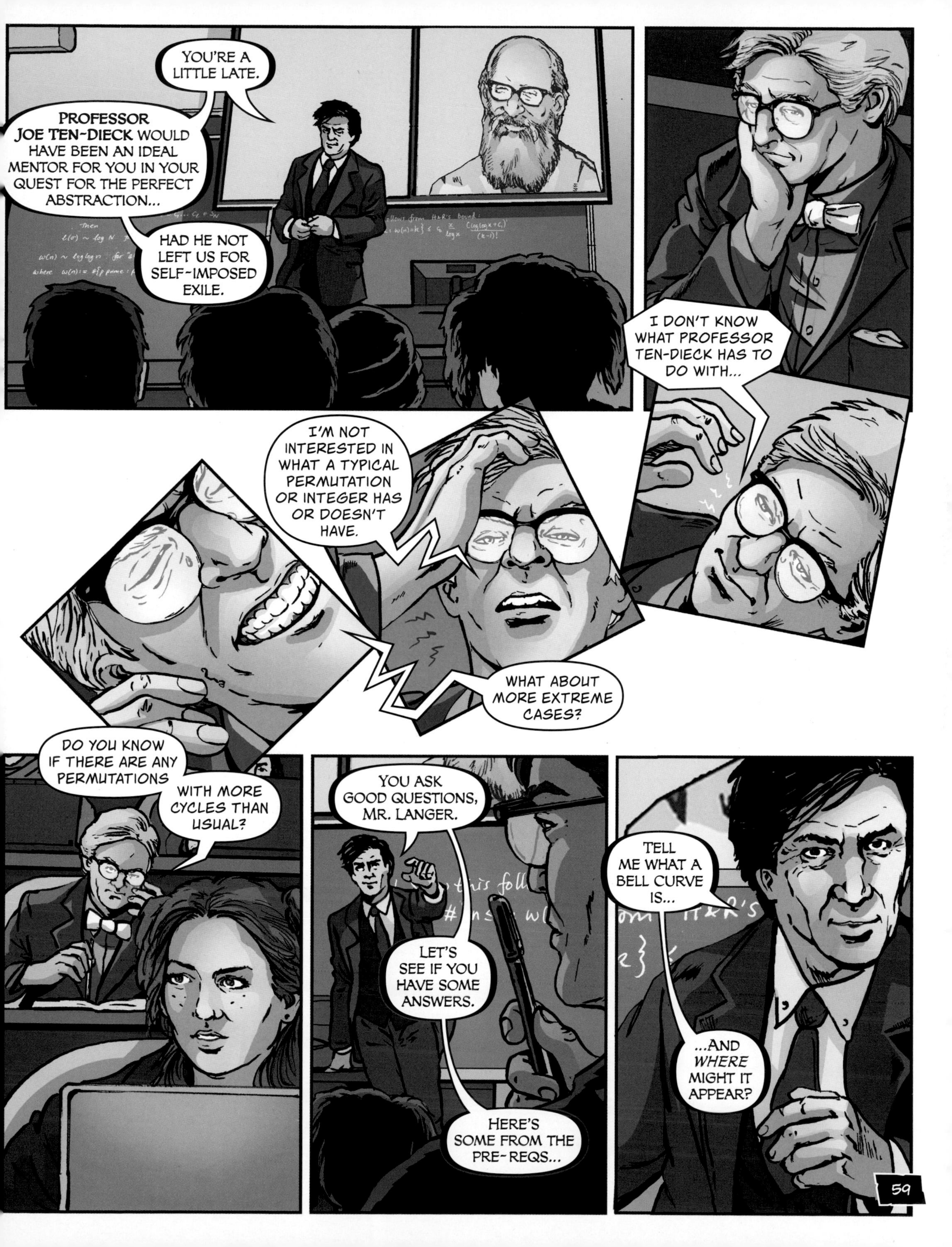
YOU'RE A LITTLE LATE.
PROFESSOR JOE TEN-DIECK WOULD HAVE BEEN AN IDEAL MENTOR FOR YOU IN YOUR QUEST FOR THE PERFECT ABSTRACTION...
HAD HE NOT LEFT US FOR SELF-IMPOSED EXILE.
I DON'T KNOW WHAT PROFESSOR TEN-DIECK HAS TO DO WITH...
I'M NOT INTERESTED IN WHAT A TYPICAL PERMUTATION OR INTEGER HAS OR DOESN'T HAVE.
WHAT ABOUT MORE EXTREME CASES?
DO YOU KNOW IF THERE ARE ANY PERMUTATIONS
WITH MORE CYCLES THAN USUAL?
YOU ASK GOOD QUESTIONS, MR. LANGER.
LET'S SEE IF YOU HAVE SOME ANSWERS.
HERE'S SOME FROM THE PRE-REQS...
TELL ME WHAT A BELL CURVE IS...
...AND WHERE MIGHT IT APPEAR?
59

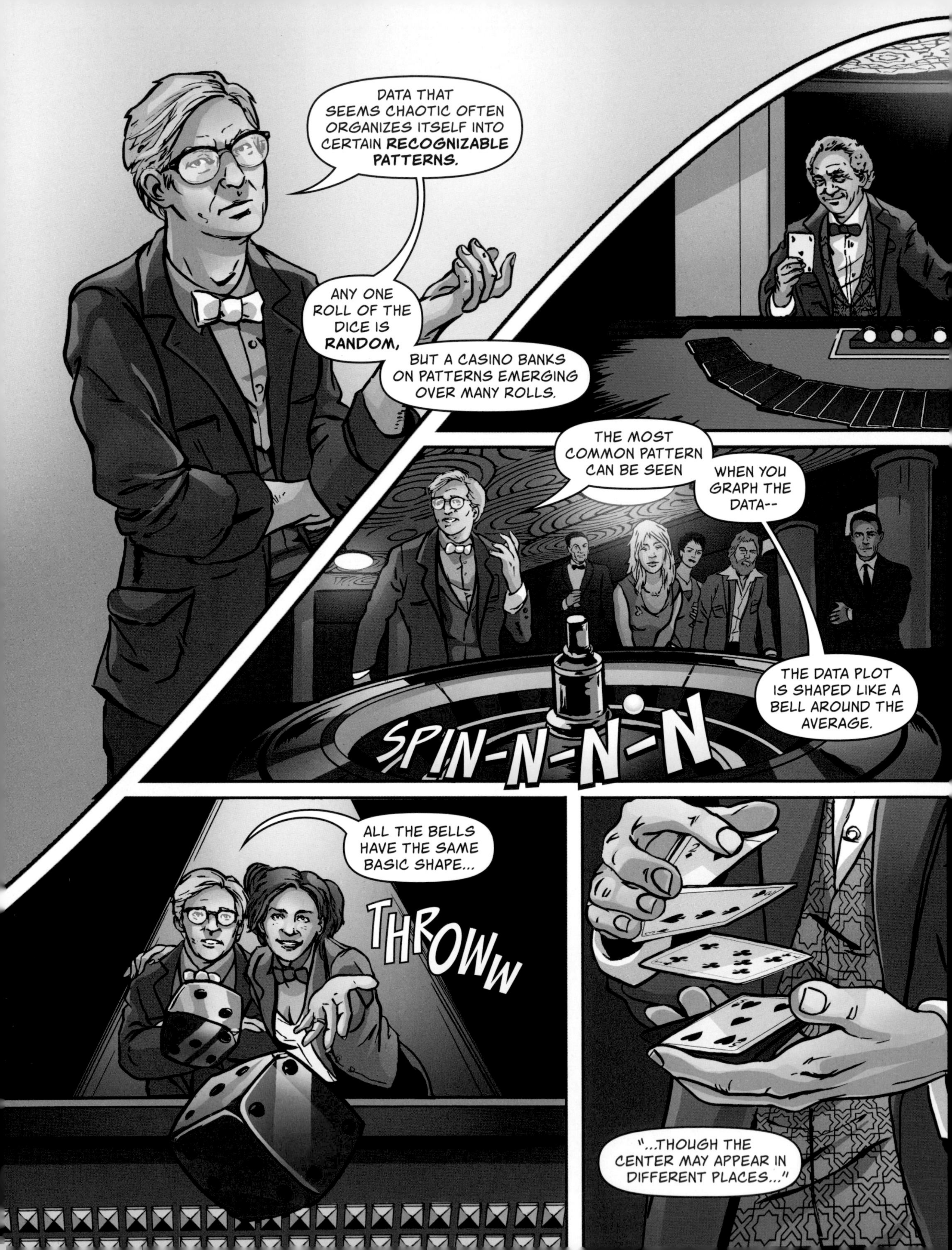
DATA THAT SEEMS CHAOTIC OFTEN ORGANIZES ITSELF INTO CERTAIN **RECOGNIZABLE PATTERNS.**
ANY ONE ROLL OF THE DICE IS **RANDOM,**
BUT A CASINO BANKS ON PATTERNS EMERGING OVER MANY ROLLS.
THE MOST COMMON PATTERN CAN BE SEEN
WHEN YOU GRAPH THE DATA--
THE DATA PLOT IS SHAPED LIKE A BELL AROUND THE AVERAGE.
SPIN-N-N-N
ALL THE BELLS HAVE THE SAME BASIC SHAPE...
THROWW
"...THOUGH THE CENTER MAY APPEAR IN DIFFERENT PLACES..."

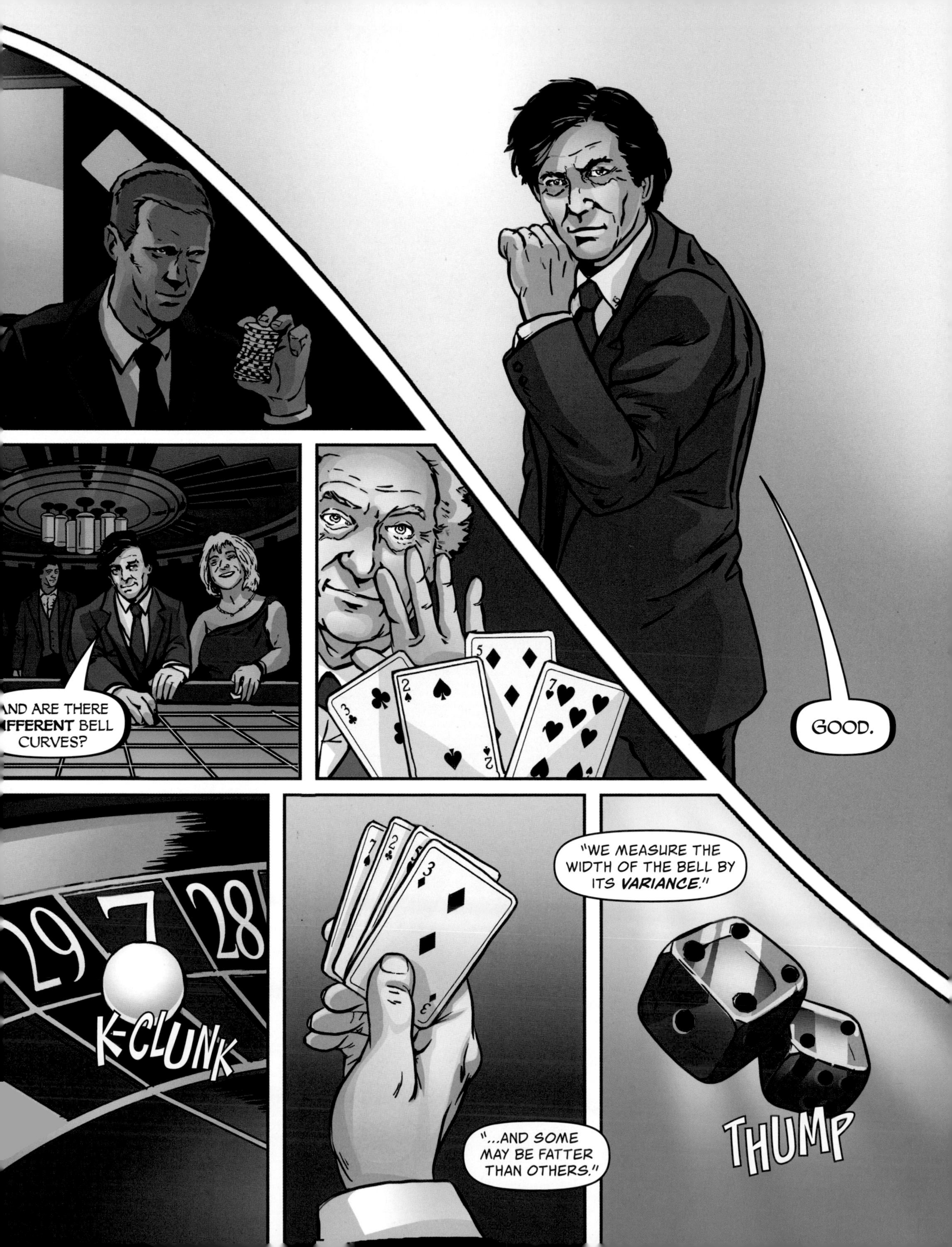

ND ARE THERE FFERENT BELL CURVES?
GOOD.
K-CLUNK
"WE MEASURE THE WIDTH OF THE BELL BY ITS VARIANCE."
THUMP
"...AND SOME MAY BE FATTER THAN OTHERS."

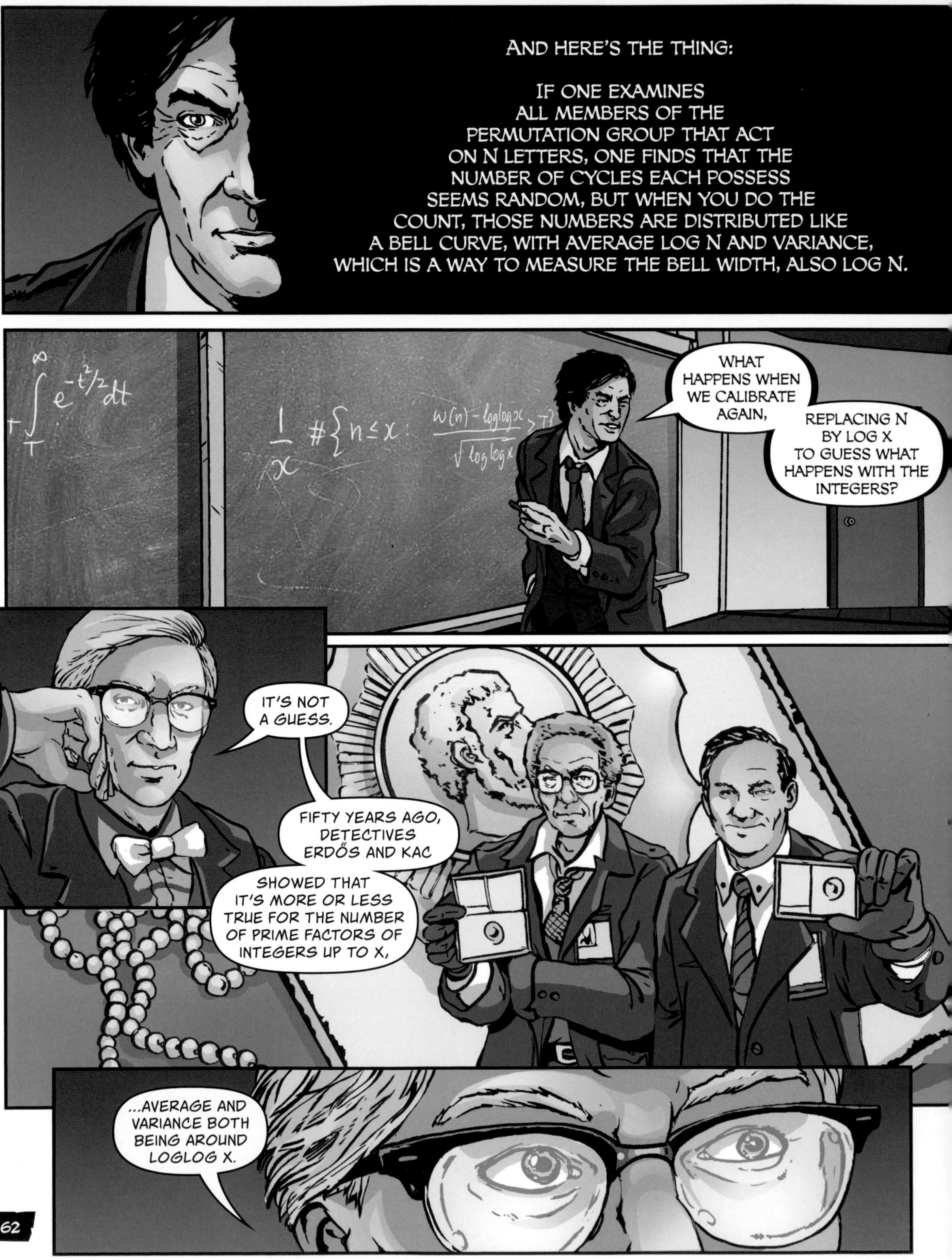
AND HERE'S THE THING:
IF ONE EXAMINES ALL MEMBERS OF THE PERMUTATION GROUP THAT ACT ON N LETTERS, ONE FINDS THAT THE NUMBER OF CYCLES EACH POSSESS SEEMS RANDOM, BUT WHEN YOU DO THE COUNT, THOSE NUMBERS ARE DISTRIBUTED LIKE A BELL CURVE, WITH AVERAGE LOG N AND VARIANCE, WHICH IS A WAY TO MEASURE THE BELL WIDTH, ALSO LOG N.
WHAT HAPPENS WHEN WE CALIBRATE AGAIN,
REPLACING N BY LOG X TO GUESS WHAT HAPPENS WITH THE INTEGERS?
IT'S NOT A GUESS.
FIFTY YEARS AGO, DETECTIVES ERDŐS AND KAC
SHOWED THAT IT'S MORE OR LESS TRUE FOR THE NUMBER OF PRIME FACTORS OF INTEGERS UP TO X,
...AVERAGE AND VARIANCE BOTH BEING AROUND LOGLOG X.

THANK YOU ONCE AGAIN FOR YOUR HELP, MR. LANGER.

I HARDLY KNOW WHERE I WOULD BE WITHOUT YOU...

DO YOU REMEMBER WHY ELSE I MENTIONED DETECTIVE KAC?

AAARGH, I'M NOT GOING TO FIND IT.

HIS REPORT, "CAN ONE HEAR THE SHAPE OF A DRUM?"

EXACTLY! SOMEONE WAS LISTENING.

ONLY TWO WEEKS AGO, I MENTIONED THAT THIS SHOULD BE REQUIRED READING FOR ALL FIRST-YEAR FORENSIC MATH STUDENTS...

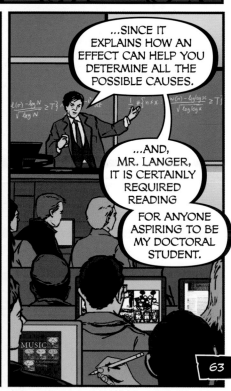

...SINCE IT EXPLAINS HOW AN EFFECT CAN HELP YOU DETERMINE ALL THE POSSIBLE CAUSES.

...AND, MR. LANGER, IT IS CERTAINLY REQUIRED READING FOR ANYONE ASPIRING TO BE MY DOCTORAL STUDENT.

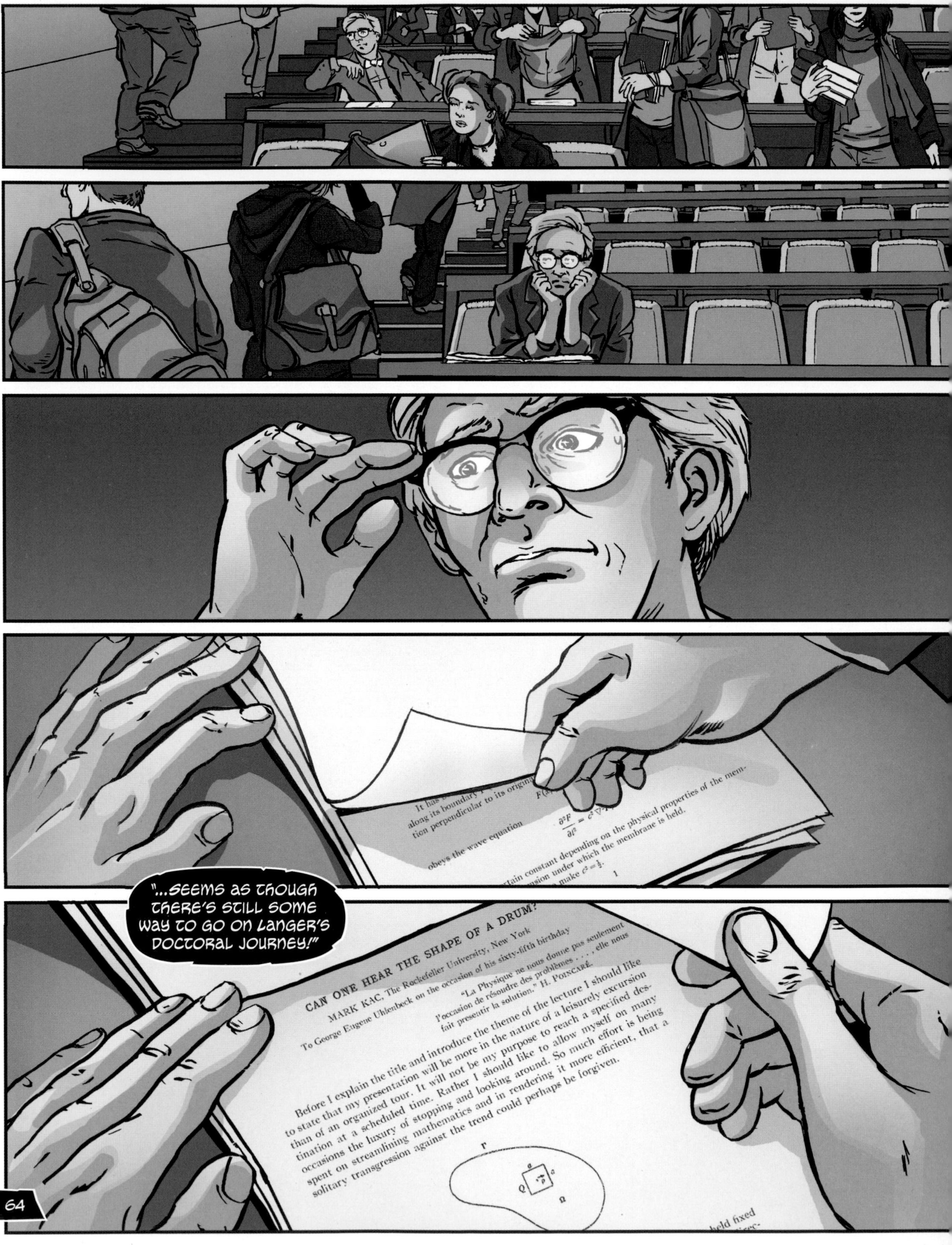
"...SEEMS AS THOUGH THERE'S STILL SOME WAY TO GO ON LANGER'S DOCTORAL JOURNEY!"
CAN ONE HEAR THE SHAPE OF A DRUM?
MARK KAC, The Rockefeller University, New York
To George Eugene Uhlenbeck on the occasion of his sixty-fifth birthday
"La Physique ne nous donne pas seulement l'occasion de résoudre des problèmes . . . , elle nous fait preseutir la solution." H. POINCARÉ.
Before I explain the title and introduce the theme of the lecture I should like to state that my presentation will be more in the nature of a leisurely excursion than of an organized tour. It will not be my purpose to reach a specified destination at a scheduled time. Rather I should like to allow myself on many occasions the luxury of stopping and looking around. So much effort is being spent on streamlining mathematics and in rendering it more efficient, that a solitary transgression against the trend could perhaps be forgiven.

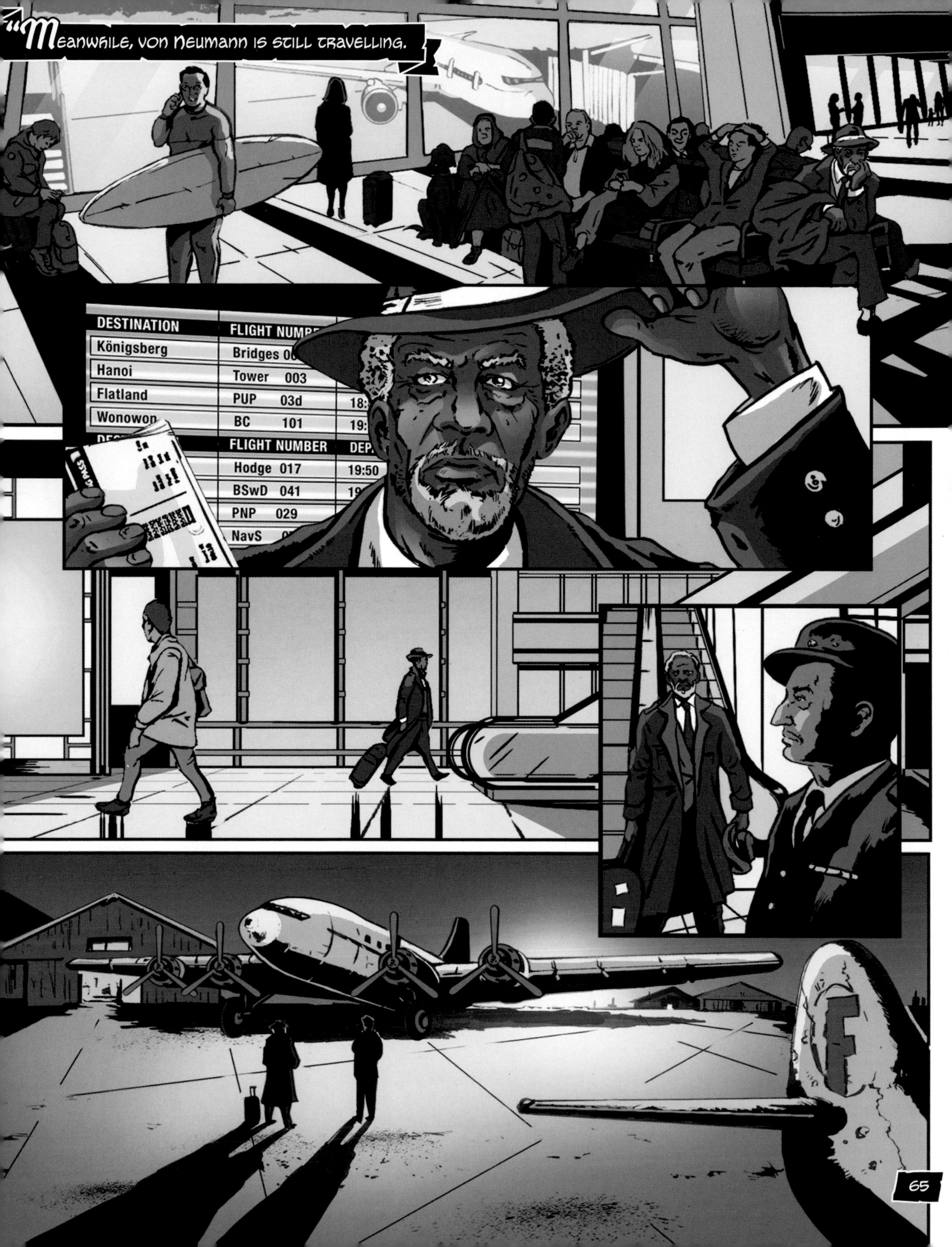
"Meanwhile, von Neumann is still travelling.
DESTINATION
FLIGHT NUMBER
Königsberg
Bridges
Hanoi
Tower 003
Flatland
PUP 03d
Wonowon
BC 101
FLIGHT NUMBER
Hodge 017
19:50
BSwD 041
PNP 029
NavS

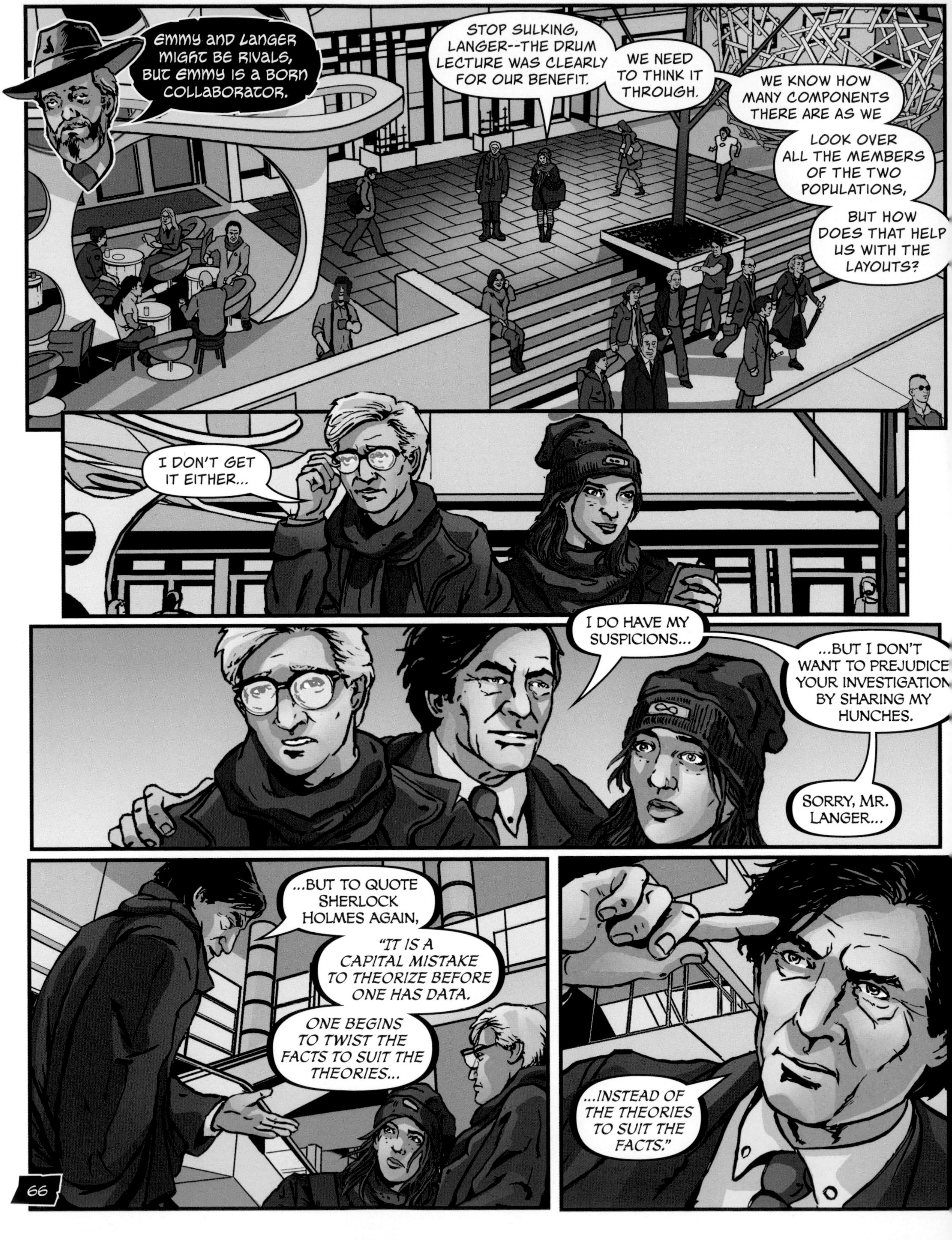
EMMY AND LANGER MIGHT BE RIVALS, BUT EMMY IS A BORN COLLABORATOR.
STOP SULKING, LANGER--THE DRUM LECTURE WAS CLEARLY FOR OUR BENEFIT.
WE NEED TO THINK IT THROUGH.
WE KNOW HOW MANY COMPONENTS THERE ARE AS WE
LOOK OVER ALL THE MEMBERS OF THE TWO POPULATIONS,
BUT HOW DOES THAT HELP US WITH THE LAYOUTS?
I DON'T GET IT EITHER...
I DO HAVE MY SUSPICIONS...
...BUT I DON'T WANT TO PREJUDICE YOUR INVESTIGATION BY SHARING MY HUNCHES.
SORRY, MR. LANGER...
...BUT TO QUOTE SHERLOCK HOLMES AGAIN,
"IT IS A CAPITAL MISTAKE TO THEORIZE BEFORE ONE HAS DATA.
ONE BEGINS TO TWIST THE FACTS TO SUIT THE THEORIES...
...INSTEAD OF THE THEORIES TO SUIT THE FACTS."

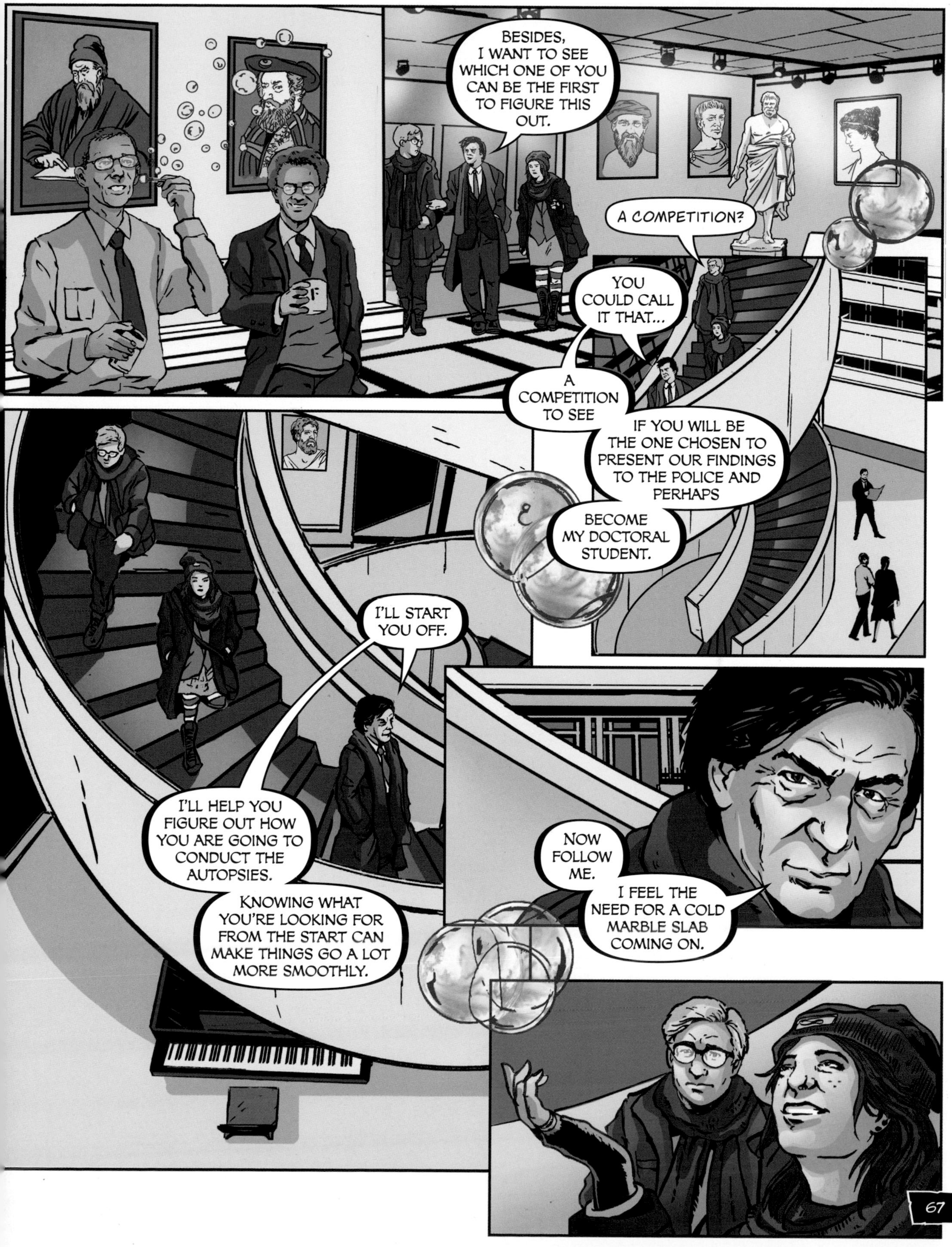
BESIDES, I WANT TO SEE WHICH ONE OF YOU CAN BE THE FIRST TO FIGURE THIS OUT.
A COMPETITION?
YOU COULD CALL IT THAT...
A COMPETITION TO SEE
IF YOU WILL BE THE ONE CHOSEN TO PRESENT OUR FINDINGS TO THE POLICE AND PERHAPS
BECOME MY DOCTORAL STUDENT.
I'LL START YOU OFF.
I'LL HELP YOU FIGURE OUT HOW YOU ARE GOING TO CONDUCT THE AUTOPSIES.
KNOWING WHAT YOU'RE LOOKING FOR FROM THE START CAN MAKE THINGS GO A LOT MORE SMOOTHLY.
NOW FOLLOW ME.
I FEEL THE NEED FOR A COLD MARBLE SLAB COMING ON.

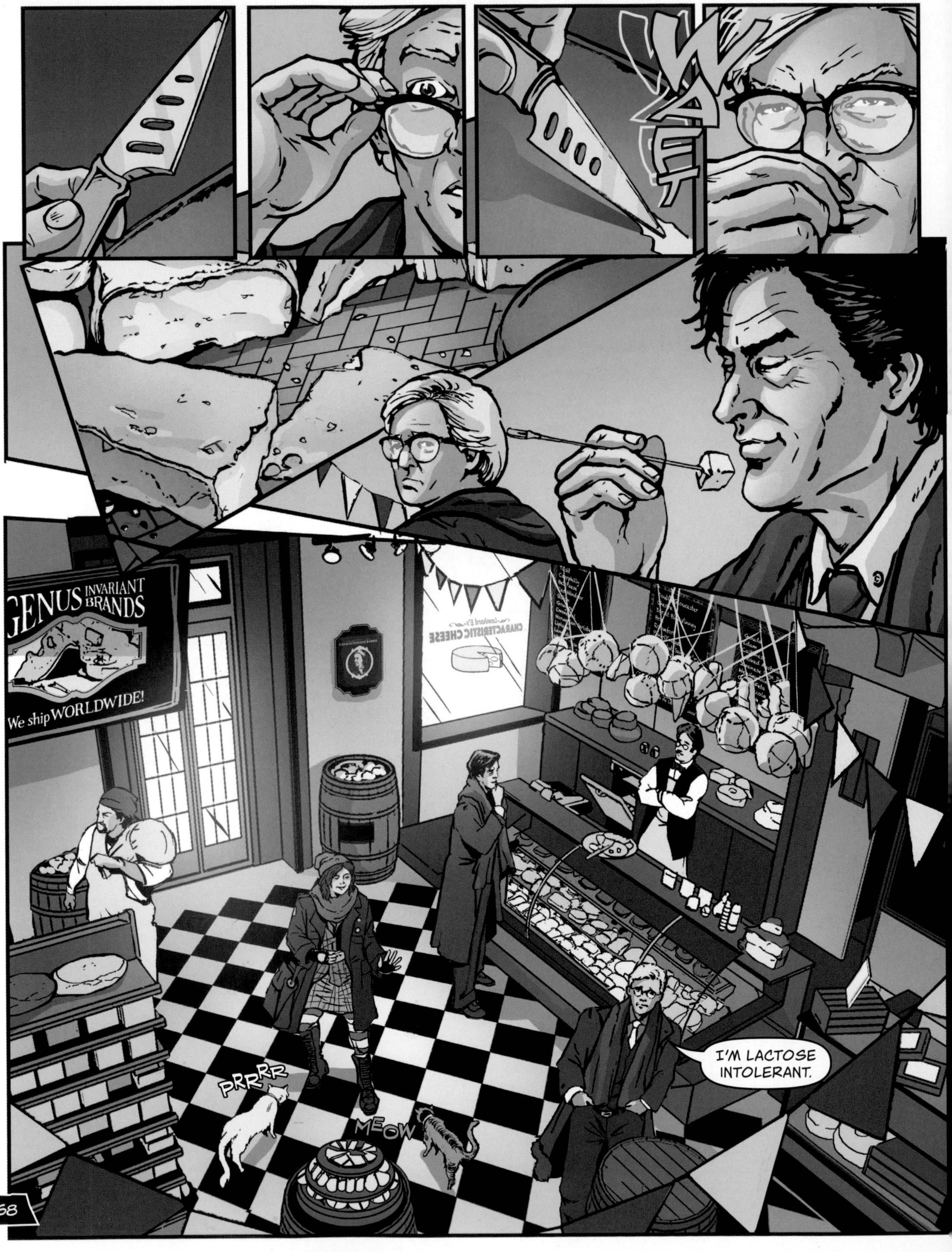
WAFT
GENUS INVARIANT BRANDS
We ship WORLDWIDE!
I'M LACTOSE INTOLERANT.
PRRRR
MEOW

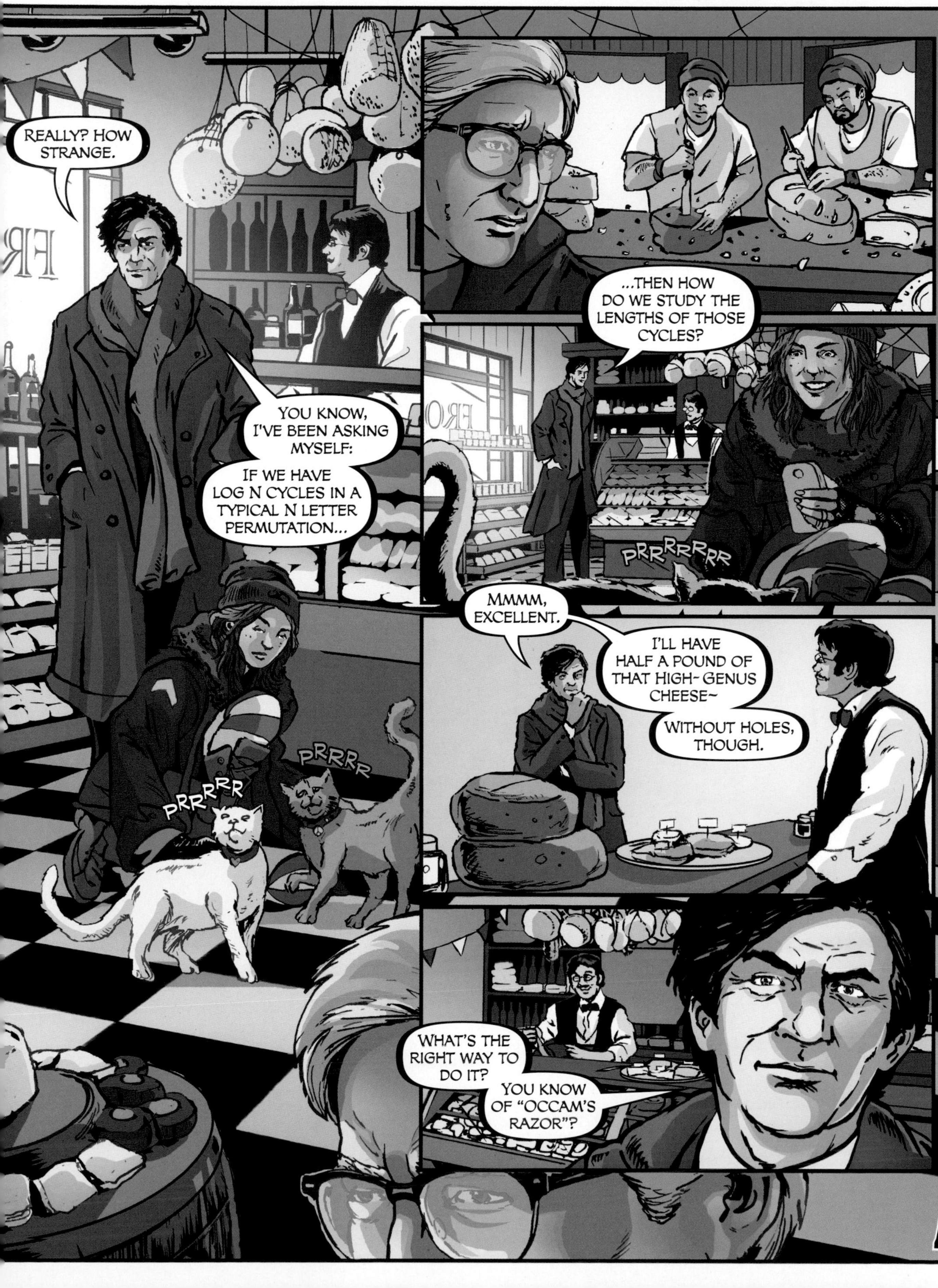
REALLY? HOW STRANGE.
YOU KNOW, I'VE BEEN ASKING MYSELF:
IF WE HAVE LOG N CYCLES IN A TYPICAL N LETTER PERMUTATION...
PRRRR
PRRRRR
...THEN HOW DO WE STUDY THE LENGTHS OF THOSE CYCLES?
PRRRRRRR
MMMM, EXCELLENT.
I'LL HAVE HALF A POUND OF THAT HIGH-GENUS CHEESE-
WITHOUT HOLES, THOUGH.
WHAT'S THE RIGHT WAY TO DO IT?
YOU KNOW OF "OCCAM'S RAZOR"?

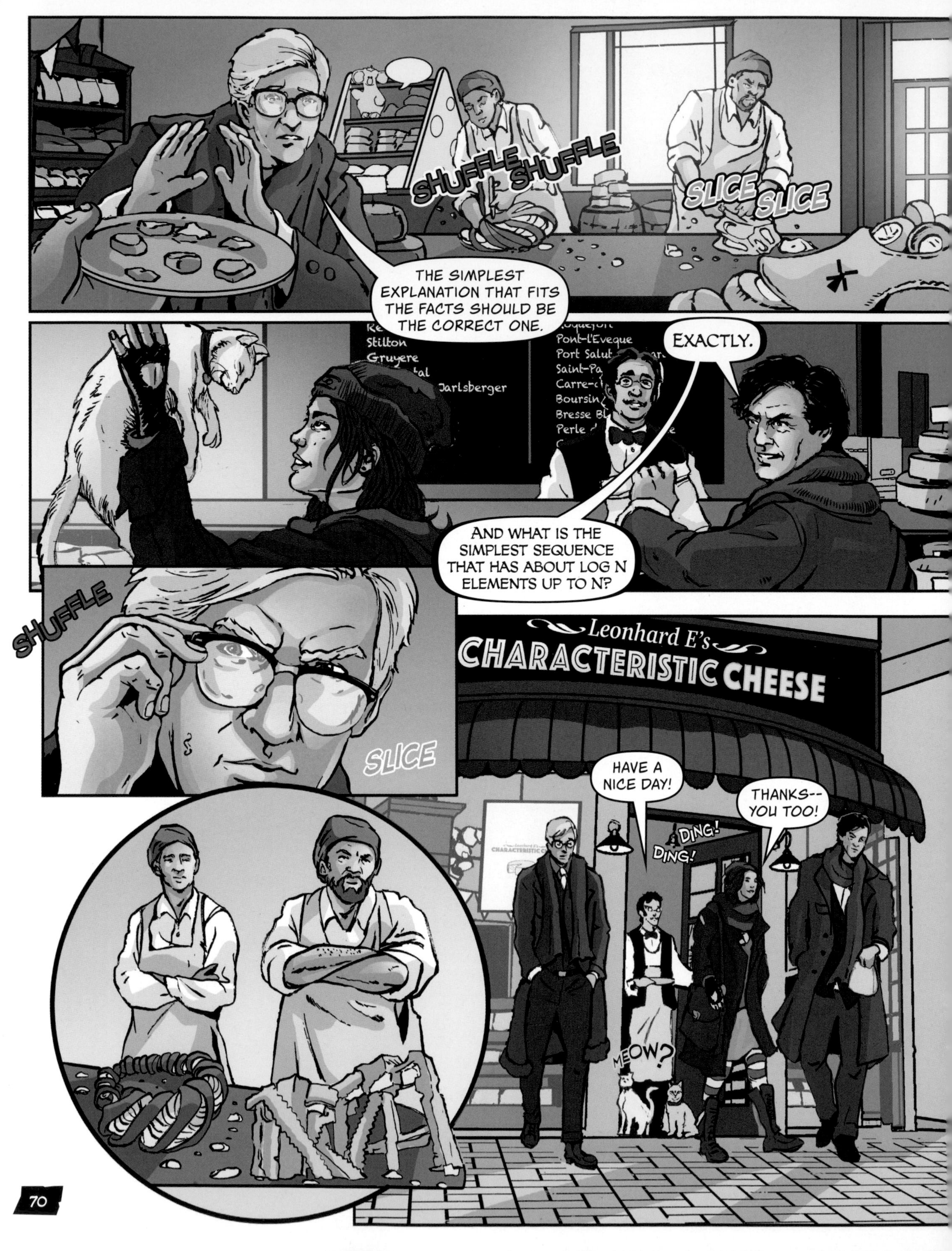
SHUFFLE SHUFFLE
SLICE SLICE
THE SIMPLEST EXPLANATION THAT FITS THE FACTS SHOULD BE THE CORRECT ONE.
Stilton
Gruyere
Jarlsberger
Pont-l'Eveque
Port Salut
Boursin
EXACTLY.
AND WHAT IS THE SIMPLEST SEQUENCE THAT HAS ABOUT LOG N ELEMENTS UP TO N?
SHUFFLE
SLICE
Leonhard E's
CHARACTERISTIC CHEESE
HAVE A NICE DAY!
THANKS-- YOU TOO!
DING! DING!
MEOW?

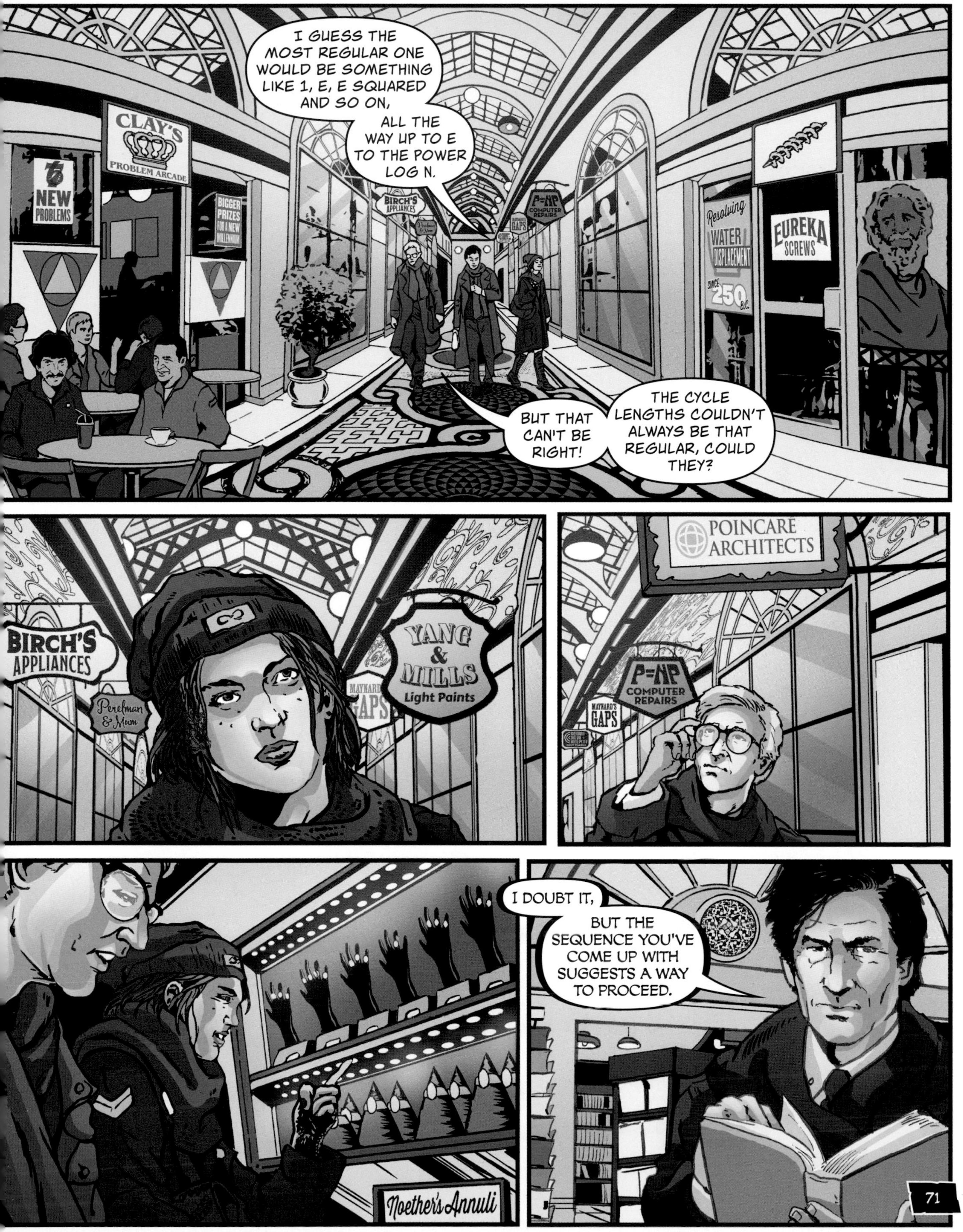
I GUESS THE MOST REGULAR ONE WOULD BE SOMETHING LIKE 1, E, E SQUARED AND SO ON,
ALL THE WAY UP TO E TO THE POWER LOG N.
CLAY'S PROBLEM ARCADE
7 NEW PROBLEMS
BIGGER PRIZES FOR A NEW MILLENNIUM
BIRCH'S APPLIANCES
P=NP COMPUTER REPAIRS
Resolving WATER DISPLACEMENT SINCE 250 B.C.
EUREKA SCREWS
BUT THAT CAN'T BE RIGHT!
THE CYCLE LENGTHS COULDN'T ALWAYS BE THAT REGULAR, COULD THEY?
BIRCH'S APPLIANCES
Perelman & Mum
MAYNARD'S GAPS
YANG & MILLS Light Paints
POINCARE ARCHITECTS
P=NP COMPUTER REPAIRS
MAYNARD'S GAPS
Noether's Annuli
I DOUBT IT,
BUT THE SEQUENCE YOU'VE COME UP WITH SUGGESTS A WAY TO PROCEED.

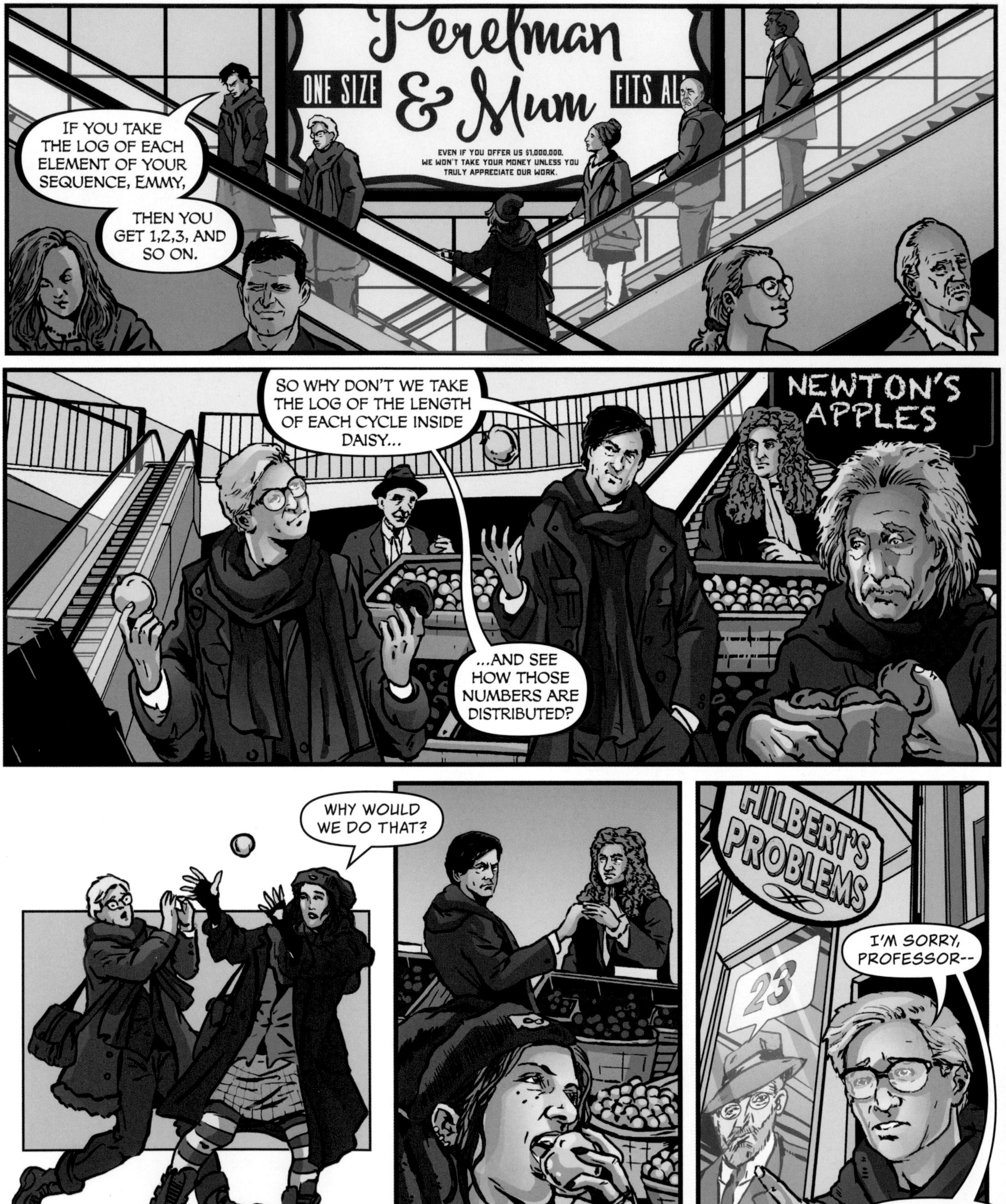
Perelman & Mum
ONE SIZE
FITS AL
EVEN IF YOU OFFER US $1,000,000, WE WON'T TAKE YOUR MONEY UNLESS YOU TRULY APPRECIATE OUR WORK.
IF YOU TAKE THE LOG OF EACH ELEMENT OF YOUR SEQUENCE, EMMY,
THEN YOU GET 1,2,3, AND SO ON.
NEWTON'S APPLES
SO WHY DON'T WE TAKE THE LOG OF THE LENGTH OF EACH CYCLE INSIDE DAISY...
...AND SEE HOW THOSE NUMBERS ARE DISTRIBUTED?
WHY WOULD WE DO THAT?
HILBERT'S PROBLEMS
23
I'M SORRY, PROFESSOR--
I CAN'T RESIST A CHALLENGE...
AN YOU
SOLVE

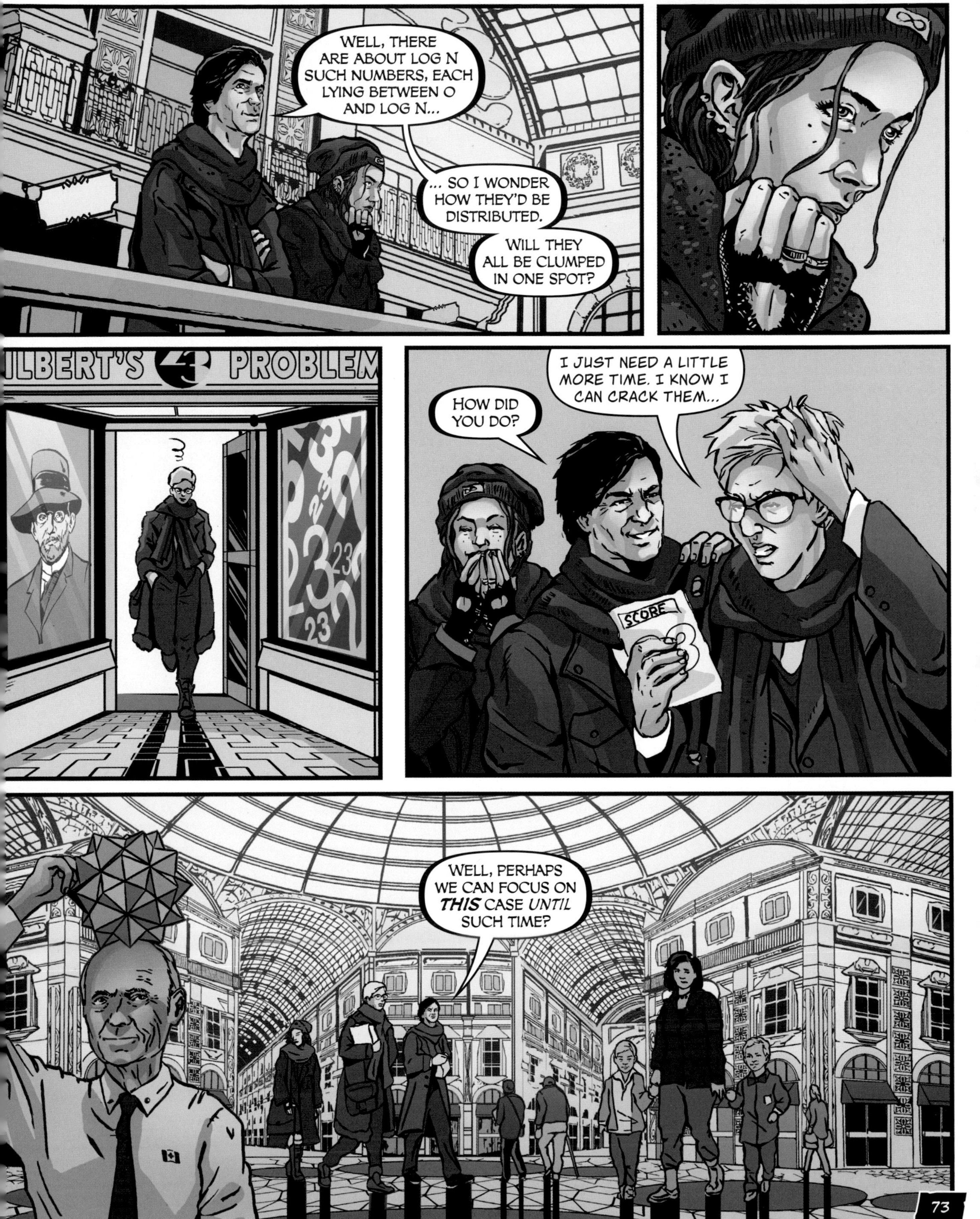
WELL, THERE ARE ABOUT LOG N SUCH NUMBERS, EACH LYING BETWEEN O AND LOG N...
... SO I WONDER HOW THEY'D BE DISTRIBUTED.
WILL THEY ALL BE CLUMPED IN ONE SPOT?
ILBERT'S 23 PROBLEM
HOW DID YOU DO?
I JUST NEED A LITTLE MORE TIME. I KNOW I CAN CRACK THEM...
SCORE
WELL, PERHAPS WE CAN FOCUS ON *THIS* CASE *UNTIL* SUCH TIME?

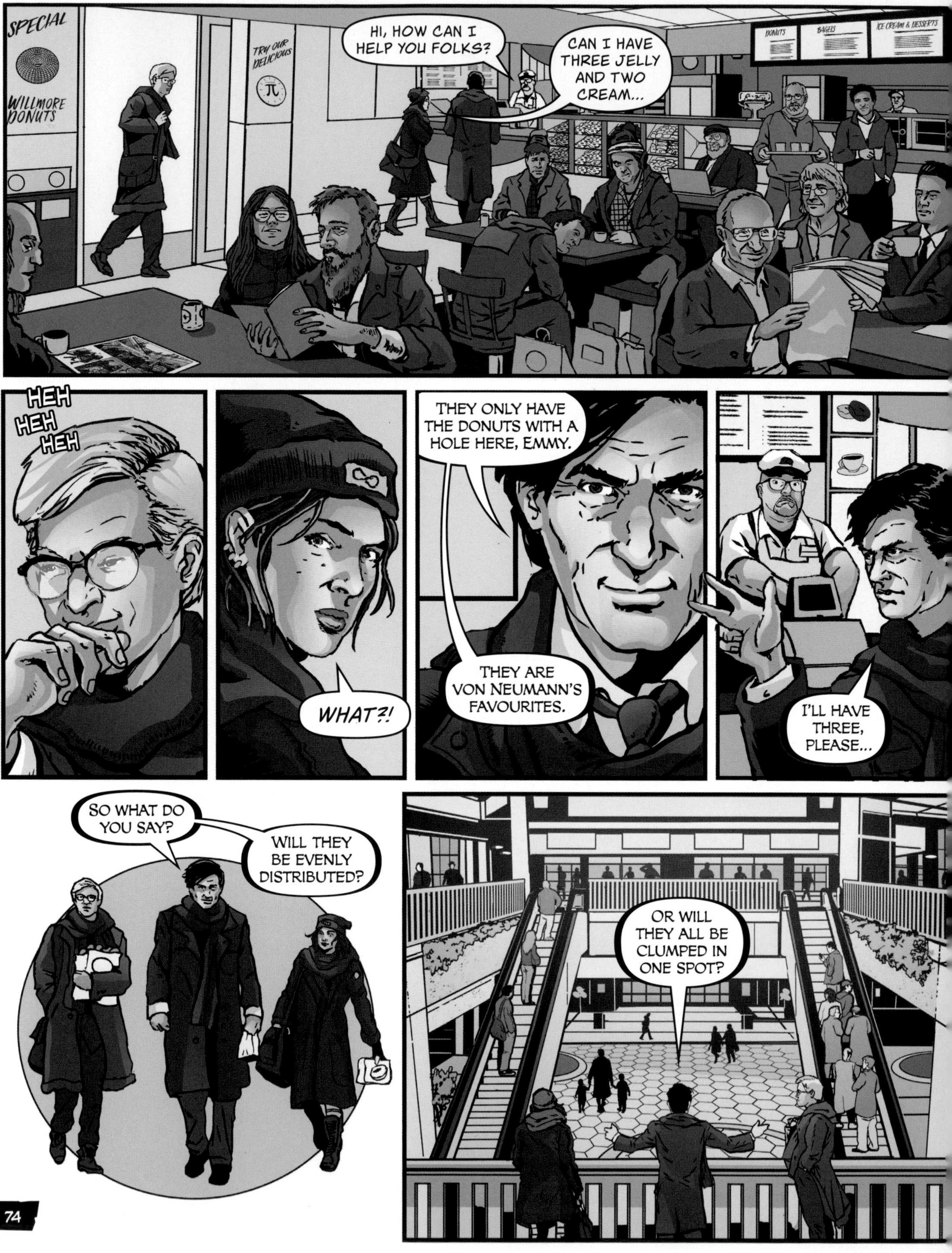
SPECIAL
WILLMORE DONUTS
TRY OUR DELICIOUS
HI, HOW CAN I HELP YOU FOLKS?
CAN I HAVE THREE JELLY AND TWO CREAM...
DONUTS
BAGELS
ICE CREAM & DESSERTS
HEH HEH HEH
WHAT?!
THEY ONLY HAVE THE DONUTS WITH A HOLE HERE, EMMY.
THEY ARE VON NEUMANN'S FAVOURITES.
I'LL HAVE THREE, PLEASE...
SO WHAT DO YOU SAY?
WILL THEY BE EVENLY DISTRIBUTED?
OR WILL THEY ALL BE CLUMPED IN ONE SPOT?

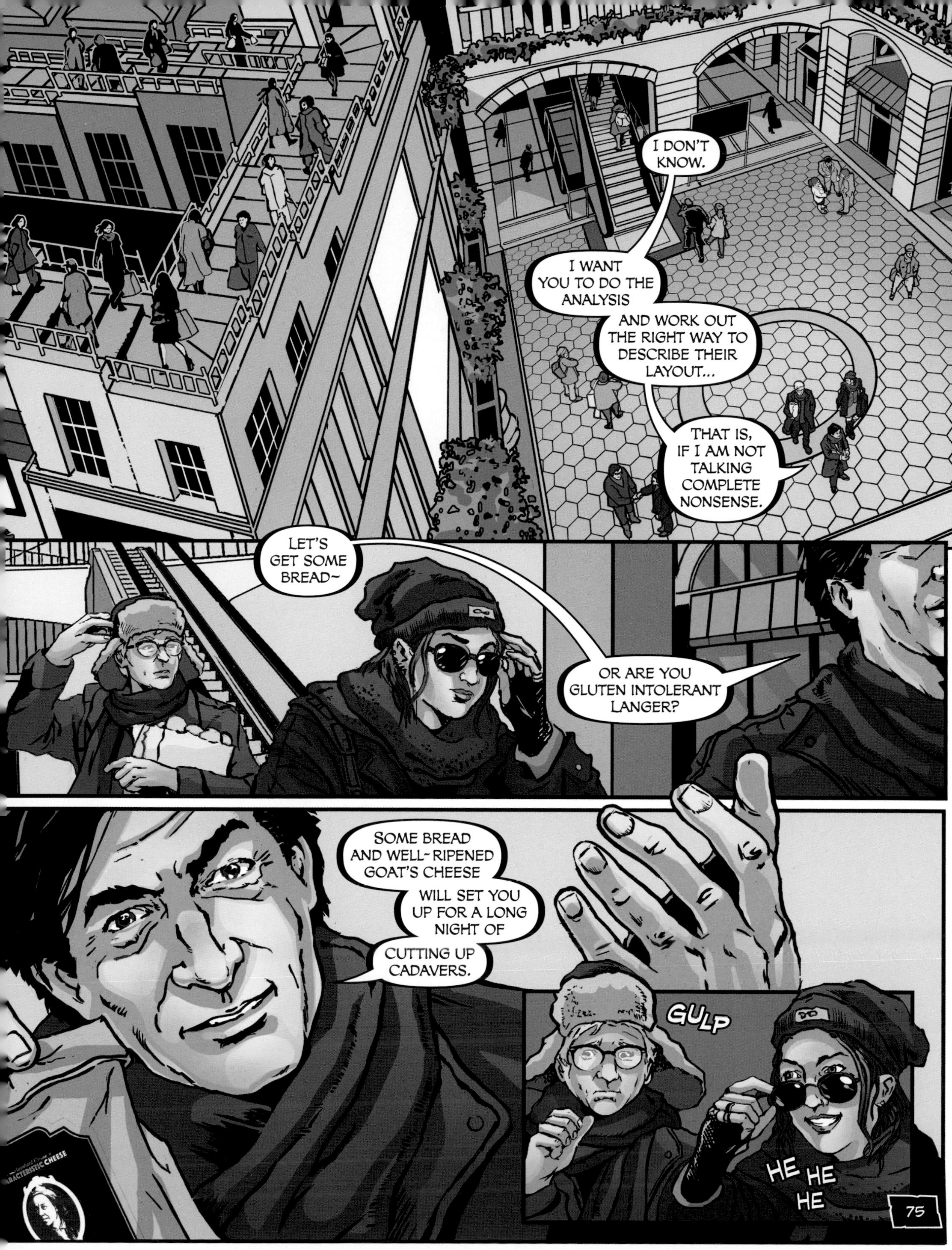
I DON'T KNOW.
I WANT YOU TO DO THE ANALYSIS
AND WORK OUT THE RIGHT WAY TO DESCRIBE THEIR LAYOUT...
THAT IS, IF I AM NOT TALKING COMPLETE NONSENSE.
LET'S GET SOME BREAD~
OR ARE YOU GLUTEN INTOLERANT LANGER?
SOME BREAD AND WELL-RIPENED GOAT'S CHEESE
WILL SET YOU UP FOR A LONG NIGHT OF
CUTTING UP CADAVERS.
GULP
HE HE HE HE

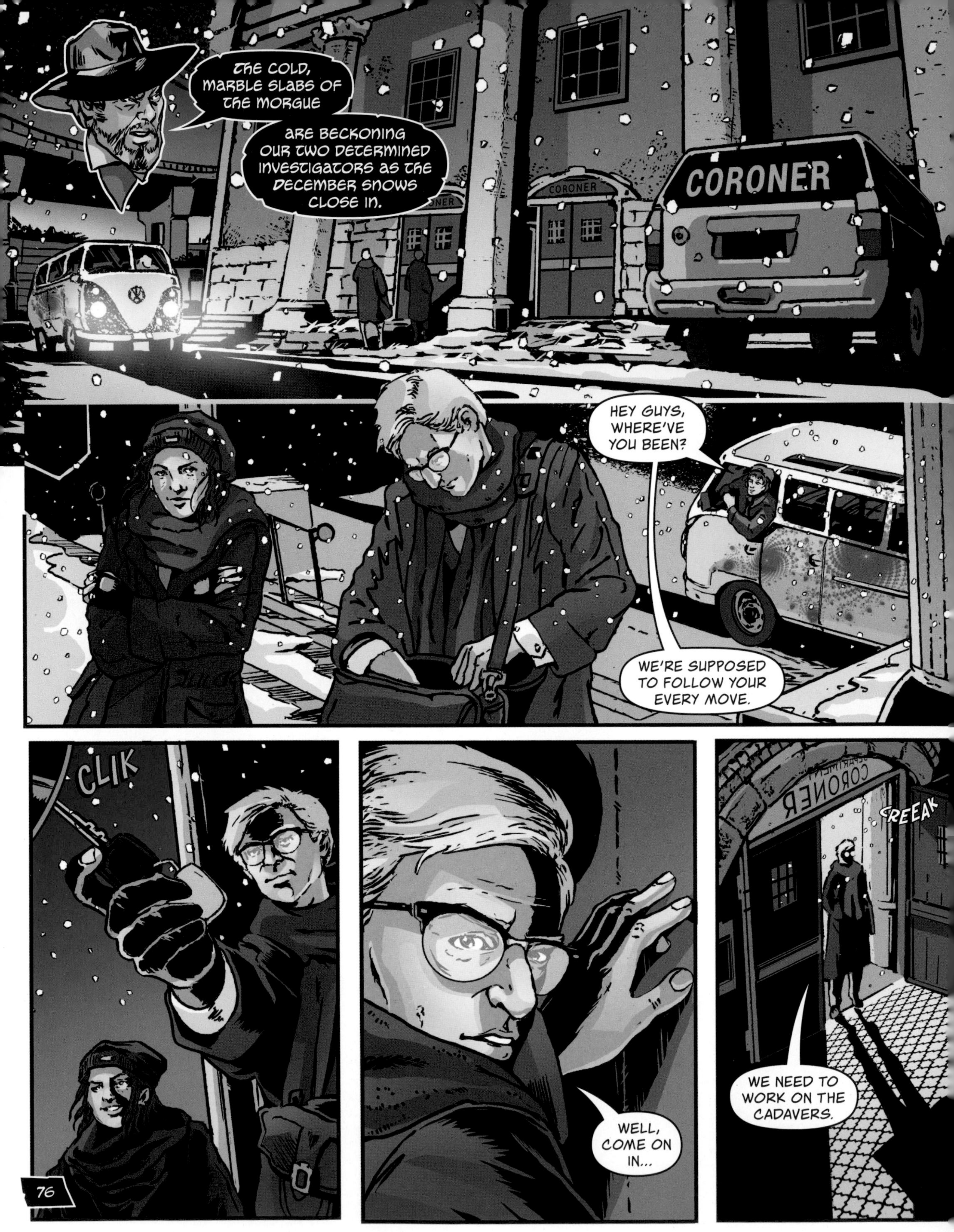
THE COLD, MARBLE SLABS OF THE MORGUE
ARE BECKONING OUR TWO DETERMINED INVESTIGATORS AS THE DECEMBER SNOWS CLOSE IN.
CORONER
CORONER
HEY GUYS, WHERE'VE YOU BEEN?
WE'RE SUPPOSED TO FOLLOW YOUR EVERY MOVE.
CLIK
WELL, COME ON IN...
CORONER
CREEAK
WE NEED TO WORK ON THE CADAVERS.

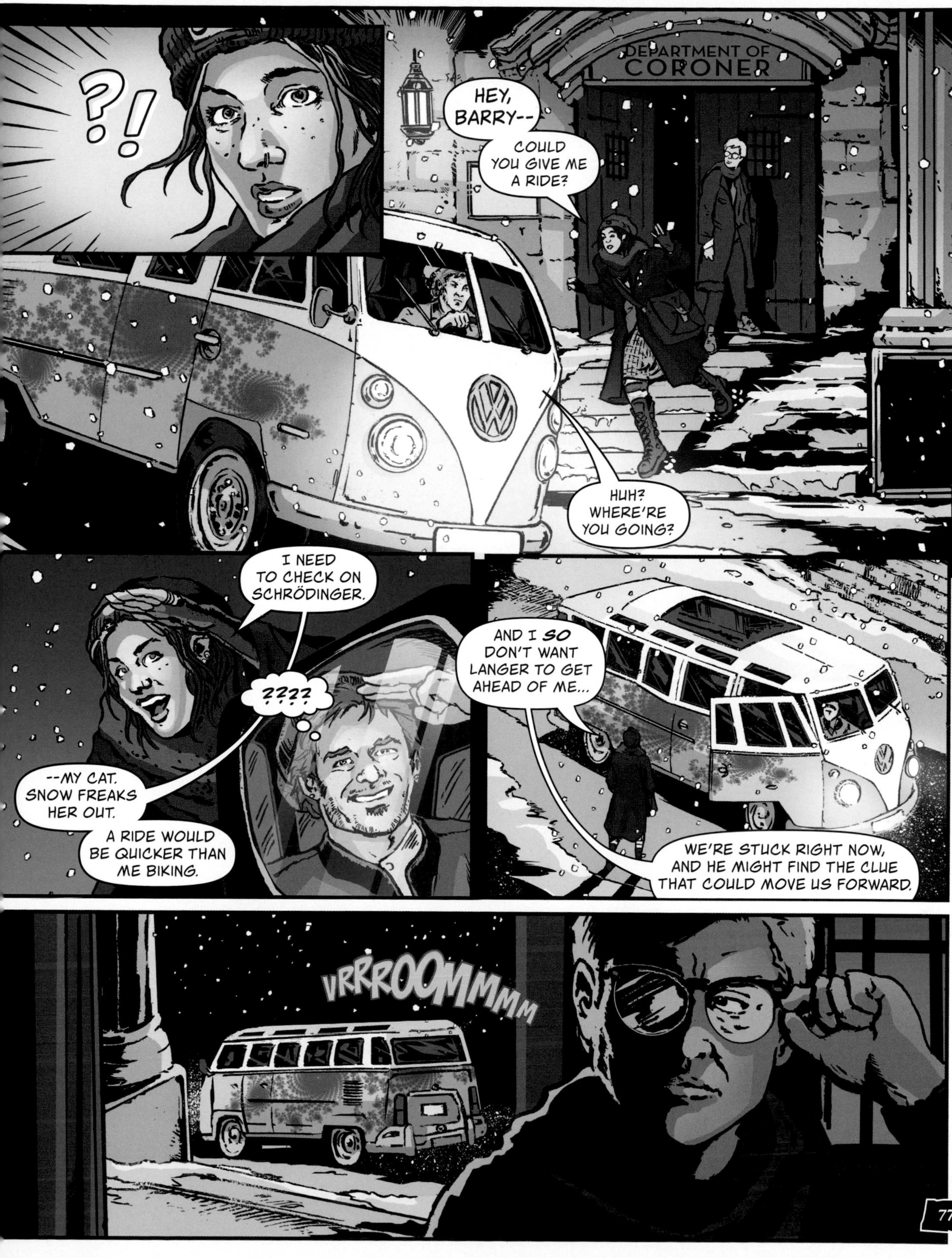
?!
DEPARTMENT OF CORONER
HEY, BARRY--
COULD YOU GIVE ME A RIDE?
HUH? WHERE'RE YOU GOING?
I NEED TO CHECK ON SCHRÖDINGER.
????
--MY CAT. SNOW FREAKS HER OUT.
A RIDE WOULD BE QUICKER THAN ME BIKING.
AND I SO DON'T WANT LANGER TO GET AHEAD OF ME...
WE'RE STUCK RIGHT NOW, AND HE MIGHT FIND THE CLUE THAT COULD MOVE US FORWARD.
VRRROOMMMM

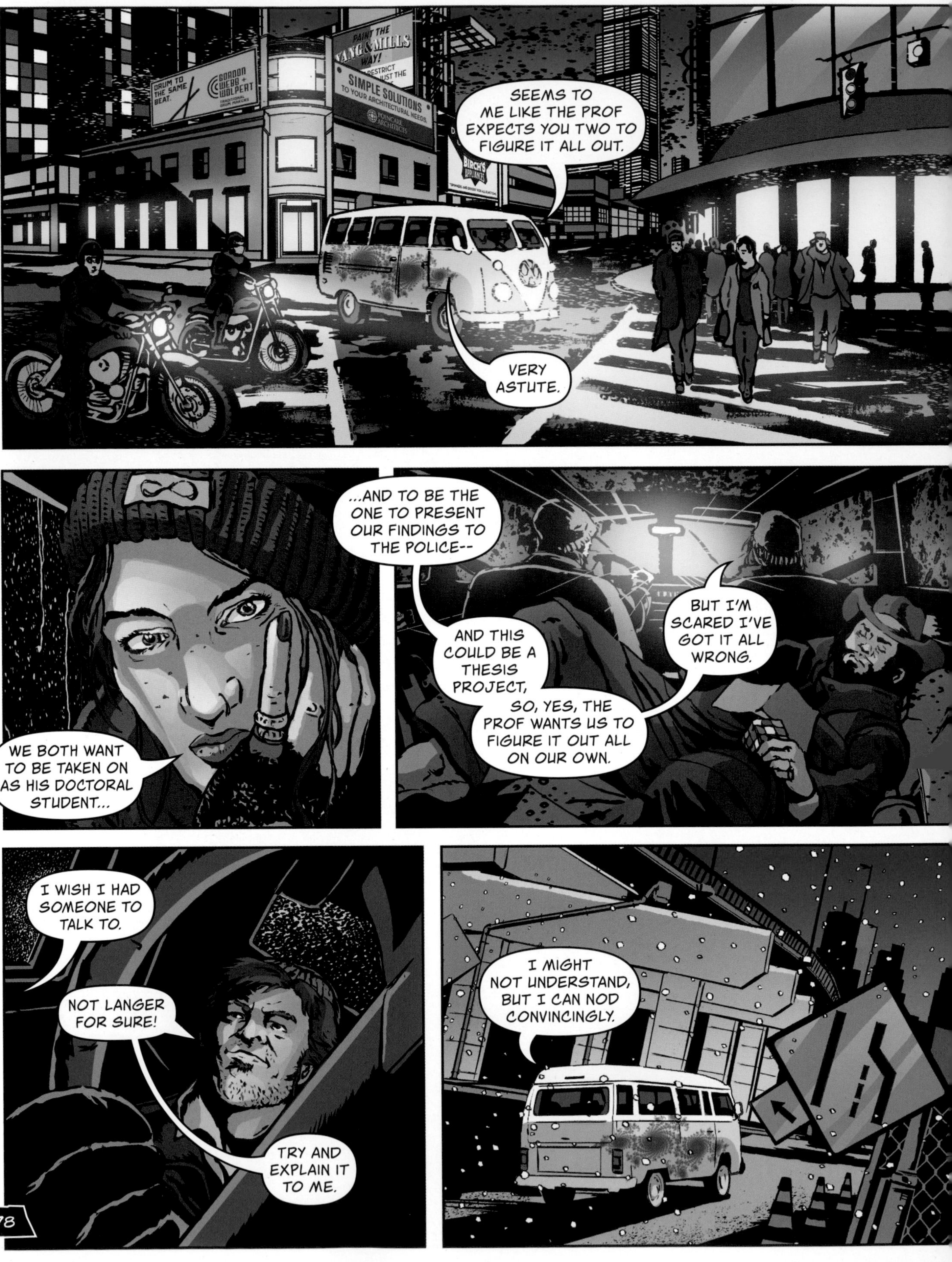
DRUM TO THE SAME BEAT.
GORDON WEBB + WOLPERT
TRADITIONAL DRUM MAKERS
PAINT THE YANG & MILLS WAY!
SIMPLE SOLUTIONS TO YOUR ARCHITECTURAL NEEDS.
POINCARE ARCHITECTS
BIRCH'S APPLIANCES
SEEMS TO ME LIKE THE PROF EXPECTS YOU TWO TO FIGURE IT ALL OUT.
VERY ASTUTE.
WE BOTH WANT TO BE TAKEN ON AS HIS DOCTORAL STUDENT...
...AND TO BE THE ONE TO PRESENT OUR FINDINGS TO THE POLICE--
AND THIS COULD BE A THESIS PROJECT,
SO, YES, THE PROF WANTS US TO FIGURE IT OUT ALL ON OUR OWN.
BUT I'M SCARED I'VE GOT IT ALL WRONG.
I WISH I HAD SOMEONE TO TALK TO.
NOT LANGER FOR SURE!
TRY AND EXPLAIN IT TO ME.
I MIGHT NOT UNDERSTAND, BUT I CAN NOD CONVINCINGLY.

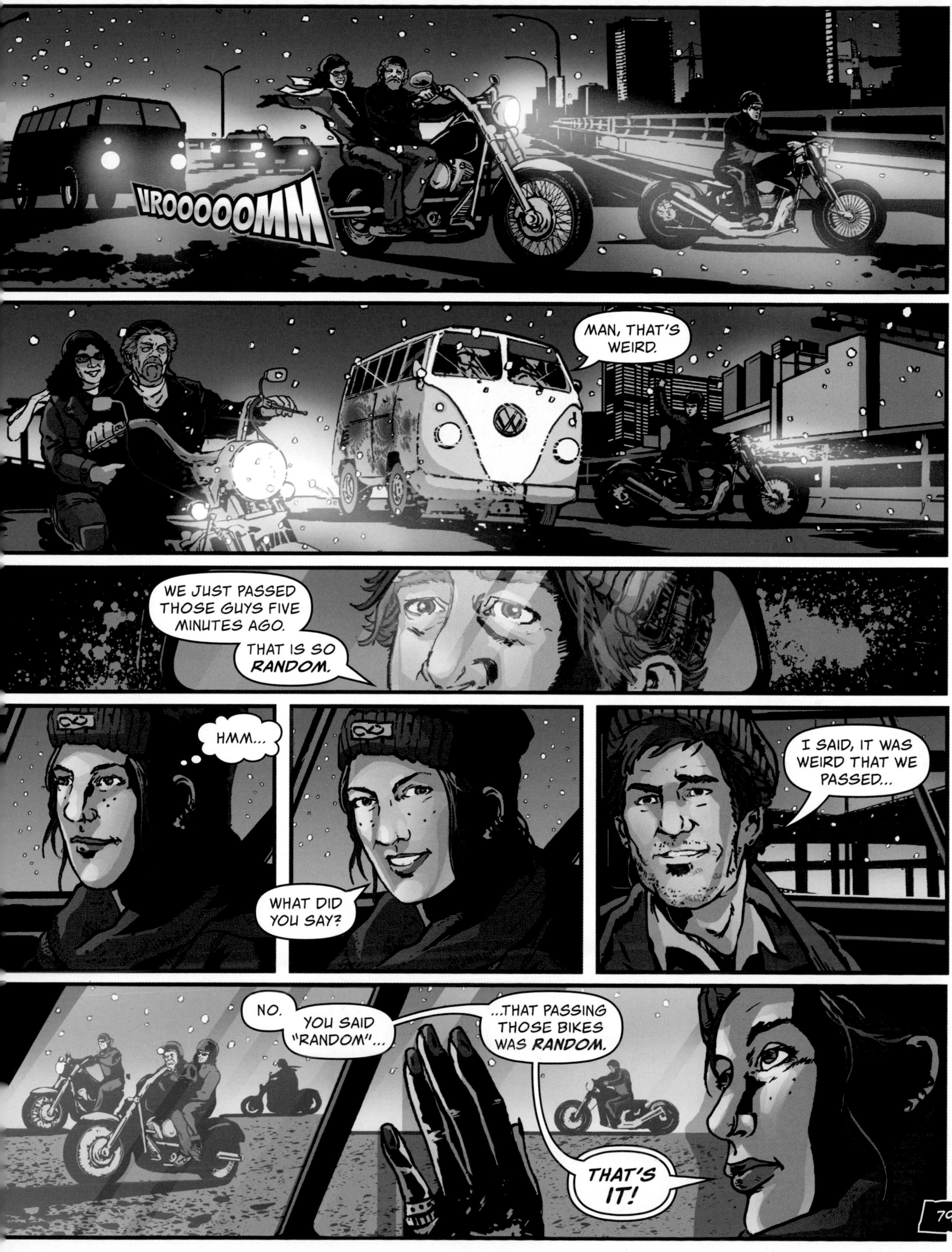

VROOOOOOMM
MAN, THAT'S WEIRD.
WE JUST PASSED THOSE GUYS FIVE MINUTES AGO.
THAT IS SO RANDOM.
HMM...
WHAT DID YOU SAY?
I SAID, IT WAS WEIRD THAT WE PASSED...
NO.
YOU SAID "RANDOM"...
...THAT PASSING THOSE BIKES WAS RANDOM.
THAT'S IT!

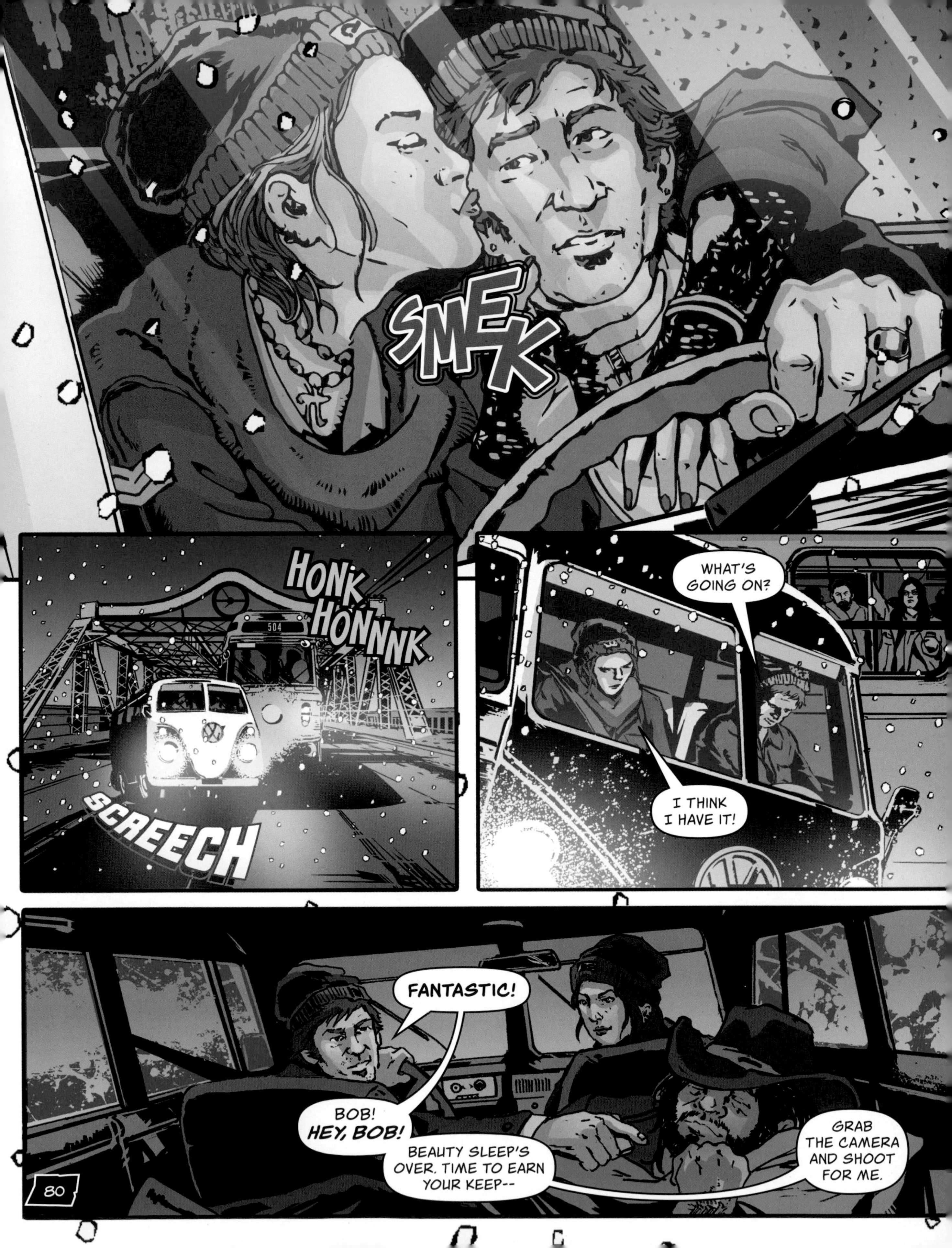
SMEK
HONK HONNNK
SCREECH
504
WHAT'S GOING ON?
I THINK I HAVE IT!
FANTASTIC!
BOB! HEY, BOB!
BEAUTY SLEEP'S OVER. TIME TO EARN YOUR KEEP--
GRAB THE CAMERA AND SHOOT FOR ME.

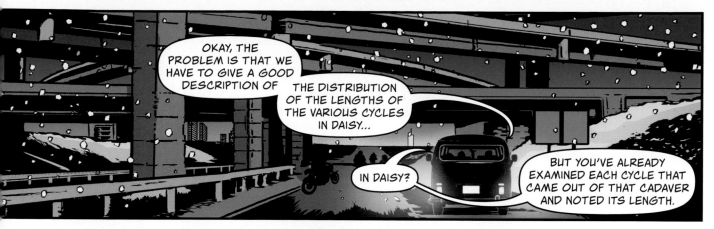

OKAY, THE PROBLEM IS THAT WE HAVE TO GIVE A GOOD DESCRIPTION OF THE DISTRIBUTION OF THE LENGTHS OF THE VARIOUS CYCLES IN DAISY...

IN DAISY?

BUT YOU'VE ALREADY EXAMINED EACH CYCLE THAT CAME OUT OF THAT CADAVER AND NOTED ITS LENGTH.

RIGHT.

SO IF WE TAKE ANY MEMBER OF DAISY PERMUTATION'S GROUP--

NOT JUST DAISY, BUT ANY OLD JANE RHO--

THE CYCLE LENGTHS ARE ALL BETWEEN 1 AND N.

THAT MEANS THAT THE LOGARITHMS OF THESE LENGTHS MUST ALL BE BETWEEN ZERO AND LOG N,

AND THERE ARE ROUGHLY LOG N OF THESE NUMBERS.

SO THE QUESTION IS TO DETERMINE HOW THEY'RE SPREAD OUT.

GOT IT SO FAR?

BUT WHAT I'M NOT GETTING IS, WHAT HAS THIS GOT TO DO WITH "RANDOM"?

PATIENCE.

WHEN WE TOOK OUR CLASS IN TRAFFIC CONTROL,

WE LEARNED THAT, AFTER A WHILE,

CARS -OR BIKES- ON A FREEWAY SPREAD OUT "RANDOMLY."

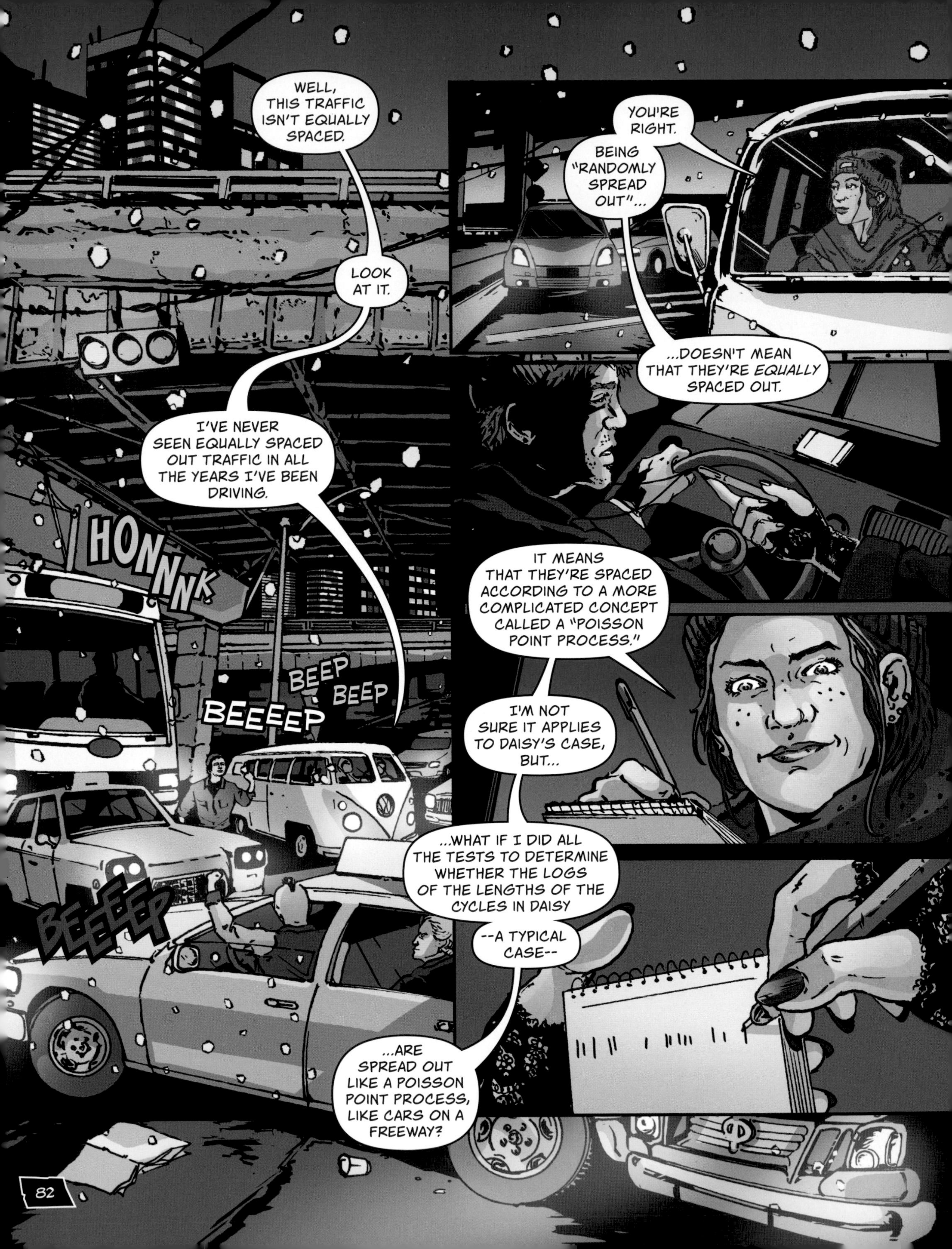
WELL, THIS TRAFFIC ISN'T EQUALLY SPACED.
LOOK AT IT.
YOU'RE RIGHT.
BEING "RANDOMLY SPREAD OUT"...
...DOESN'T MEAN THAT THEY'RE EQUALLY SPACED OUT.
I'VE NEVER SEEN EQUALLY SPACED OUT TRAFFIC IN ALL THE YEARS I'VE BEEN DRIVING.
HONNNNK
BEEP BEEP
BEEEEP
BEEEEP
IT MEANS THAT THEY'RE SPACED ACCORDING TO A MORE COMPLICATED CONCEPT CALLED A "POISSON POINT PROCESS."
I'M NOT SURE IT APPLIES TO DAISY'S CASE, BUT...
...WHAT IF I DID ALL THE TESTS TO DETERMINE WHETHER THE LOGS OF THE LENGTHS OF THE CYCLES IN DAISY
--A TYPICAL CASE--
...ARE SPREAD OUT LIKE A POISSON POINT PROCESS, LIKE CARS ON A FREEWAY?

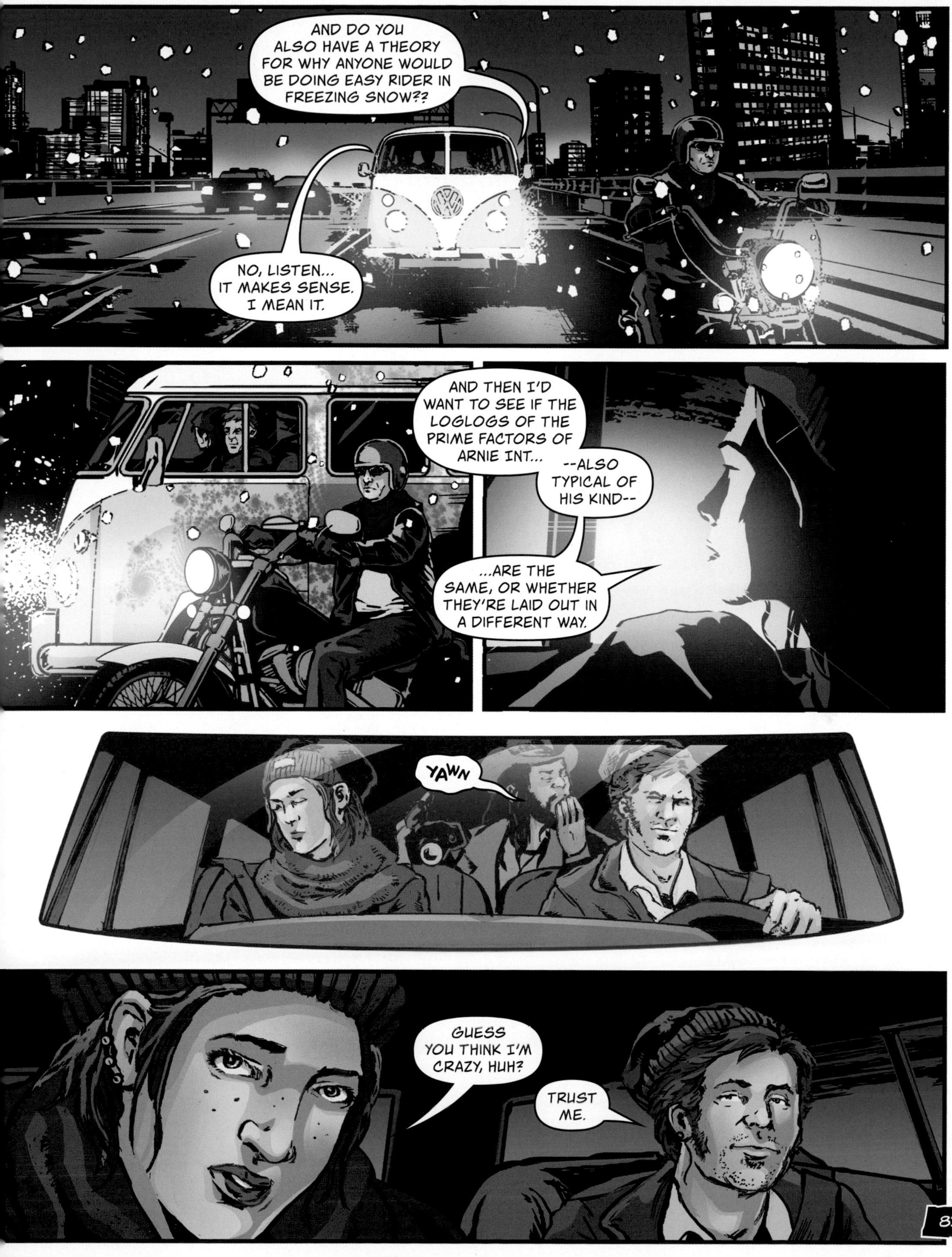
AND DO YOU ALSO HAVE A THEORY FOR WHY ANYONE WOULD BE DOING EASY RIDER IN FREEZING SNOW??
NO, LISTEN... IT MAKES SENSE. I MEAN IT.
AND THEN I'D WANT TO SEE IF THE LOGLOGS OF THE PRIME FACTORS OF ARNIE INT...
--ALSO TYPICAL OF HIS KIND--
...ARE THE SAME, OR WHETHER THEY'RE LAID OUT IN A DIFFERENT WAY.
YAWN
GUESS YOU THINK I'M CRAZY, HUH?
TRUST ME.

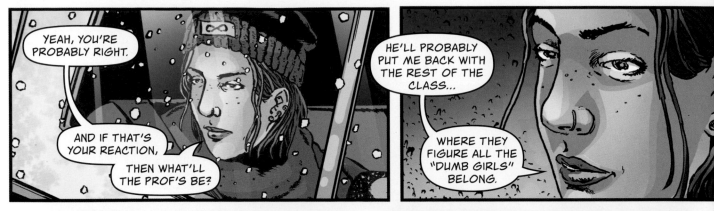

YEAH, YOU'RE PROBABLY RIGHT.

AND IF THAT'S YOUR REACTION,

THEN WHAT'LL THE PROF'S BE?

HE'LL PROBABLY PUT ME BACK WITH THE REST OF THE CLASS...

WHERE THEY FIGURE ALL THE "DUMB GIRLS" BELONG.

AAAEEEEYAAAAHHHHHHHA'AAEEEEHHHH!!!!

WHAT WAS THAT?

SOME KIND OF WEIRD MATH YELL?

SNAP

I'M SORRY. I JUST GET SO FRUSTRATED.

I'VE GOT SO MUCH TO PROVE.

GO BACK TO SLEEP, DAMN YOU.

NO WAY.

THIS IS THE KIND OF DRAMA WE'RE HERE TO GET.

JUST COOL IT.

DIDN'T YOU GIVE ME AN INTERVIEW TELLING ME HOW OLE C. F. IS LIKE THE GREATEST MATH DUDE THERE EVER WAS?

DID HE GET THAT WAY WITHOUT TAKING A FEW RISKS AND SOUNDING A LITTLE CRAZY FROM TIME TO TIME?

IF THAT'S YOUR BEST SHOT, YOU SHOULD GO FOR IT.

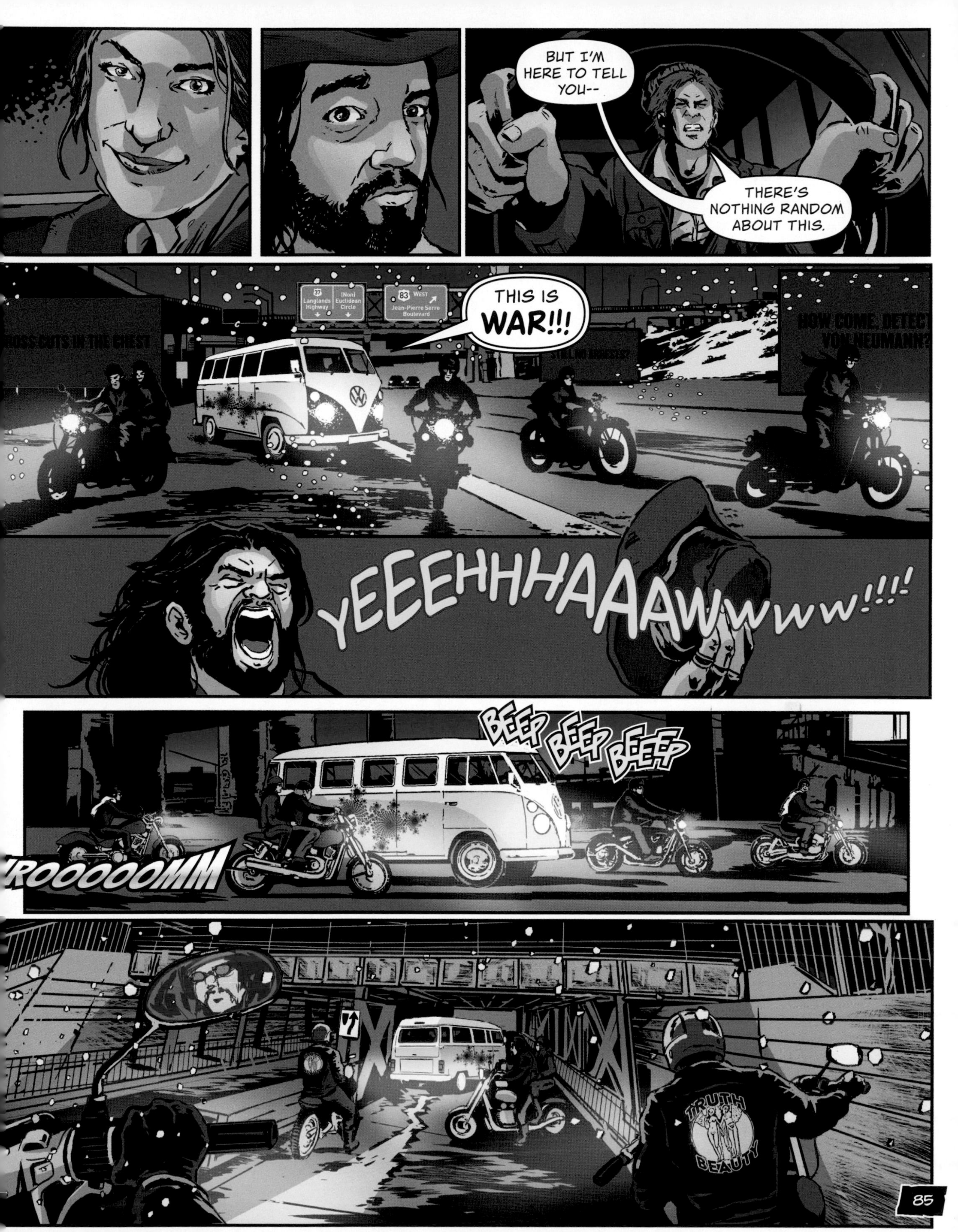

BUT I'M HERE TO TELL YOU--
THERE'S NOTHING RANDOM ABOUT THIS.
37 Langlands Highway
(Non) Euclidean Circle
83 West Jean-Pierre Serre Boulevard
THIS IS WAR!!!
YEEEHHHAAAAWwww!!!!
BEEP BEEP BEEEP
ROOOOOOMM
TRUTH BEAUTY

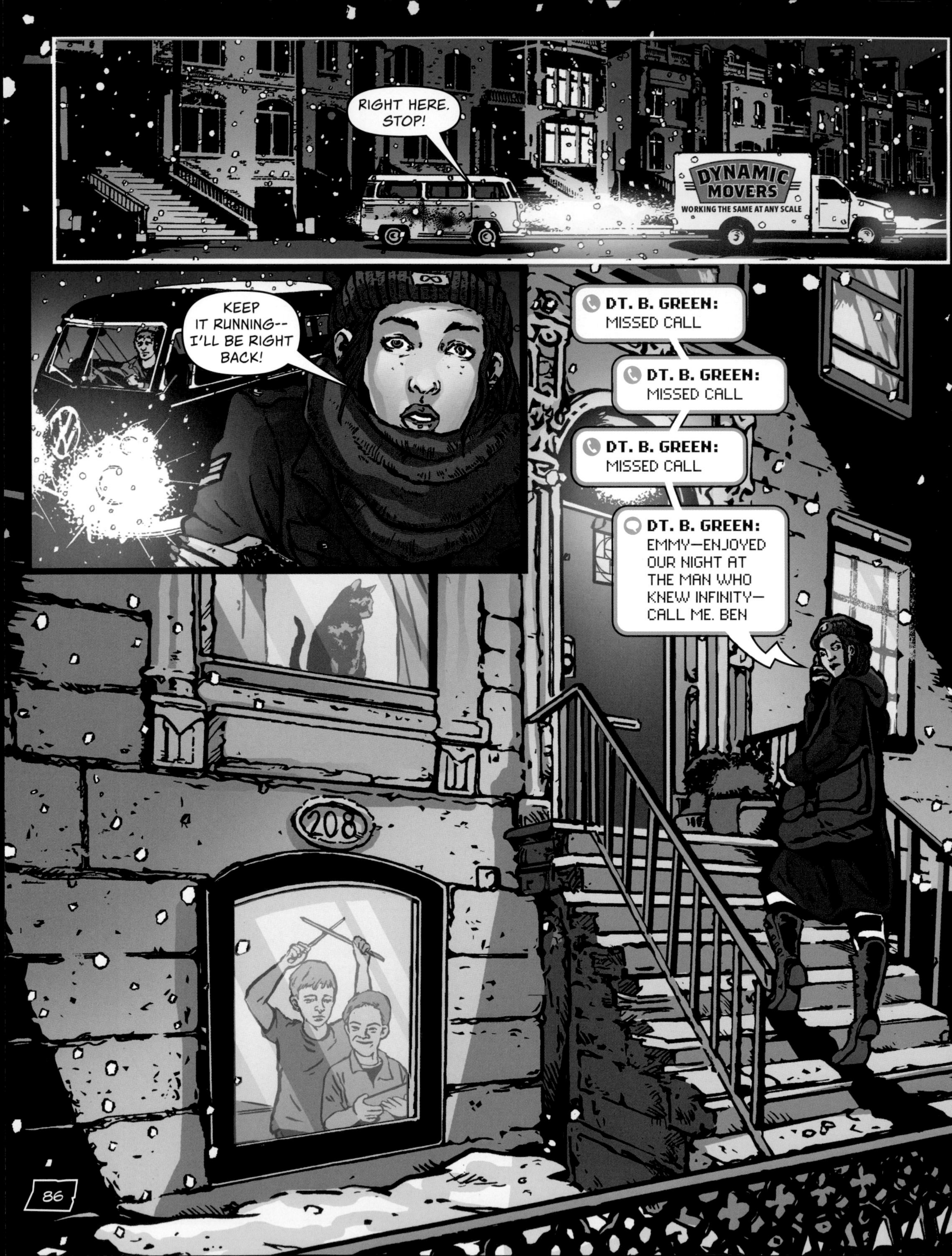
RIGHT HERE. STOP!
DYNAMIC MOVERS
WORKING THE SAME AT ANY SCALE
KEEP IT RUNNING-- I'LL BE RIGHT BACK!
DT. B. GREEN: MISSED CALL
DT. B. GREEN: MISSED CALL
DT. B. GREEN: MISSED CALL
DT. B. GREEN: EMMY—ENJOYED OUR NIGHT AT THE MAN WHO KNEW INFINITY—CALL ME. BEN
208

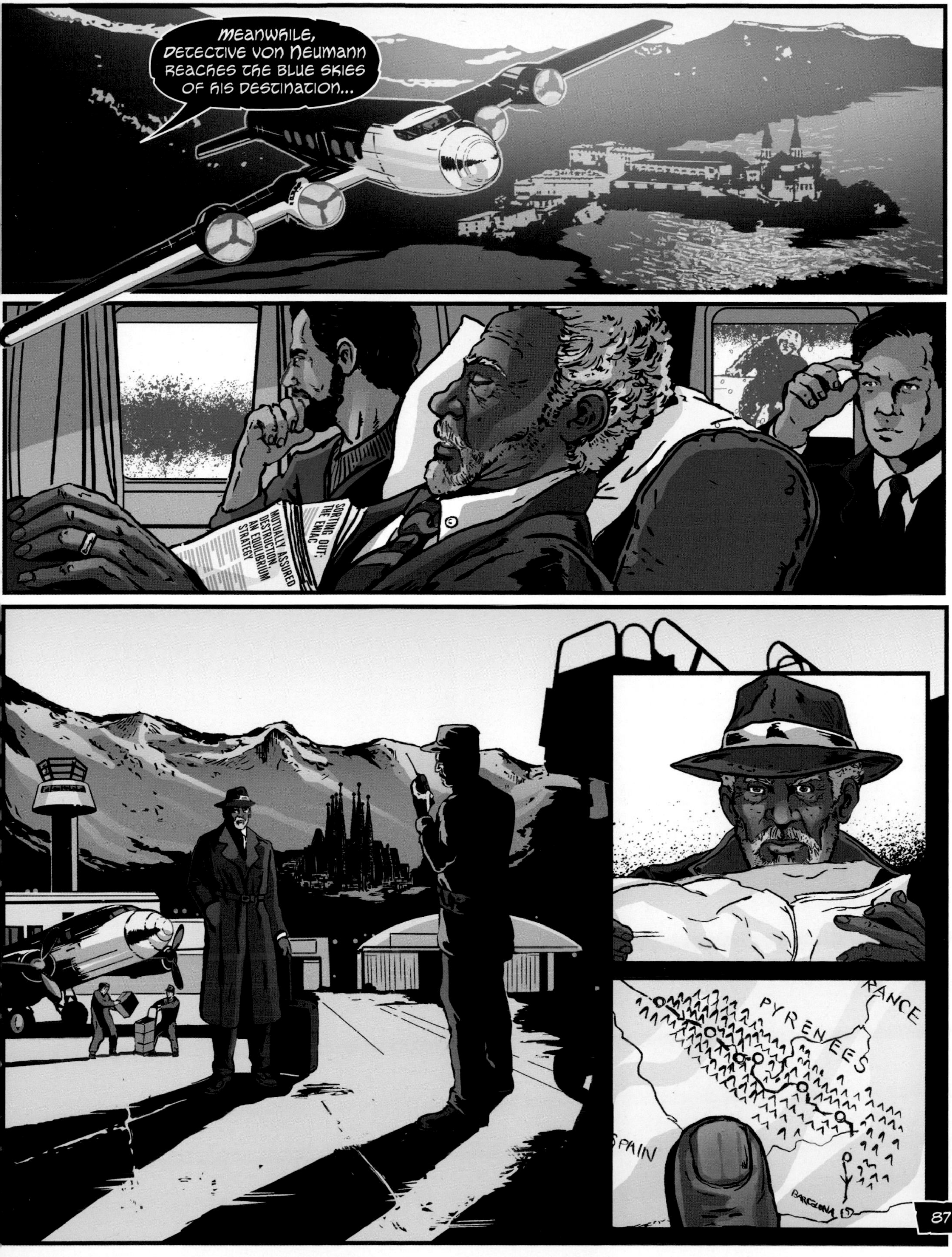
MEANWHILE, DETECTIVE VON NEUMANN REACHES THE BLUE SKIES OF HIS DESTINATION...
SORTING OUT: THE ENIAC
MUTUALLY ASSURED DESTRUCTION: AN EQUILIBRIUM STRATEGY
PYRENEES
RANCE
SPAIN
BARCELONA

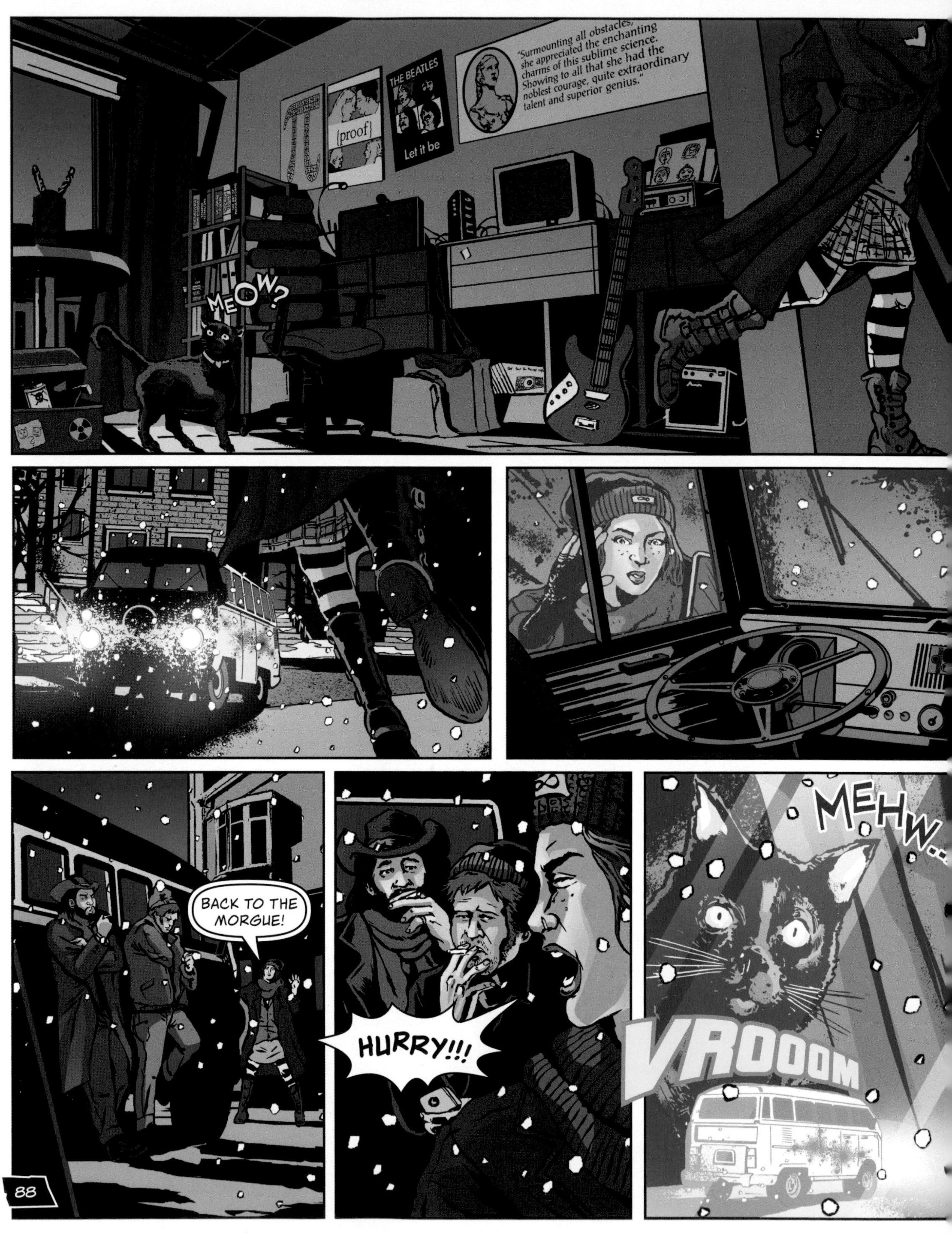
"Surmounting all obstacles, she appreciated the enchanting charms of this sublime science. Showing to all that she had the noblest courage, quite extraordinary talent and superior genius."
{proof}
THE BEATLES
Let it be
MEOW?
BACK TO THE MORGUE!
HURRY!!!
MEHW...
VROOOM

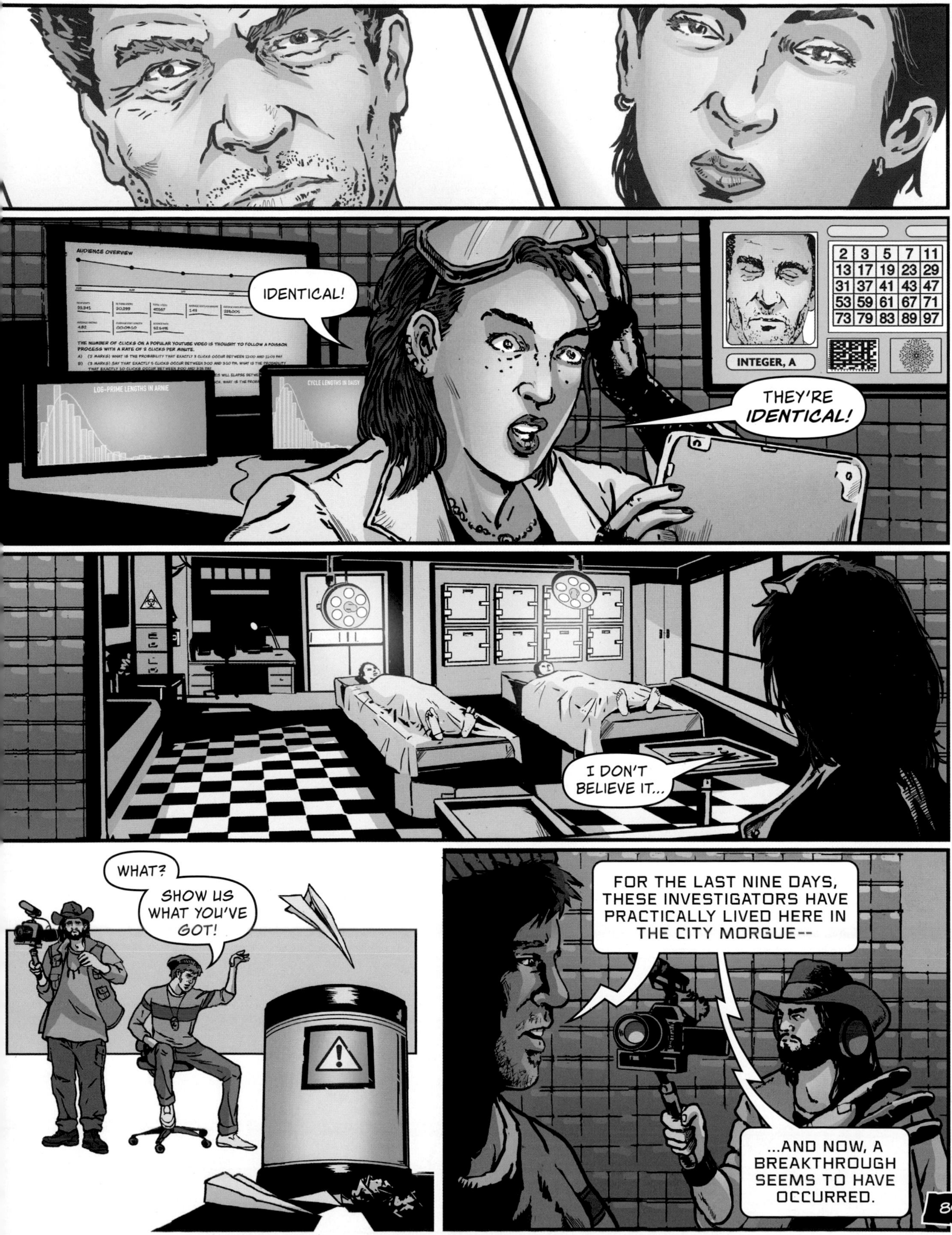

AUDIENCE OVERVIEW
THE NUMBER OF CLICKS ON A POPULAR YOUTUBE VIDEO IS THOUGHT TO FOLLOW A POISSON PROCESS WITH A RATE OF 2 CLICKS PER MINUTE.
LOG-PRIME LENGTHS IN ARNIE
CYCLE LENGTHS IN DAISY
IDENTICAL!
2 3 5 7 11
13 17 19 23 29
31 37 41 43 47
53 59 61 67 71
73 79 83 89 97
INTEGER, A
THEY'RE *IDENTICAL!*
I DON'T BELIEVE IT...
WHAT?
SHOW US WHAT YOU'VE GOT!
FOR THE LAST NINE DAYS, THESE INVESTIGATORS HAVE PRACTICALLY LIVED HERE IN THE CITY MORGUE--
...AND NOW, A BREAKTHROUGH SEEMS TO HAVE OCCURRED.

HEY, LANGER, EMMY SEEMS TO HAVE FOUND SOMETHING.

AREN'T YOU INTERESTED?

NO MORE THAN I EXPECTED...

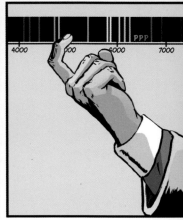

I SUSPECT EMMY HAS DISCOVERED THAT THE SIZES OF THE COMPONENTS

ARE LAID OUT LIKE A POISSON POINT PROCESS?

UBIQUITY OF PPP?

PRETTY WELL UBIQUITOUS IN SUCH SITUATIONS.

PERHAPS EMMY WOULD LIKE TO STOP AND REMEMBER FORENSIC PROBABILITY THEORY...

...BEFORE GETTING QUITE SO EXCITED?

1-0-1

OF COURSE, IF SHE EVER READ MY EMAILS,

SHE'D KNOW THAT I'D CC'D HER ON AN EMAIL TO THE PROF,

PROPOSING EXACTLY THIS IDEA.

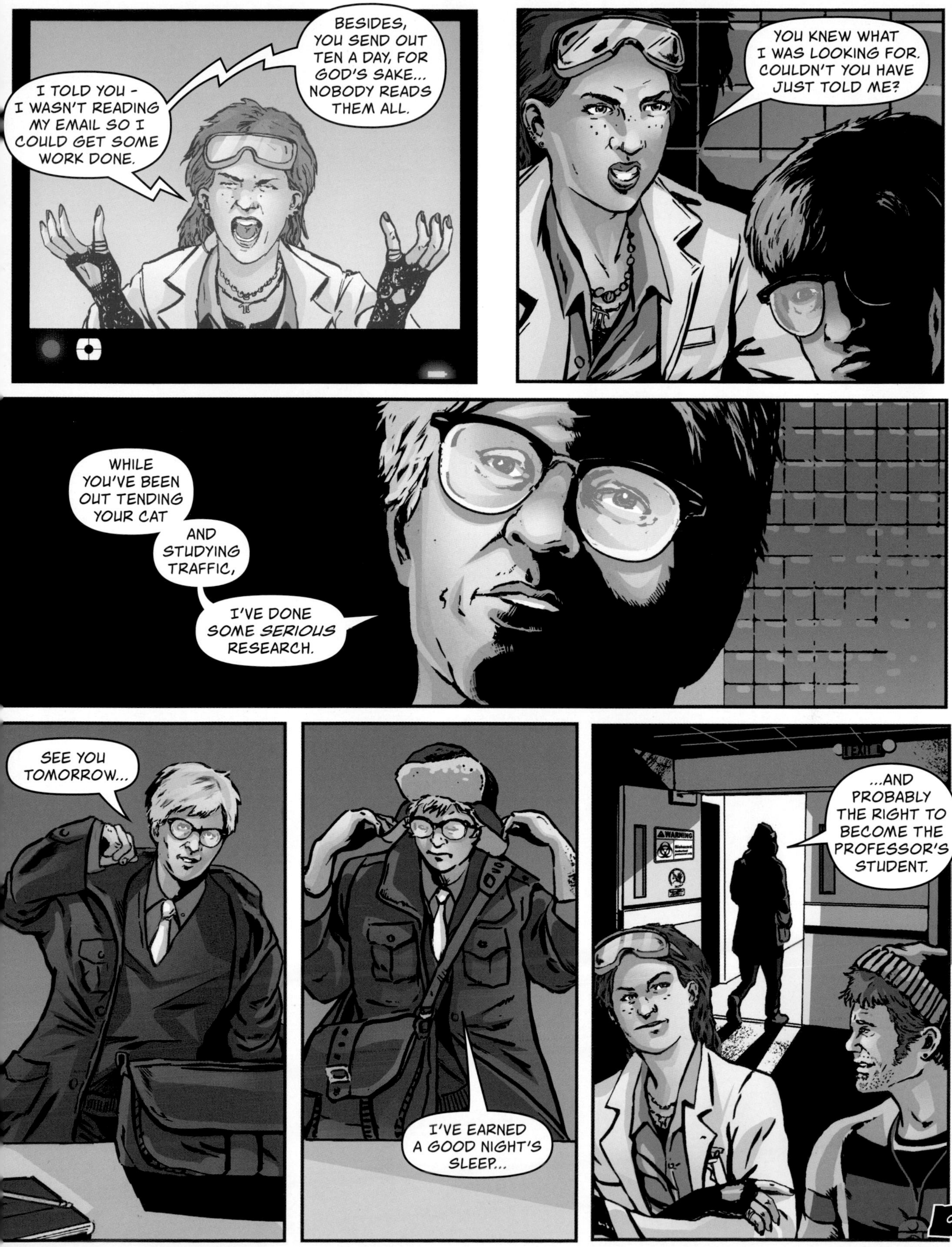
I TOLD YOU - I WASN'T READING MY EMAIL SO I COULD GET SOME WORK DONE.
BESIDES, YOU SEND OUT TEN A DAY, FOR GOD'S SAKE... NOBODY READS THEM ALL.
YOU KNEW WHAT I WAS LOOKING FOR. COULDN'T YOU HAVE JUST TOLD ME?
WHILE YOU'VE BEEN OUT TENDING YOUR CAT
AND STUDYING TRAFFIC,
I'VE DONE SOME SERIOUS RESEARCH.
SEE YOU TOMORROW...
I'VE EARNED A GOOD NIGHT'S SLEEP...
...AND PROBABLY THE RIGHT TO BECOME THE PROFESSOR'S STUDENT.
EXIT
WARNING

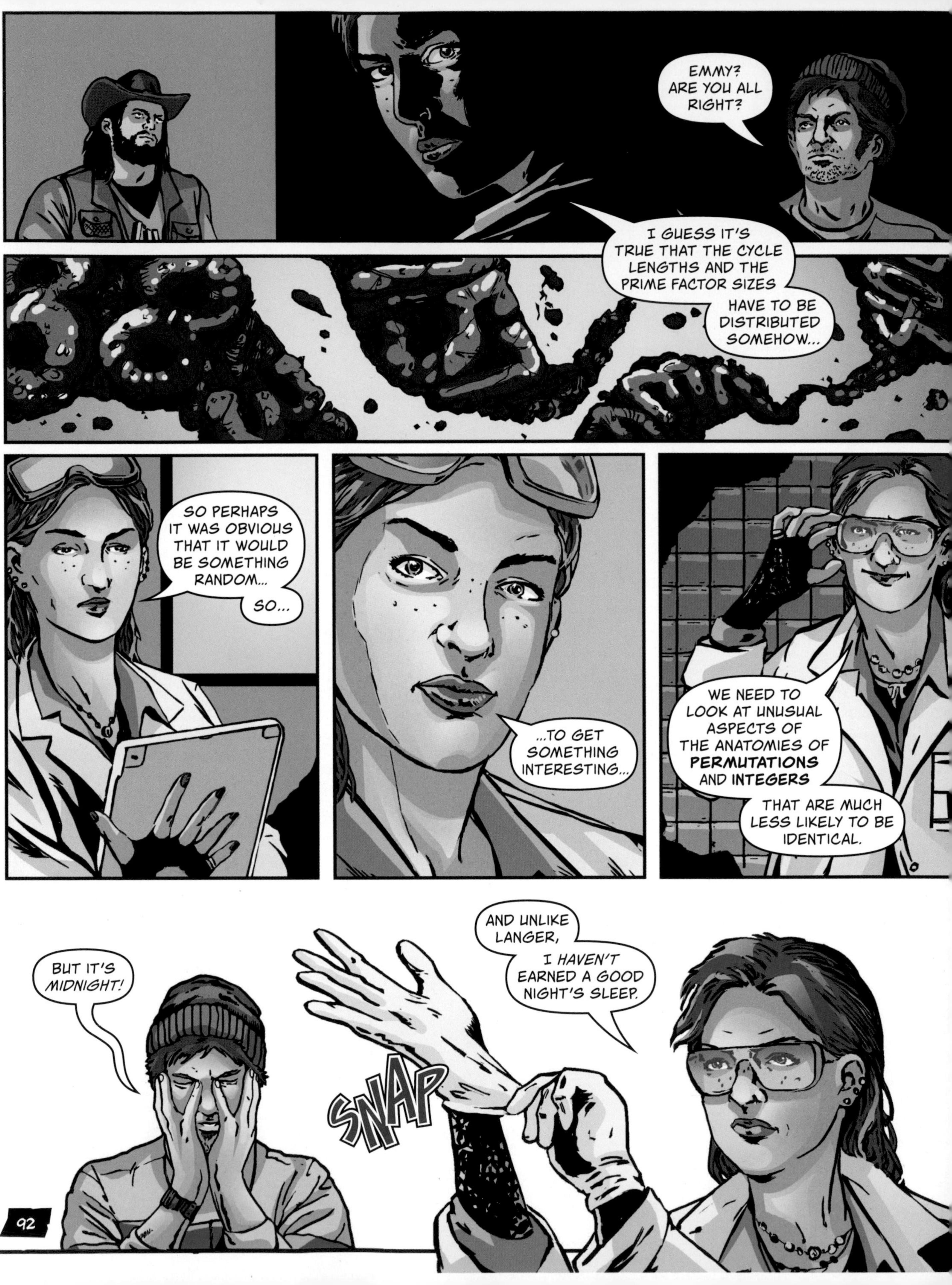
EMMY? ARE YOU ALL RIGHT?
I GUESS IT'S TRUE THAT THE CYCLE LENGTHS AND THE PRIME FACTOR SIZES
HAVE TO BE DISTRIBUTED SOMEHOW...
SO PERHAPS IT WAS OBVIOUS THAT IT WOULD BE SOMETHING RANDOM...
SO...
...TO GET SOMETHING INTERESTING...
WE NEED TO LOOK AT UNUSUAL ASPECTS OF THE ANATOMIES OF **PERMUTATIONS** AND **INTEGERS**
THAT ARE MUCH LESS LIKELY TO BE IDENTICAL.
BUT IT'S *MIDNIGHT!*
AND UNLIKE LANGER,
I *HAVEN'T* EARNED A GOOD NIGHT'S SLEEP.
SNAP

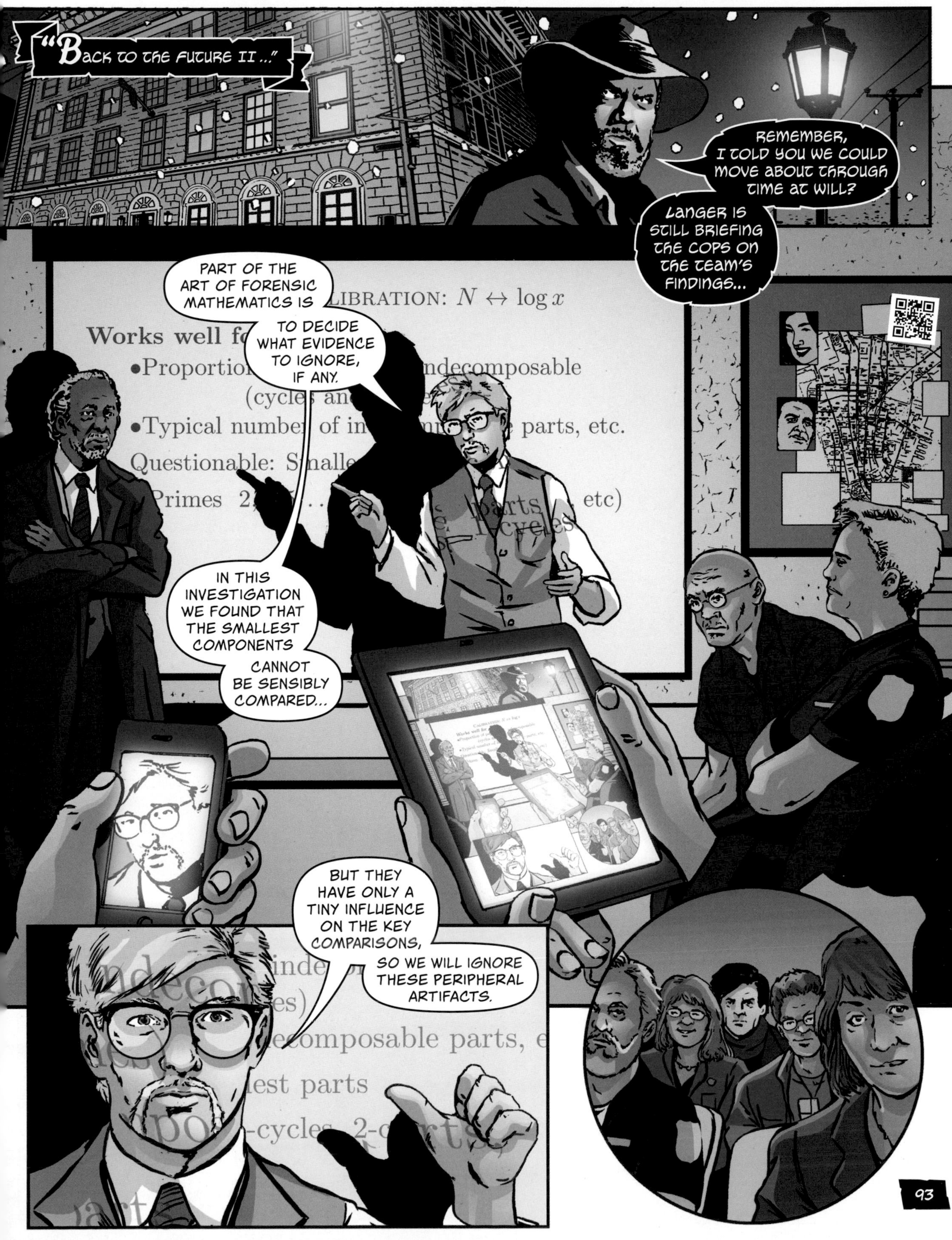
"Back to the future II ..."
Remember, I told you we could move about through time at will?
Langer is still briefing the cops on the team's findings...
Part of the art of forensic mathematics is
to decide what evidence to ignore, if any.
In this investigation we found that the smallest components
cannot be sensibly compared...
But they have only a tiny influence on the key comparisons,
so we will ignore these peripheral artifacts.

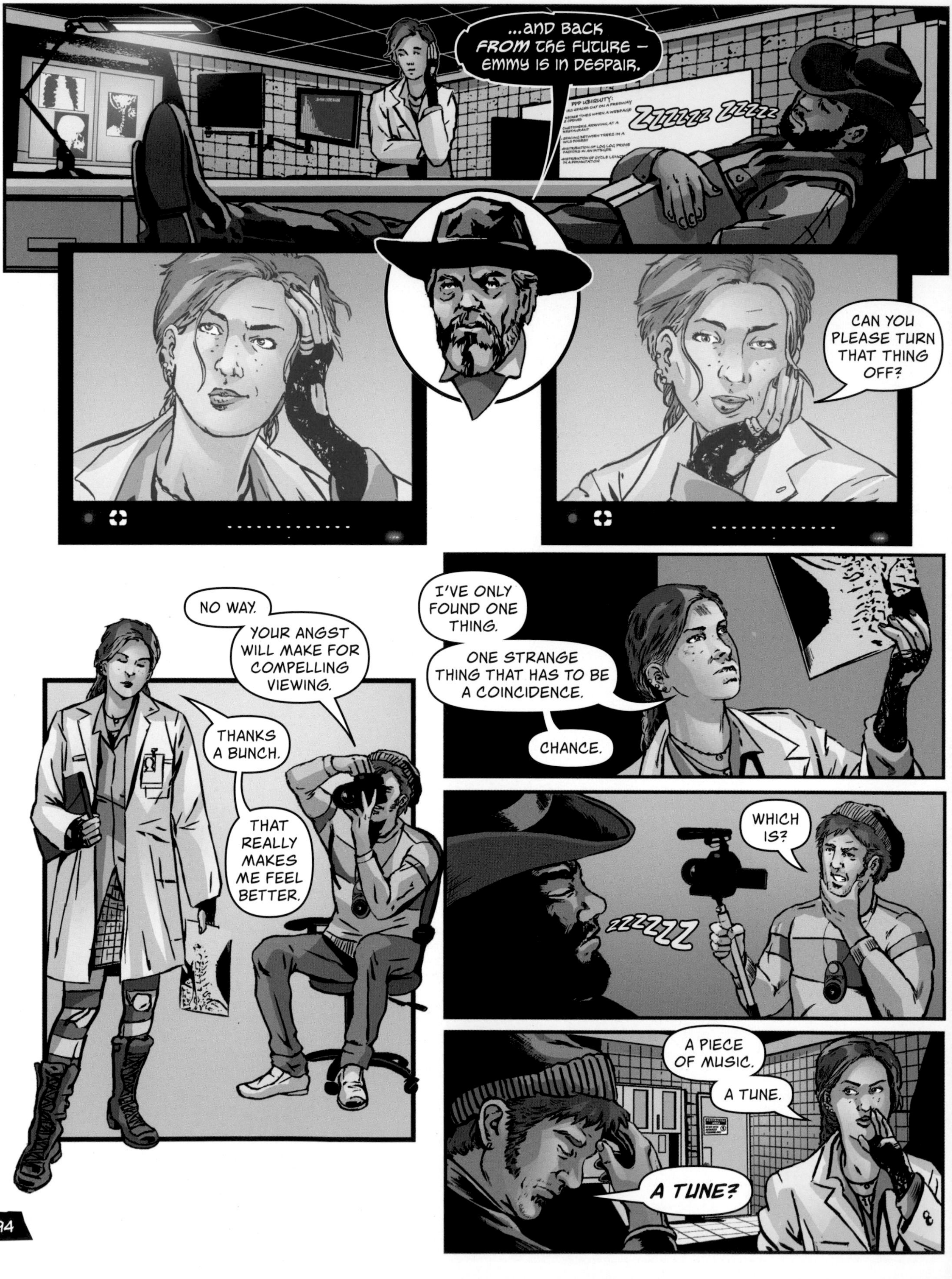
...and back *FROM* the future – Emmy is in despair.
ZZZZZZZ ZZZZZ
CAN YOU PLEASE TURN THAT THING OFF?
NO WAY.
YOUR ANGST WILL MAKE FOR COMPELLING VIEWING.
THANKS A BUNCH.
THAT REALLY MAKES ME FEEL BETTER.
I'VE ONLY FOUND ONE THING.
ONE STRANGE THING THAT HAS TO BE A COINCIDENCE.
CHANCE.
WHICH IS?
ZZZZZZ
A PIECE OF MUSIC.
A TUNE.
**A TUNE?**

YES, WELL, I KNOW IT SOUNDS CRAZY, JUST *CRAZY*, BUT...

IN BOTH CADAVERS...

IN THEIR HEADS...

THERE WAS THIS *TUNE*...

THIS SAME *TUNE*,

GOING ROUND AND ROUND...

SNIKT

WHAT ARE YOU *TALKING* ABOUT...?

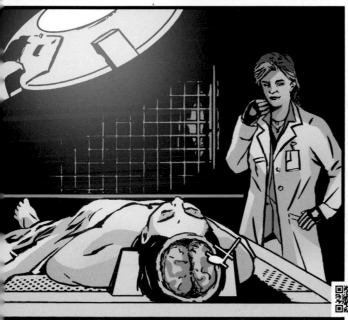

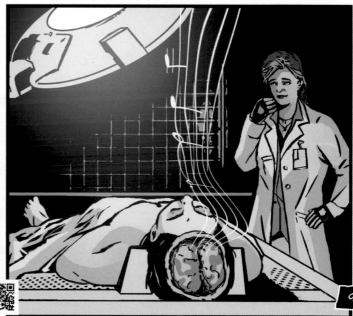

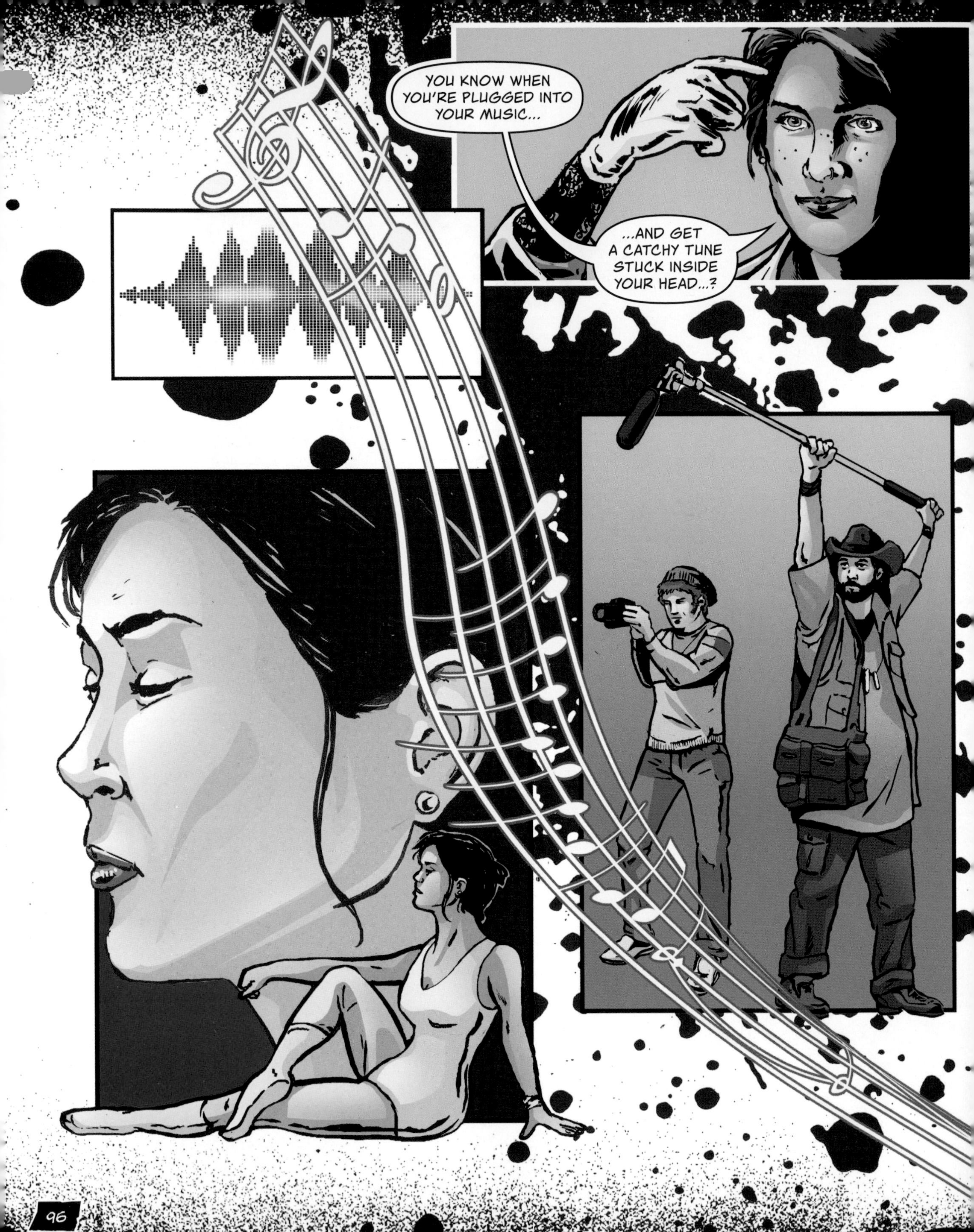
YOU KNOW WHEN YOU'RE PLUGGED INTO YOUR MUSIC...
...AND GET A CATCHY TUNE STUCK INSIDE YOUR HEAD...?

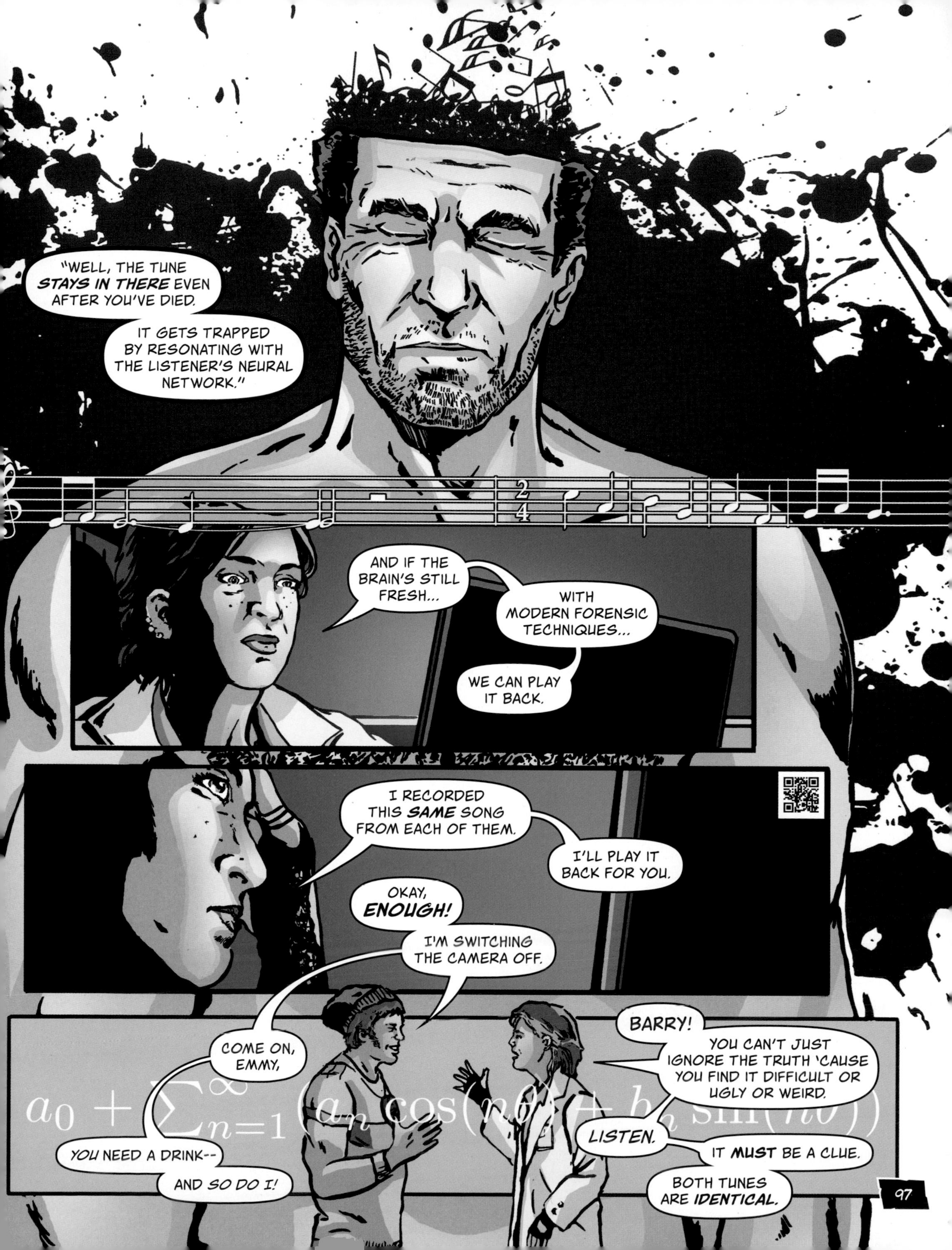
"WELL, THE TUNE STAYS IN THERE EVEN AFTER YOU'VE DIED.
IT GETS TRAPPED BY RESONATING WITH THE LISTENER'S NEURAL NETWORK."
AND IF THE BRAIN'S STILL FRESH...
WITH MODERN FORENSIC TECHNIQUES...
WE CAN PLAY IT BACK.
I RECORDED THIS SAME SONG FROM EACH OF THEM.
I'LL PLAY IT BACK FOR YOU.
OKAY, ENOUGH!
I'M SWITCHING THE CAMERA OFF.
COME ON, EMMY,
YOU NEED A DRINK--
AND SO DO I!
BARRY!
YOU CAN'T JUST IGNORE THE TRUTH 'CAUSE YOU FIND IT DIFFICULT OR UGLY OR WEIRD.
LISTEN.
IT MUST BE A CLUE.
BOTH TUNES ARE IDENTICAL.
$a_0 + \sum_{n=1}^{\infty} (a_n \cos(n\theta) + b_n \sin(n\theta))$

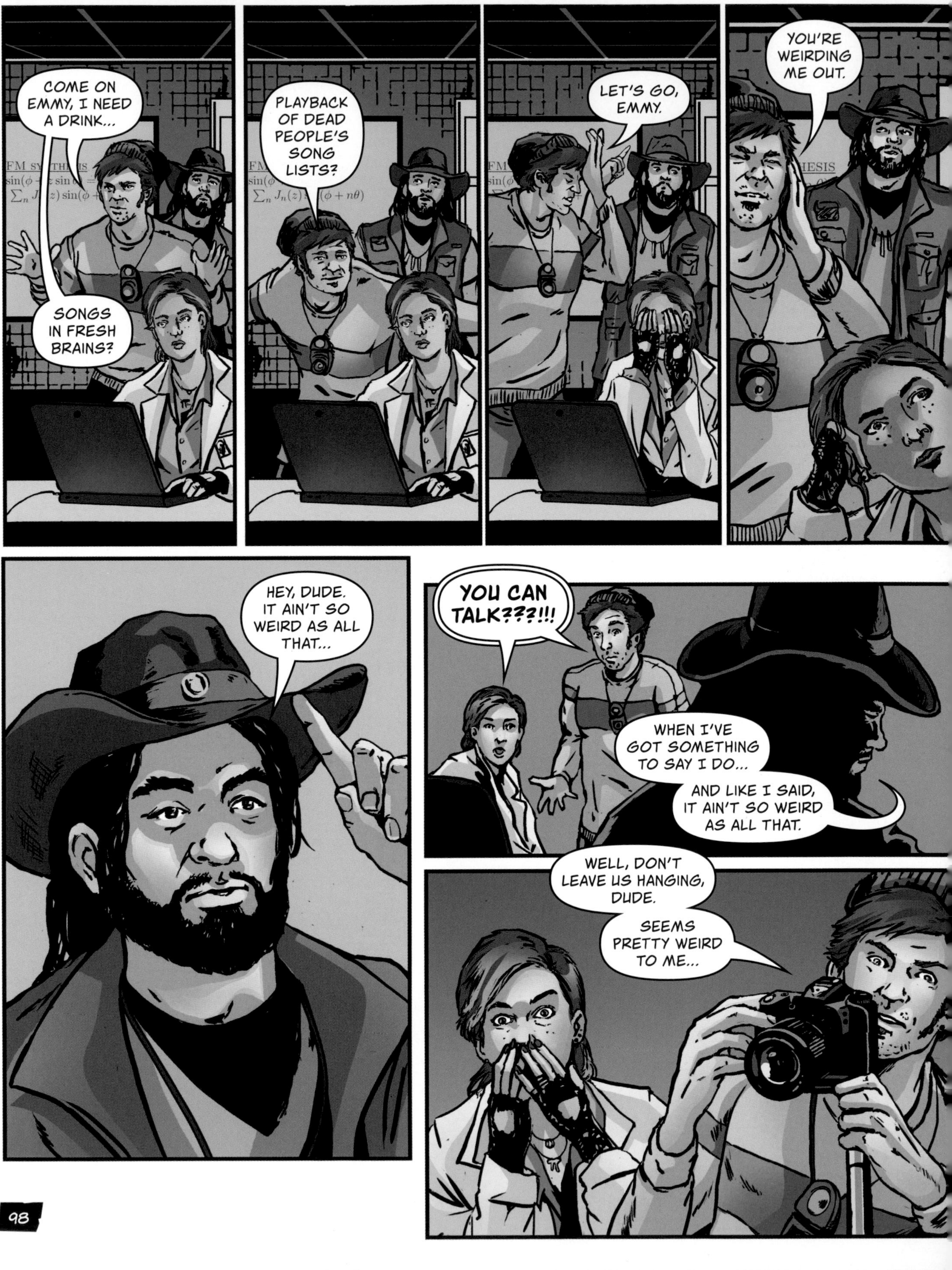

COME ON EMMY, I NEED A DRINK...
SONGS IN FRESH BRAINS?
PLAYBACK OF DEAD PEOPLE'S SONG LISTS?
LET'S GO, EMMY.
YOU'RE WEIRDING ME OUT.
HEY, DUDE. IT AIN'T SO WEIRD AS ALL THAT...
YOU CAN TALK???!!!
WHEN I'VE GOT SOMETHING TO SAY I DO...
AND LIKE I SAID, IT AIN'T SO WEIRD AS ALL THAT.
WELL, DON'T LEAVE US HANGING, DUDE.
SEEMS PRETTY WEIRD TO ME...

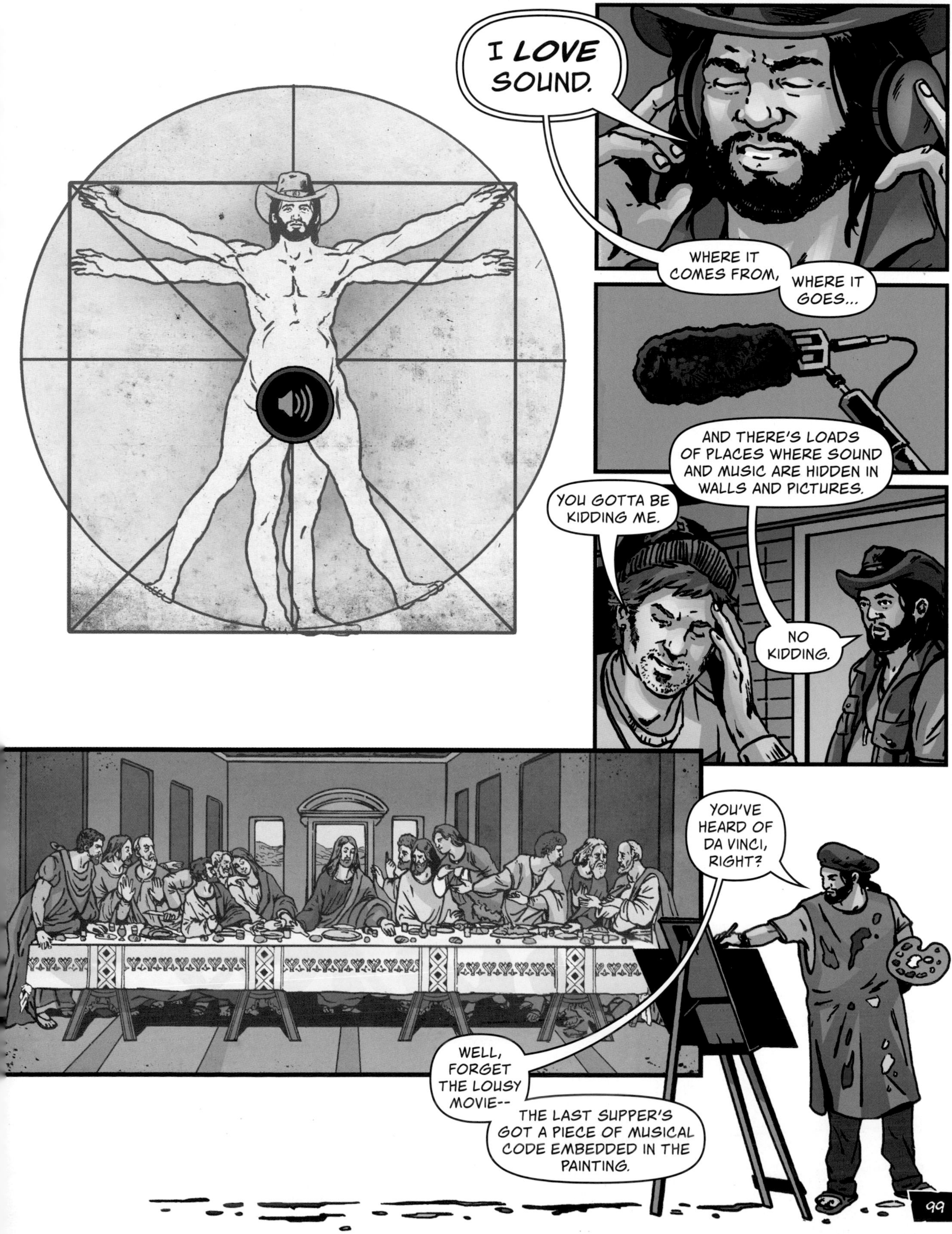
I LOVE SOUND.
WHERE IT COMES FROM,
WHERE IT GOES...
AND THERE'S LOADS OF PLACES WHERE SOUND AND MUSIC ARE HIDDEN IN WALLS AND PICTURES.
YOU GOTTA BE KIDDING ME.
NO KIDDING.
YOU'VE HEARD OF DA VINCI, RIGHT?
WELL, FORGET THE LOUSY MOVIE--
THE LAST SUPPER'S GOT A PIECE OF MUSICAL CODE EMBEDDED IN THE PAINTING.

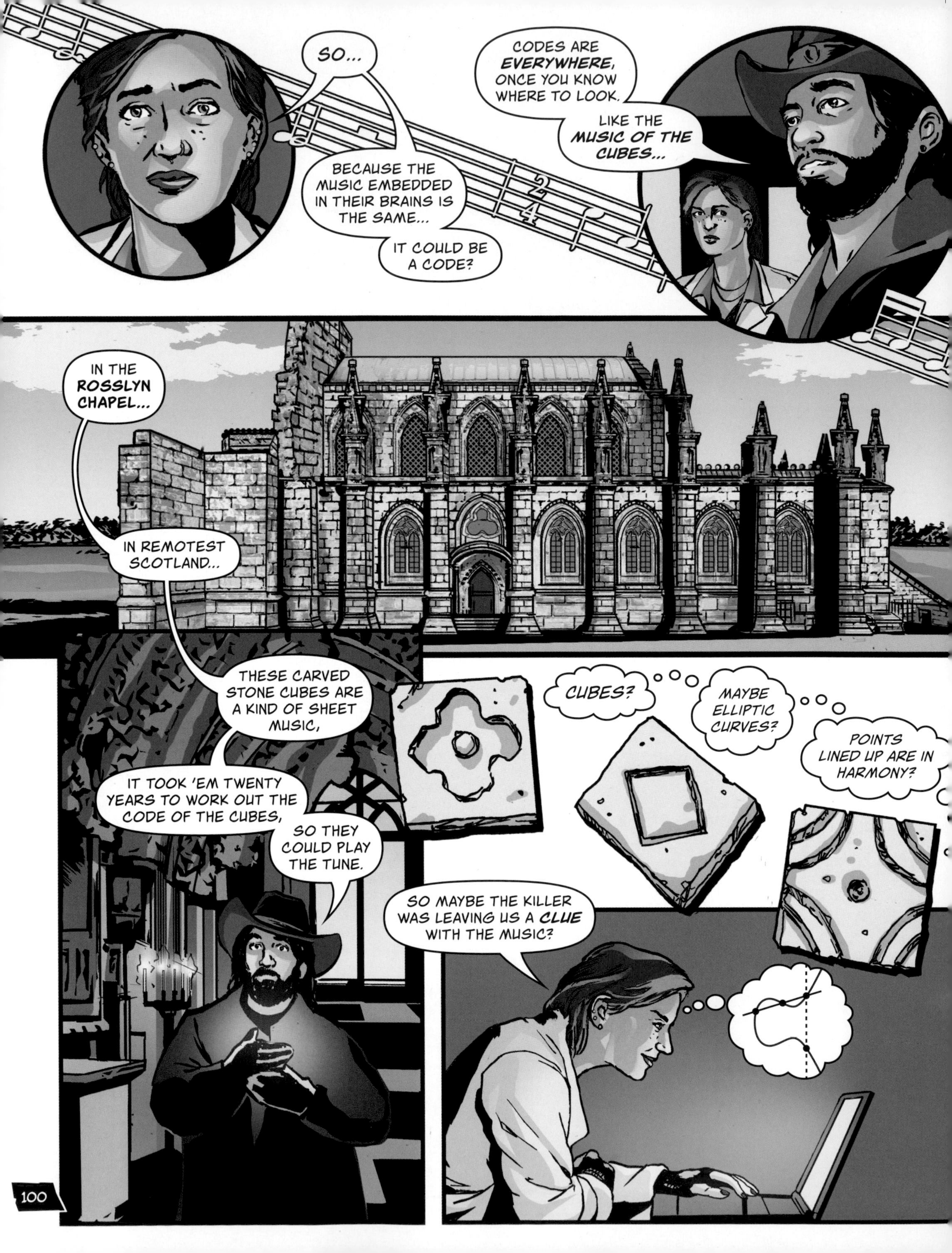
SO...
BECAUSE THE MUSIC EMBEDDED IN THEIR BRAINS IS THE SAME...
IT COULD BE A CODE?
CODES ARE EVERYWHERE, ONCE YOU KNOW WHERE TO LOOK.
LIKE THE MUSIC OF THE CUBES...
IN THE ROSSLYN CHAPEL...
IN REMOTEST SCOTLAND...
THESE CARVED STONE CUBES ARE A KIND OF SHEET MUSIC,
IT TOOK 'EM TWENTY YEARS TO WORK OUT THE CODE OF THE CUBES,
SO THEY COULD PLAY THE TUNE.
CUBES?
MAYBE ELLIPTIC CURVES?
POINTS LINED UP ARE IN HARMONY?
SO MAYBE THE KILLER WAS LEAVING US A CLUE WITH THE MUSIC?

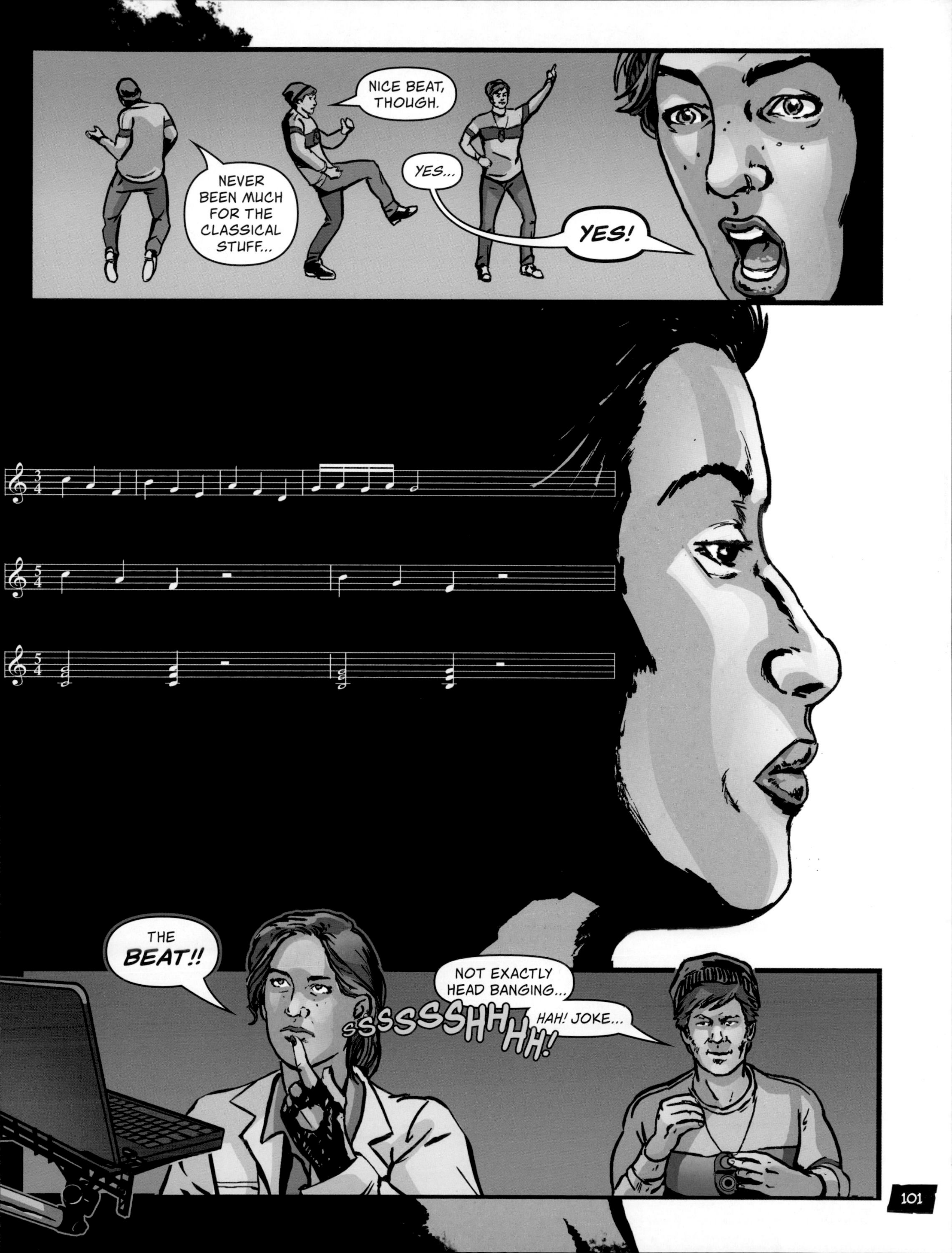

NEVER BEEN MUCH FOR THE CLASSICAL STUFF...
NICE BEAT, THOUGH.
YES...
YES!
THE BEAT!!
SSSSSSHHHHH!
NOT EXACTLY HEAD BANGING...
HAH! JOKE...

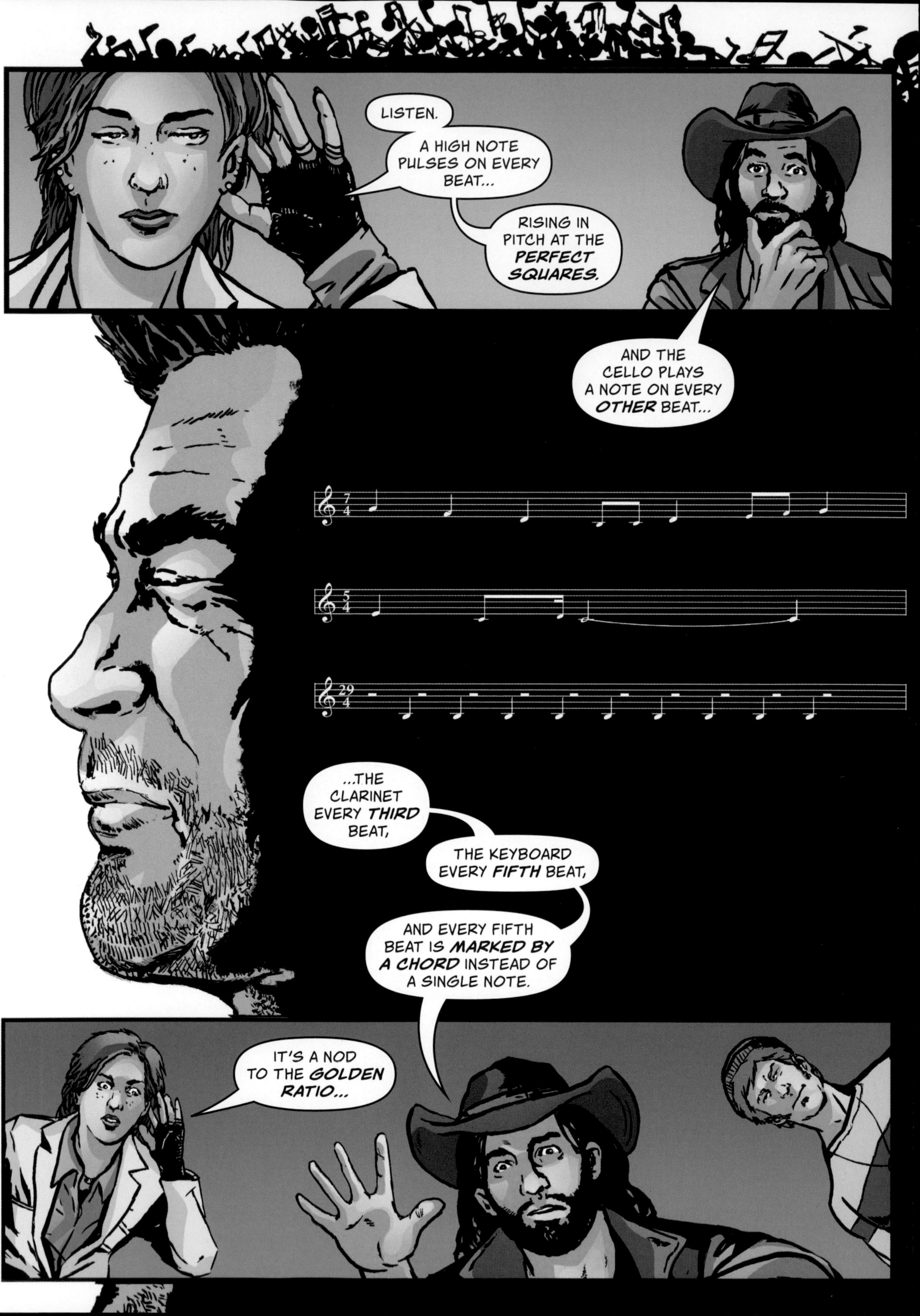
LISTEN.
A HIGH NOTE PULSES ON EVERY BEAT...
RISING IN PITCH AT THE PERFECT SQUARES.
AND THE CELLO PLAYS A NOTE ON EVERY OTHER BEAT...
...THE CLARINET EVERY THIRD BEAT,
THE KEYBOARD EVERY FIFTH BEAT,
AND EVERY FIFTH BEAT IS MARKED BY A CHORD INSTEAD OF A SINGLE NOTE.
IT'S A NOD TO THE GOLDEN RATIO...

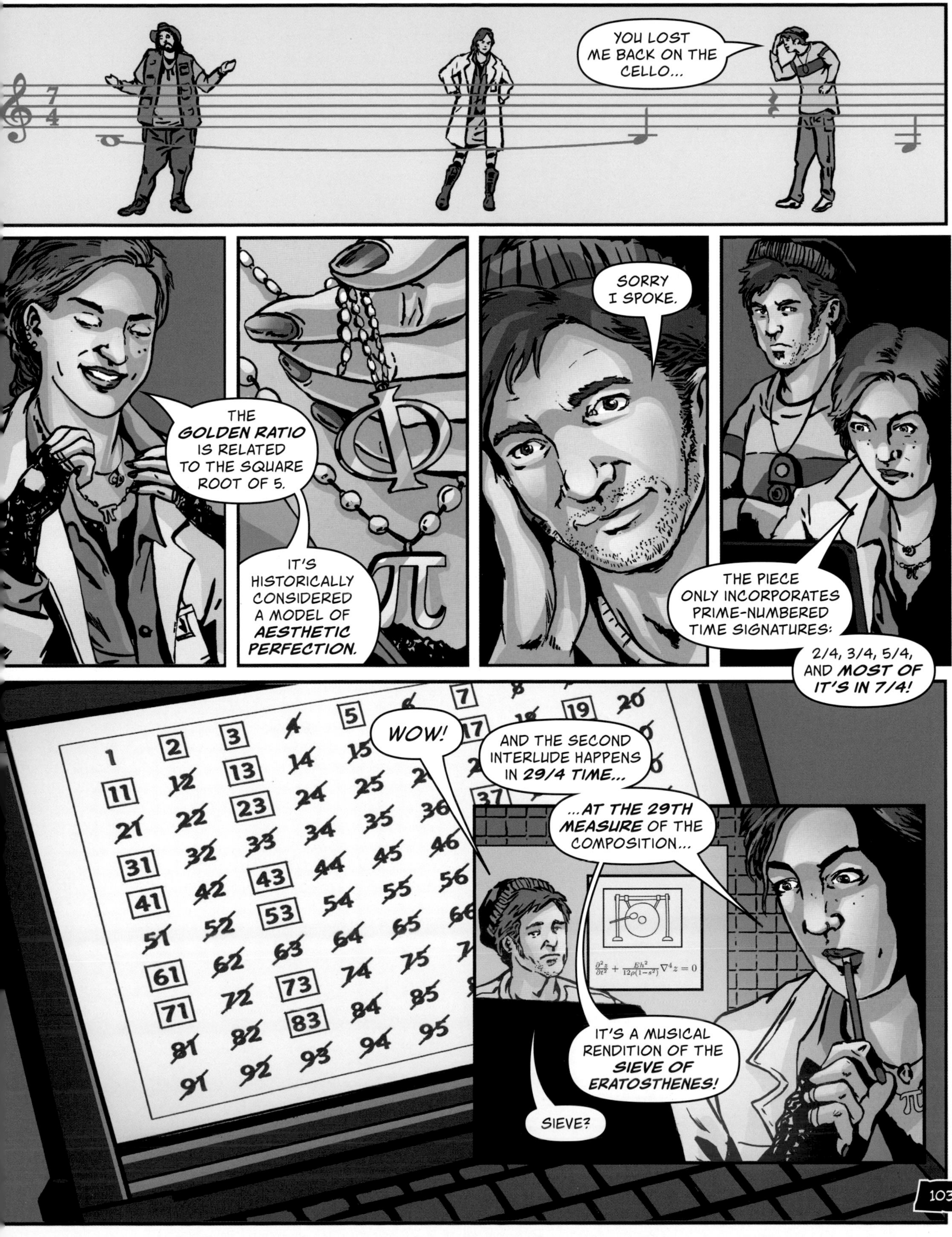

YOU LOST ME BACK ON THE CELLO...
THE GOLDEN RATIO IS RELATED TO THE SQUARE ROOT OF 5.
IT'S HISTORICALLY CONSIDERED A MODEL OF AESTHETIC PERFECTION.
SORRY I SPOKE.
THE PIECE ONLY INCORPORATES PRIME-NUMBERED TIME SIGNATURES:
2/4, 3/4, 5/4, AND MOST OF IT'S IN 7/4!
WOW!
AND THE SECOND INTERLUDE HAPPENS IN 29/4 TIME...
...AT THE 29TH MEASURE OF THE COMPOSITION...
IT'S A MUSICAL RENDITION OF THE SIEVE OF ERATOSTHENES!
SIEVE?

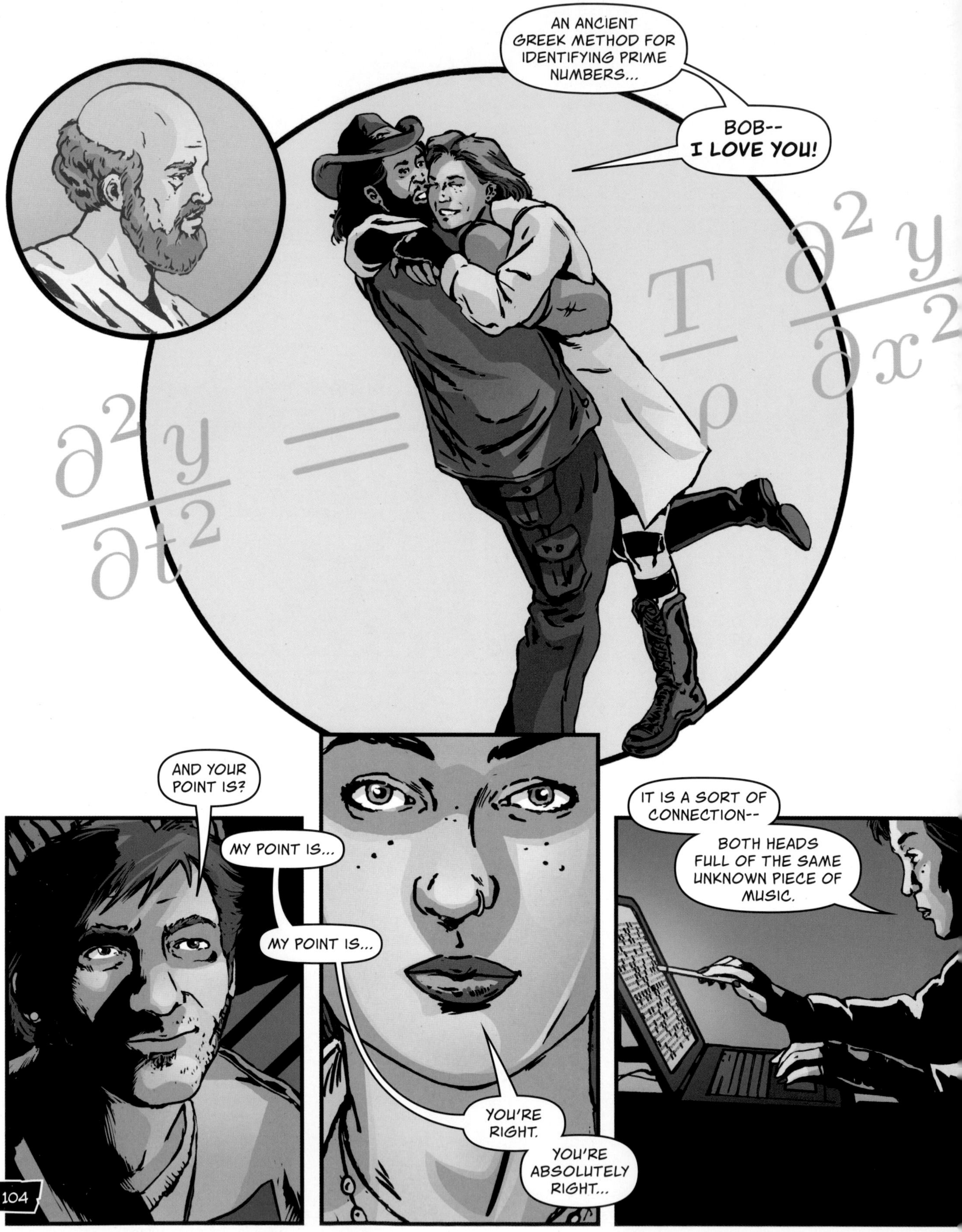
AN ANCIENT GREEK METHOD FOR IDENTIFYING PRIME NUMBERS...
BOB-- I LOVE YOU!
$$\frac{\partial^2 y}{\partial t^2} = \frac{T}{\rho}\frac{\partial^2 y}{\partial x^2}$$
AND YOUR POINT IS?
MY POINT IS...
MY POINT IS...
YOU'RE RIGHT.
YOU'RE ABSOLUTELY RIGHT...
IT IS A SORT OF CONNECTION--
BOTH HEADS FULL OF THE SAME UNKNOWN PIECE OF MUSIC.

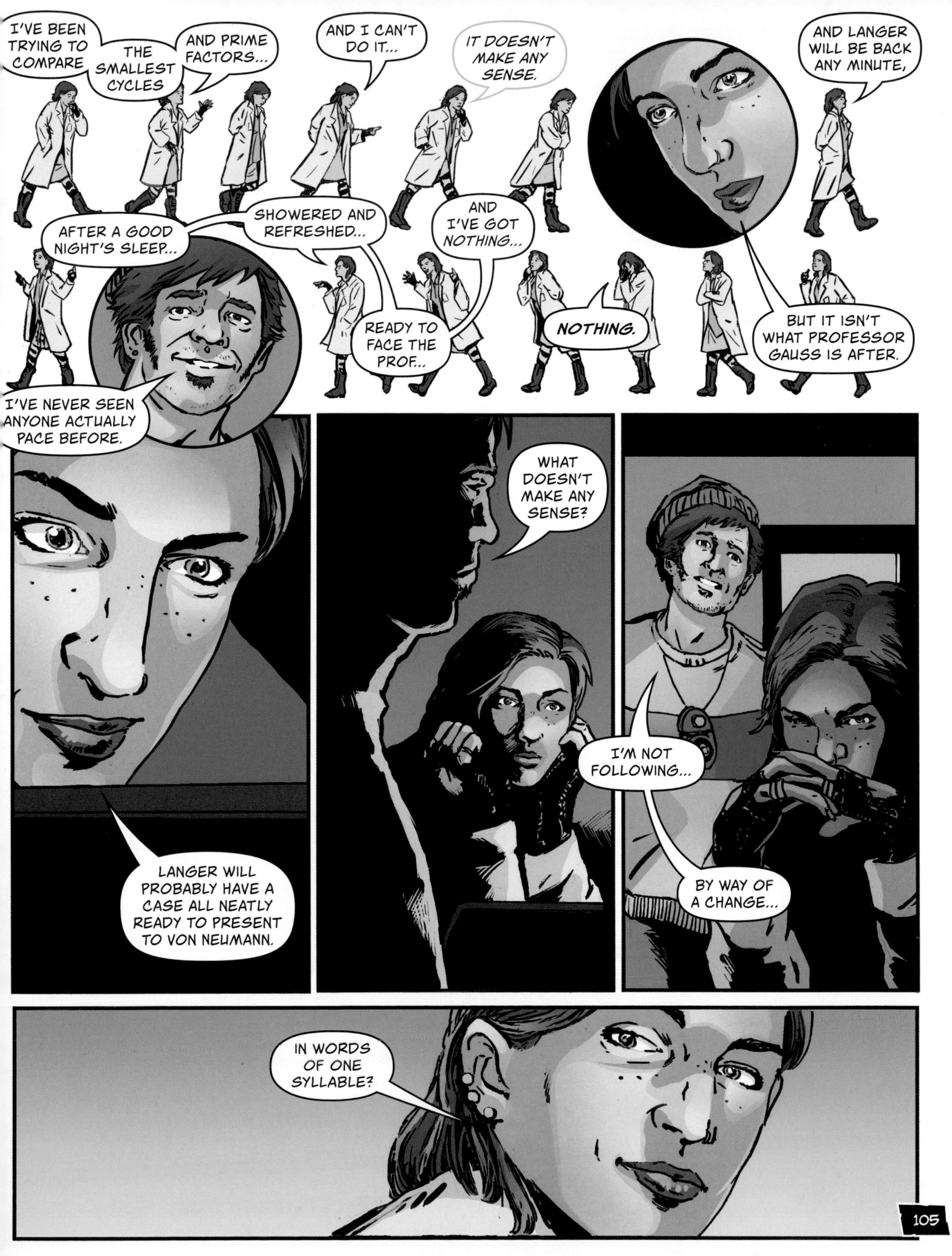

I'VE BEEN TRYING TO COMPARE
THE SMALLEST CYCLES
AND PRIME FACTORS...
AND I CAN'T DO IT...
IT DOESN'T MAKE ANY SENSE.
AND LANGER WILL BE BACK ANY MINUTE,
BUT IT ISN'T WHAT PROFESSOR GAUSS IS AFTER.
AFTER A GOOD NIGHT'S SLEEP...
SHOWERED AND REFRESHED...
READY TO FACE THE PROF...
AND I'VE GOT NOTHING...
NOTHING.
I'VE NEVER SEEN ANYONE ACTUALLY PACE BEFORE.
LANGER WILL PROBABLY HAVE A CASE ALL NEATLY READY TO PRESENT TO VON NEUMANN.
WHAT DOESN'T MAKE ANY SENSE?
I'M NOT FOLLOWING...
BY WAY OF A CHANGE...
IN WORDS OF ONE SYLLABLE?

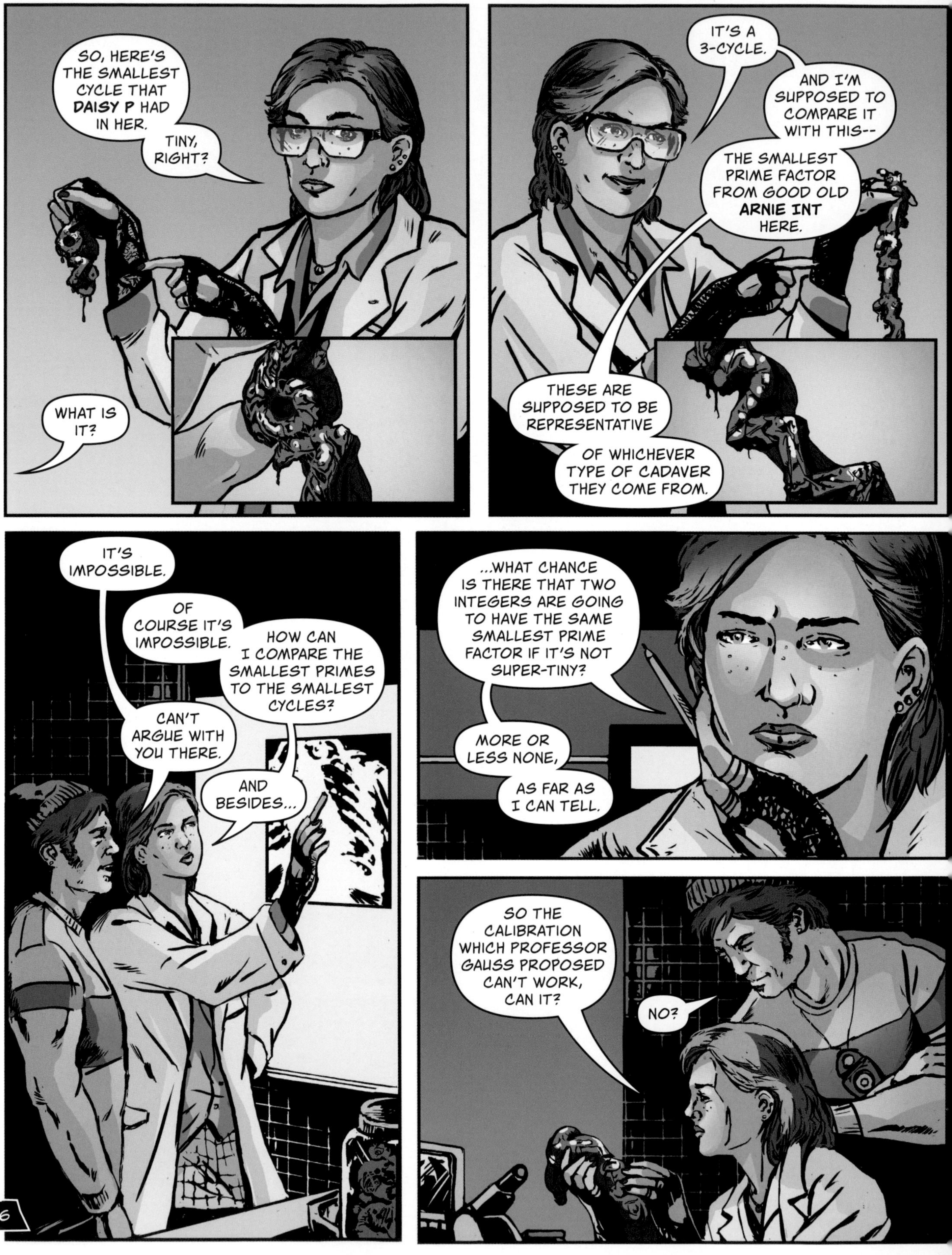
SO, HERE'S THE SMALLEST CYCLE THAT DAISY P HAD IN HER.
TINY, RIGHT?
WHAT IS IT?
IT'S A 3-CYCLE.
AND I'M SUPPOSED TO COMPARE IT WITH THIS--
THE SMALLEST PRIME FACTOR FROM GOOD OLD ARNIE INT HERE.
THESE ARE SUPPOSED TO BE REPRESENTATIVE
OF WHICHEVER TYPE OF CADAVER THEY COME FROM.
IT'S IMPOSSIBLE.
OF COURSE IT'S IMPOSSIBLE.
HOW CAN I COMPARE THE SMALLEST PRIMES TO THE SMALLEST CYCLES?
CAN'T ARGUE WITH YOU THERE.
AND BESIDES...
...WHAT CHANCE IS THERE THAT TWO INTEGERS ARE GOING TO HAVE THE SAME SMALLEST PRIME FACTOR IF IT'S NOT SUPER-TINY?
MORE OR LESS NONE,
AS FAR AS I CAN TELL.
SO THE CALIBRATION WHICH PROFESSOR GAUSS PROPOSED CAN'T WORK, CAN IT?
NO?

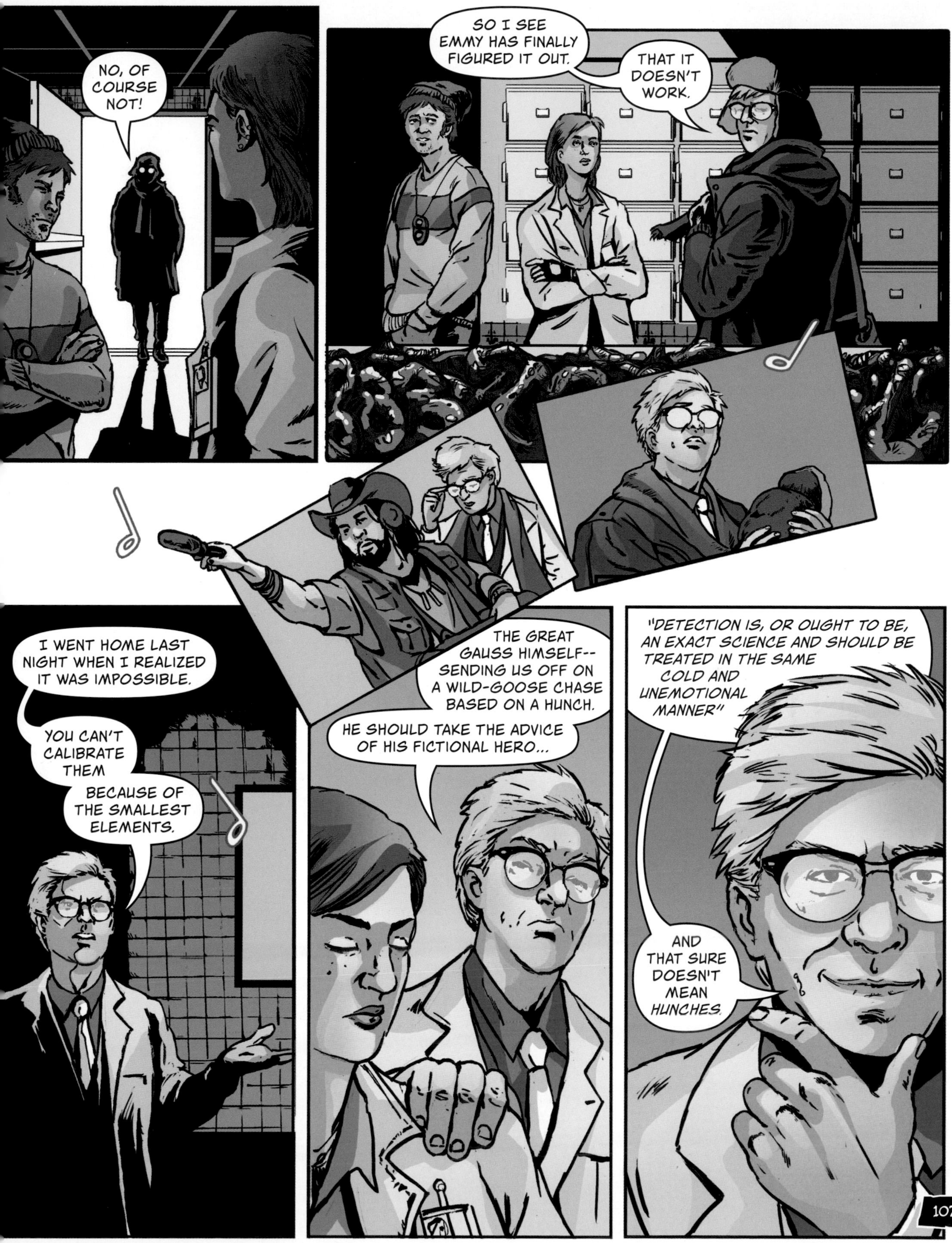
NO, OF COURSE NOT!
SO I SEE EMMY HAS FINALLY FIGURED IT OUT.
THAT IT DOESN'T WORK.
I WENT HOME LAST NIGHT WHEN I REALIZED IT WAS IMPOSSIBLE.
YOU CAN'T CALIBRATE THEM
BECAUSE OF THE SMALLEST ELEMENTS.
THE GREAT GAUSS HIMSELF-- SENDING US OFF ON A WILD-GOOSE CHASE BASED ON A HUNCH.
HE SHOULD TAKE THE ADVICE OF HIS FICTIONAL HERO...
"DETECTION IS, OR OUGHT TO BE, AN EXACT SCIENCE AND SHOULD BE TREATED IN THE SAME COLD AND UNEMOTIONAL MANNER"
AND THAT SURE DOESN'T MEAN HUNCHES.

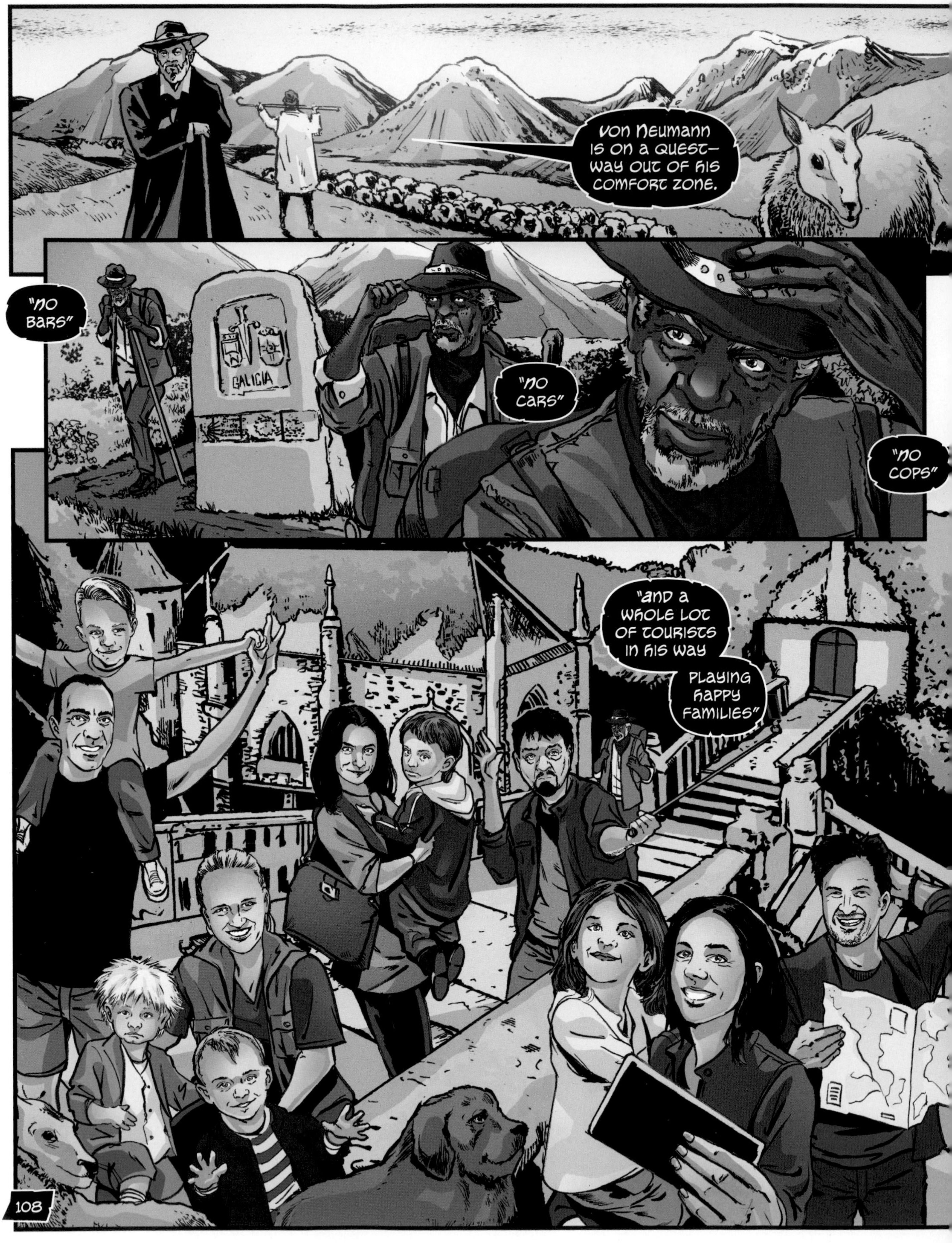
Von Neumann is on a quest—way out of his comfort zone.
"No bars"
GALICIA
"No cars"
"No cops"
"And a whole lot of tourists in his way
playing happy families"

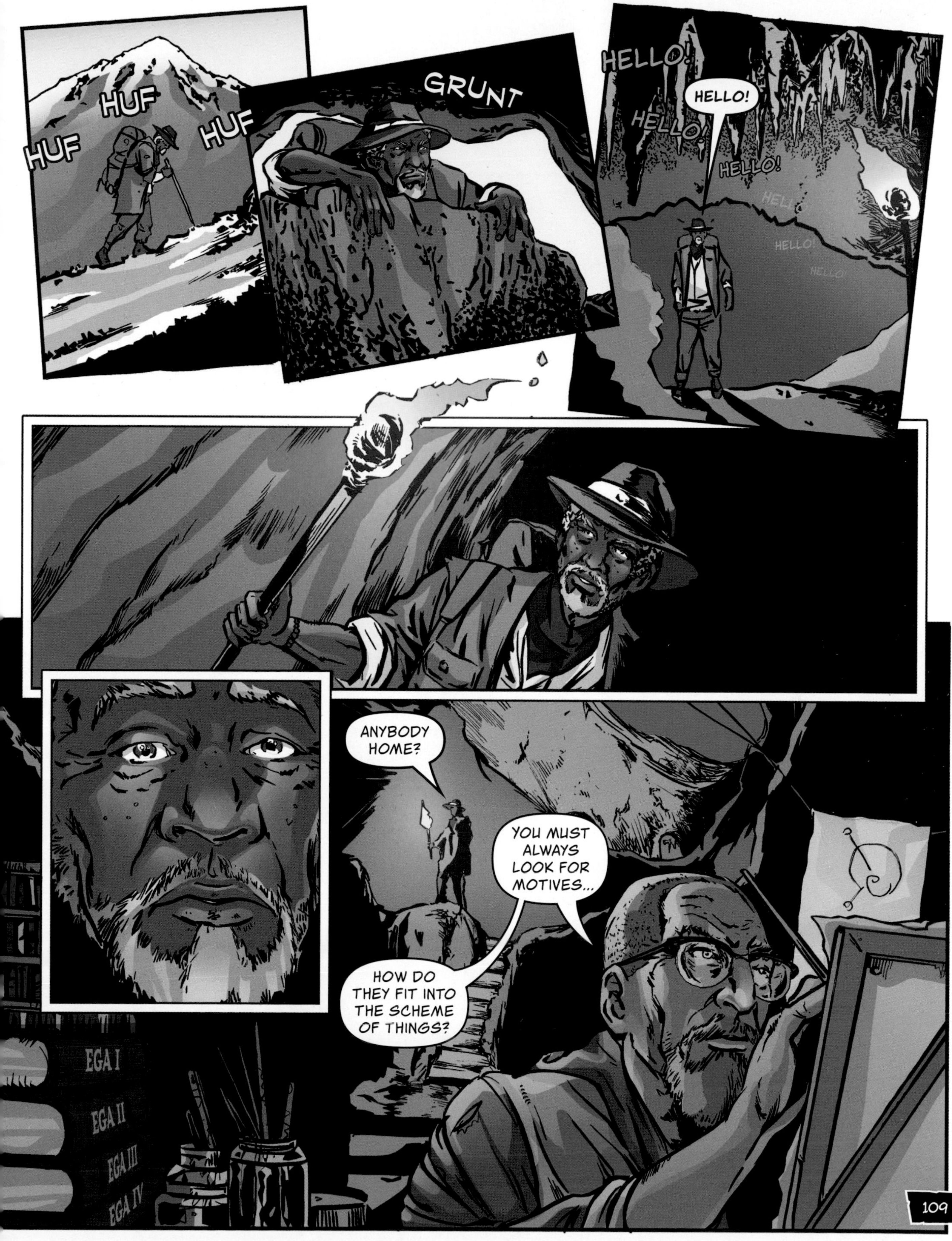

HUF
HUF
HUF
GRUNT
HELLO!
HELLO!
HELLO!
HELLO!
HELLO!
HELLO!
HELLO!
ANYBODY HOME?
YOU MUST ALWAYS LOOK FOR MOTIVES...
HOW DO THEY FIT INTO THE SCHEME OF THINGS?
EGA I
EGA II
EGA III
EGA IV

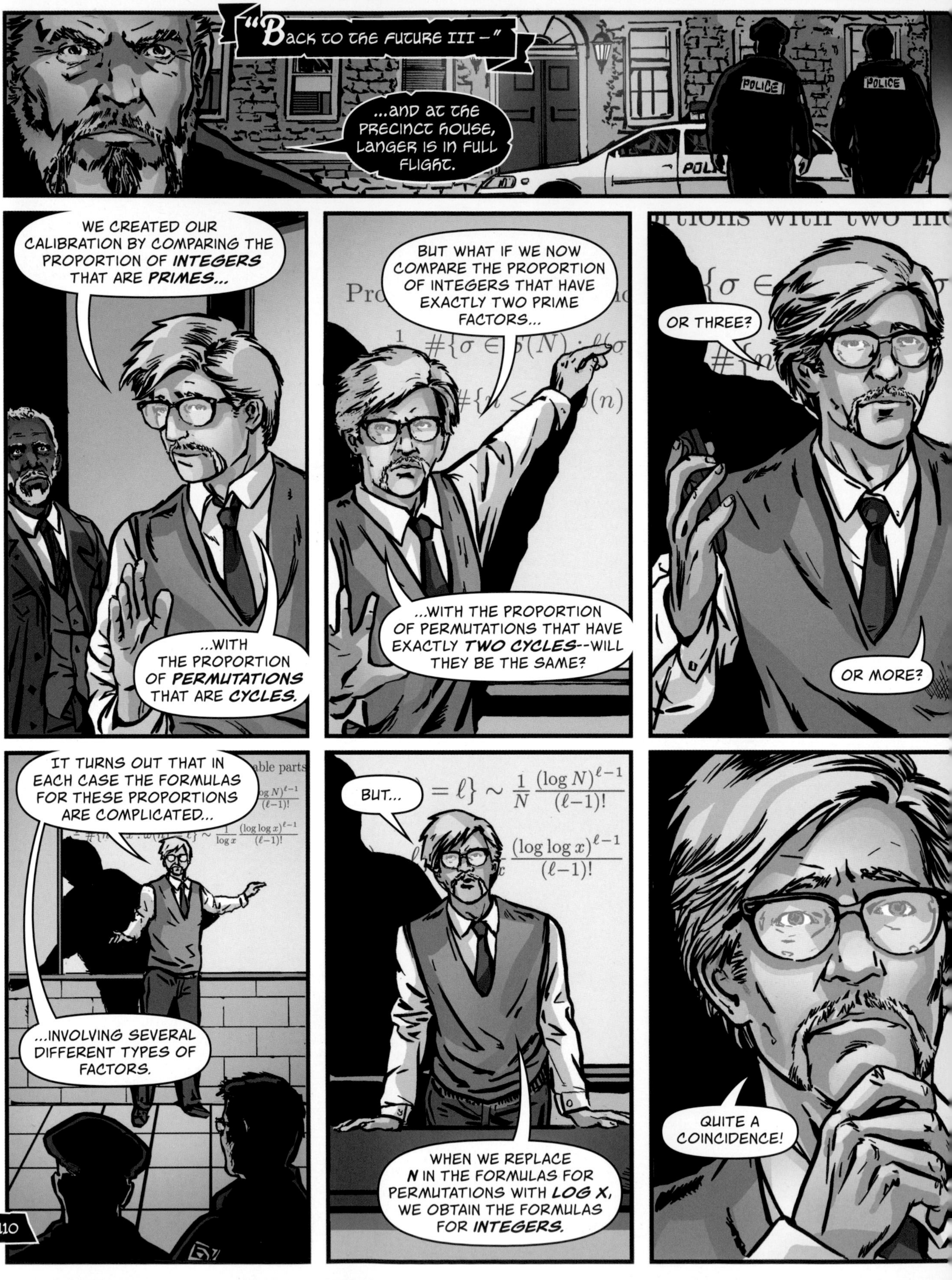
"Back to the Future III –"
...and at the precinct house, Langer is in full flight.
WE CREATED OUR CALIBRATION BY COMPARING THE PROPORTION OF INTEGERS THAT ARE PRIMES...
...WITH THE PROPORTION OF PERMUTATIONS THAT ARE CYCLES.
BUT WHAT IF WE NOW COMPARE THE PROPORTION OF INTEGERS THAT HAVE EXACTLY TWO PRIME FACTORS...
...WITH THE PROPORTION OF PERMUTATIONS THAT HAVE EXACTLY TWO CYCLES--WILL THEY BE THE SAME?
OR THREE?
OR MORE?
IT TURNS OUT THAT IN EACH CASE THE FORMULAS FOR THESE PROPORTIONS ARE COMPLICATED...
...INVOLVING SEVERAL DIFFERENT TYPES OF FACTORS.
BUT...
WHEN WE REPLACE N IN THE FORMULAS FOR PERMUTATIONS WITH LOG X, WE OBTAIN THE FORMULAS FOR INTEGERS.
QUITE A COINCIDENCE!

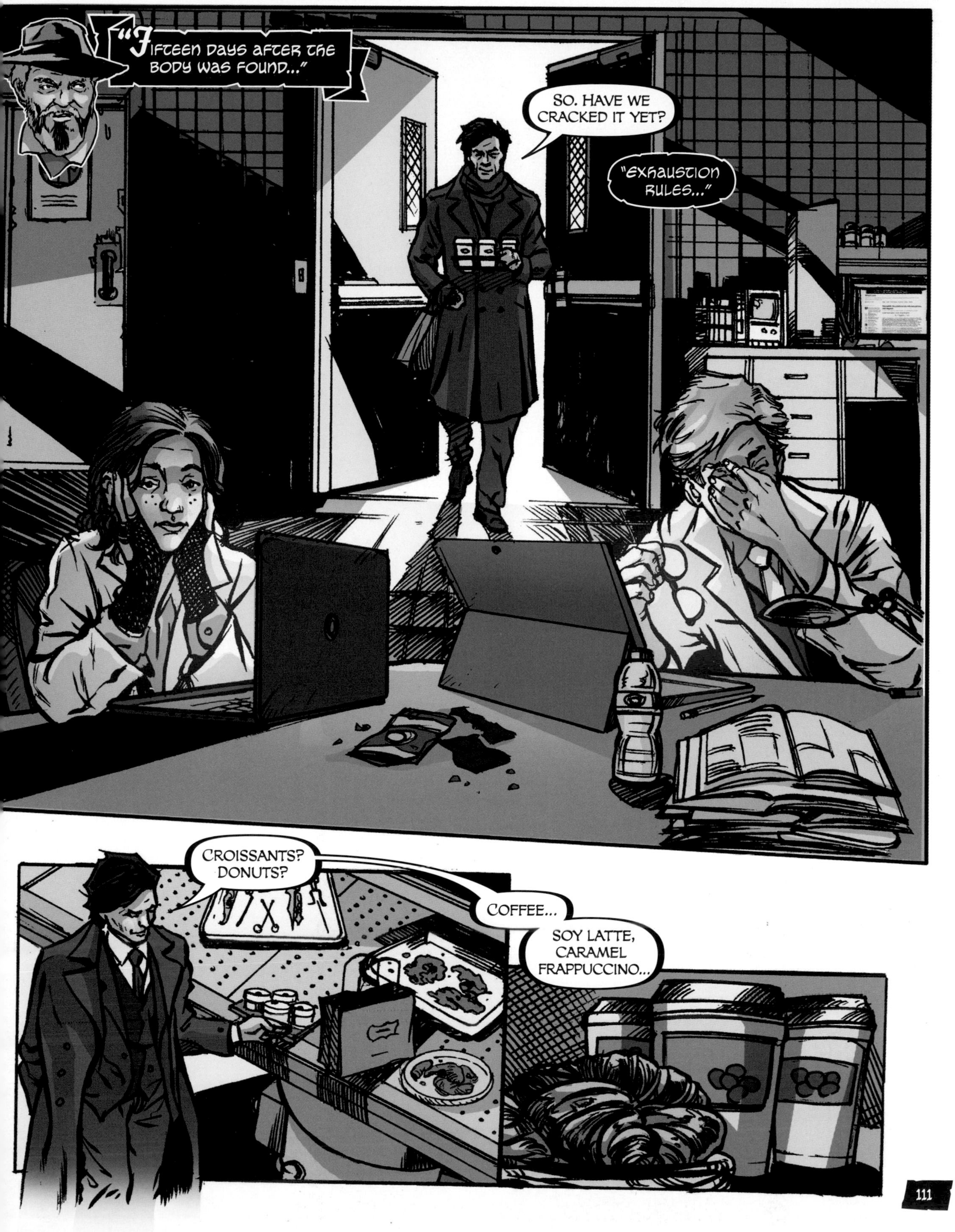
"Fifteen days after the body was found..."
SO. HAVE WE CRACKED IT YET?
"Exhaustion rules..."
CROISSANTS? DONUTS?
COFFEE...
SOY LATTE, CARAMEL FRAPPUCCINO...

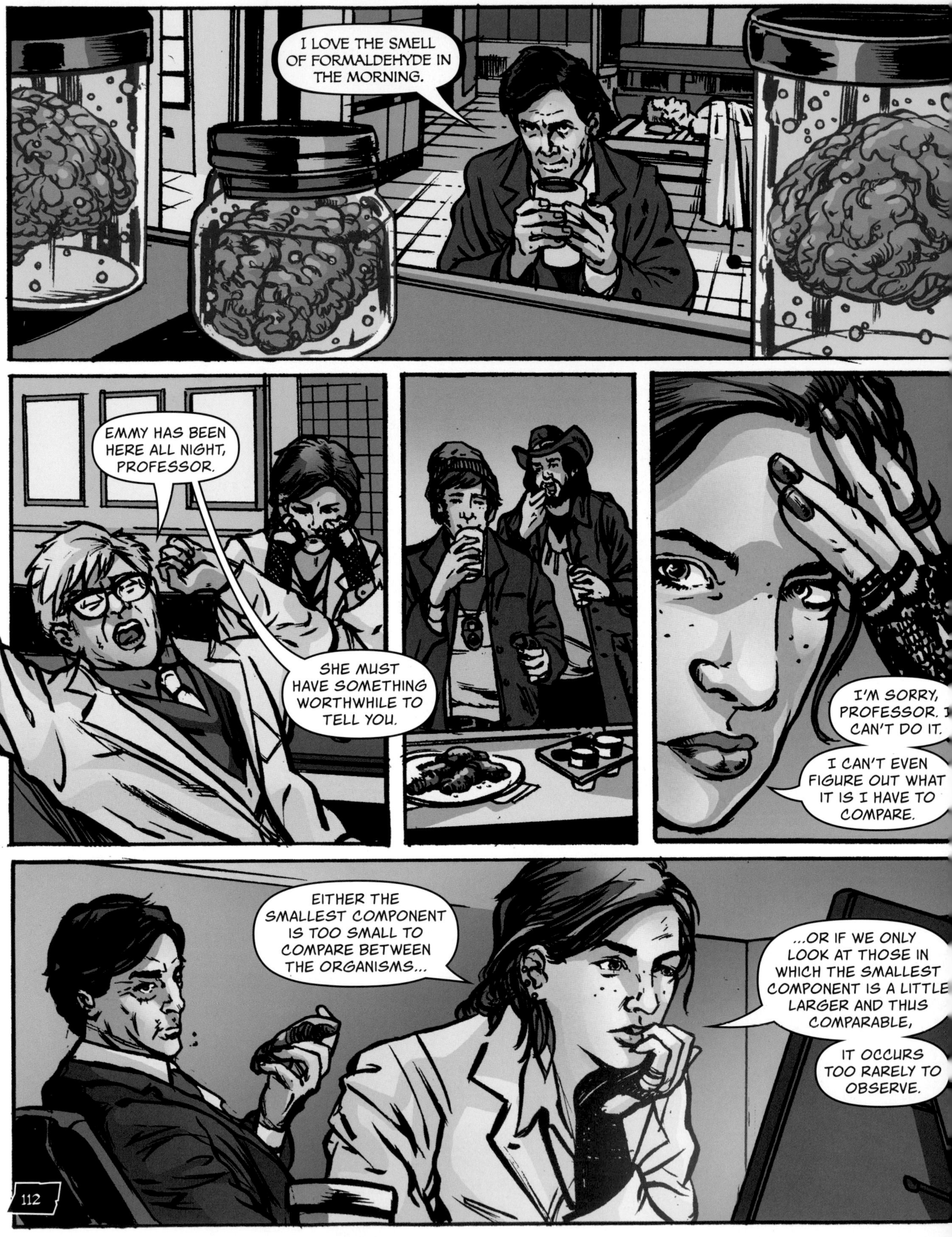
I LOVE THE SMELL OF FORMALDEHYDE IN THE MORNING.
EMMY HAS BEEN HERE ALL NIGHT, PROFESSOR.
SHE MUST HAVE SOMETHING WORTHWHILE TO TELL YOU.
I'M SORRY, PROFESSOR. I CAN'T DO IT.
I CAN'T EVEN FIGURE OUT WHAT IT IS I HAVE TO COMPARE.
EITHER THE SMALLEST COMPONENT IS TOO SMALL TO COMPARE BETWEEN THE ORGANISMS...
...OR IF WE ONLY LOOK AT THOSE IN WHICH THE SMALLEST COMPONENT IS A LITTLE LARGER AND THUS COMPARABLE,
IT OCCURS TOO RARELY TO OBSERVE.

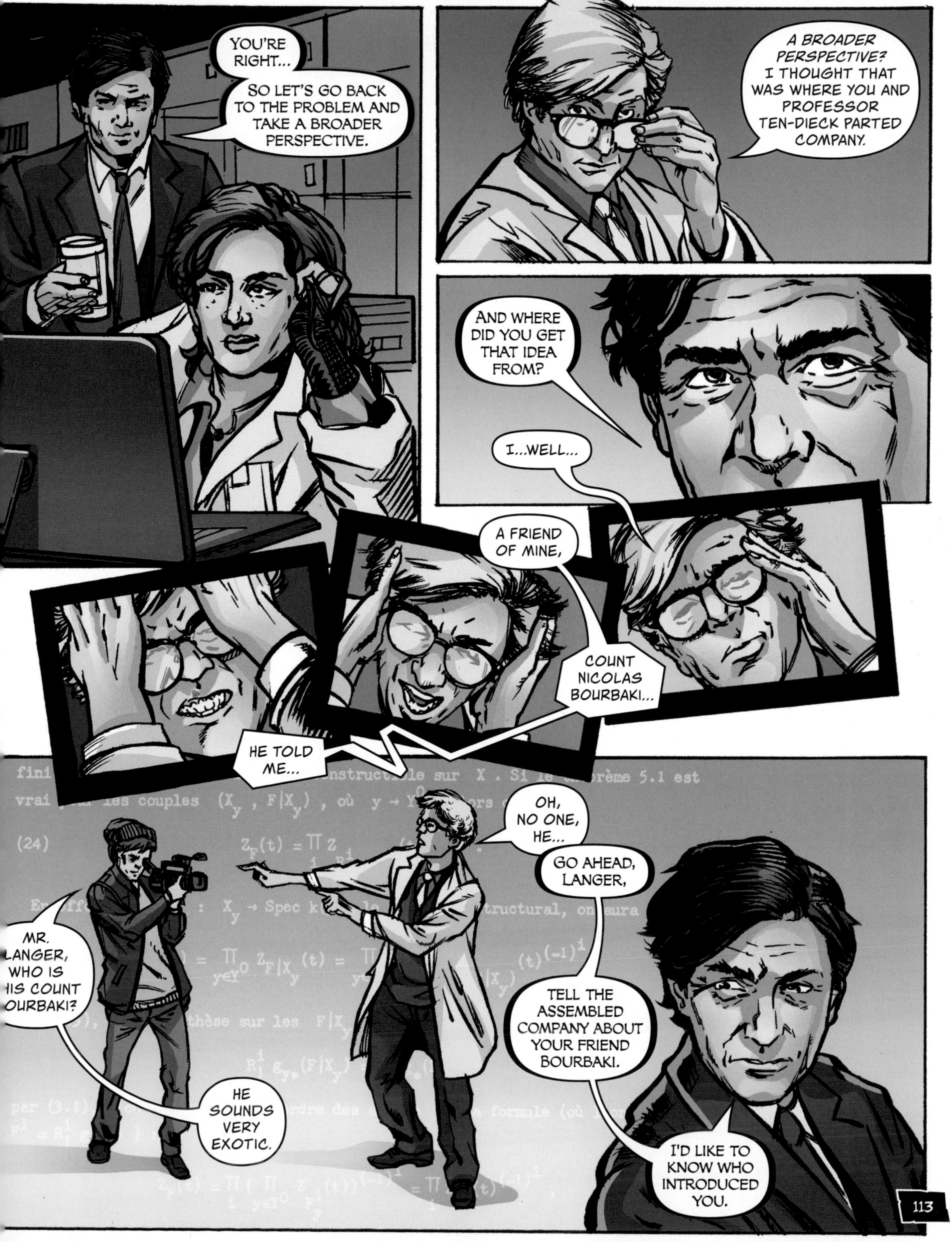
YOU'RE RIGHT...
SO LET'S GO BACK TO THE PROBLEM AND TAKE A BROADER PERSPECTIVE.
A BROADER PERSPECTIVE? I THOUGHT THAT WAS WHERE YOU AND PROFESSOR TEN-DIECK PARTED COMPANY.
AND WHERE DID YOU GET THAT IDEA FROM?
I...WELL...
A FRIEND OF MINE,
COUNT NICOLAS BOURBAKI...
HE TOLD ME...
OH, NO ONE, HE...
GO AHEAD, LANGER,
TELL THE ASSEMBLED COMPANY ABOUT YOUR FRIEND BOURBAKI.
I'D LIKE TO KNOW WHO INTRODUCED YOU.
MR. LANGER, WHO IS HIS COUNT OURBAKI?
HE SOUNDS VERY EXOTIC.

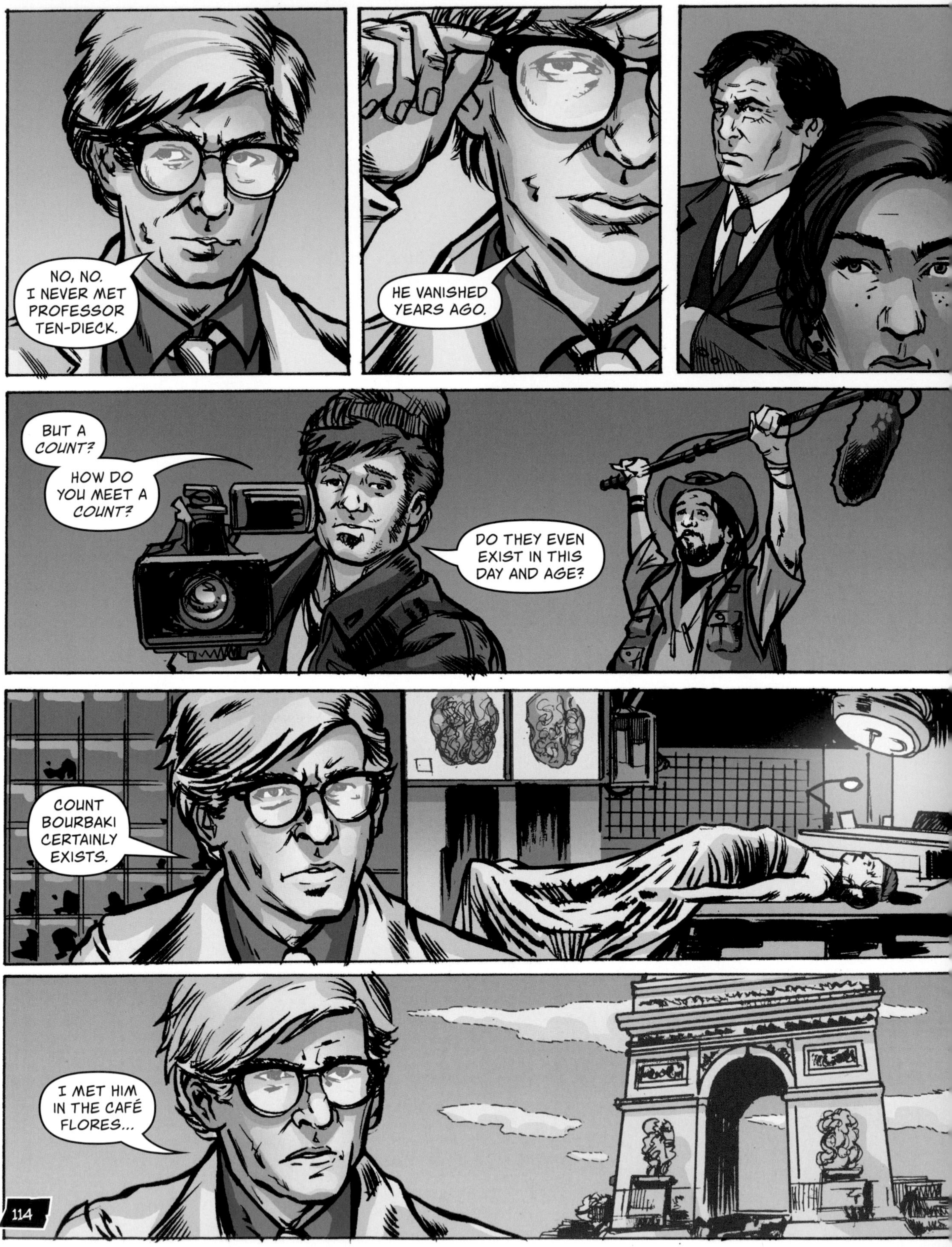
NO, NO.
I NEVER MET PROFESSOR TEN-DIECK.
HE VANISHED YEARS AGO.
BUT A COUNT?
HOW DO YOU MEET A COUNT?
DO THEY EVEN EXIST IN THIS DAY AND AGE?
COUNT BOURBAKI CERTAINLY EXISTS.
I MET HIM IN THE CAFÉ FLORES...

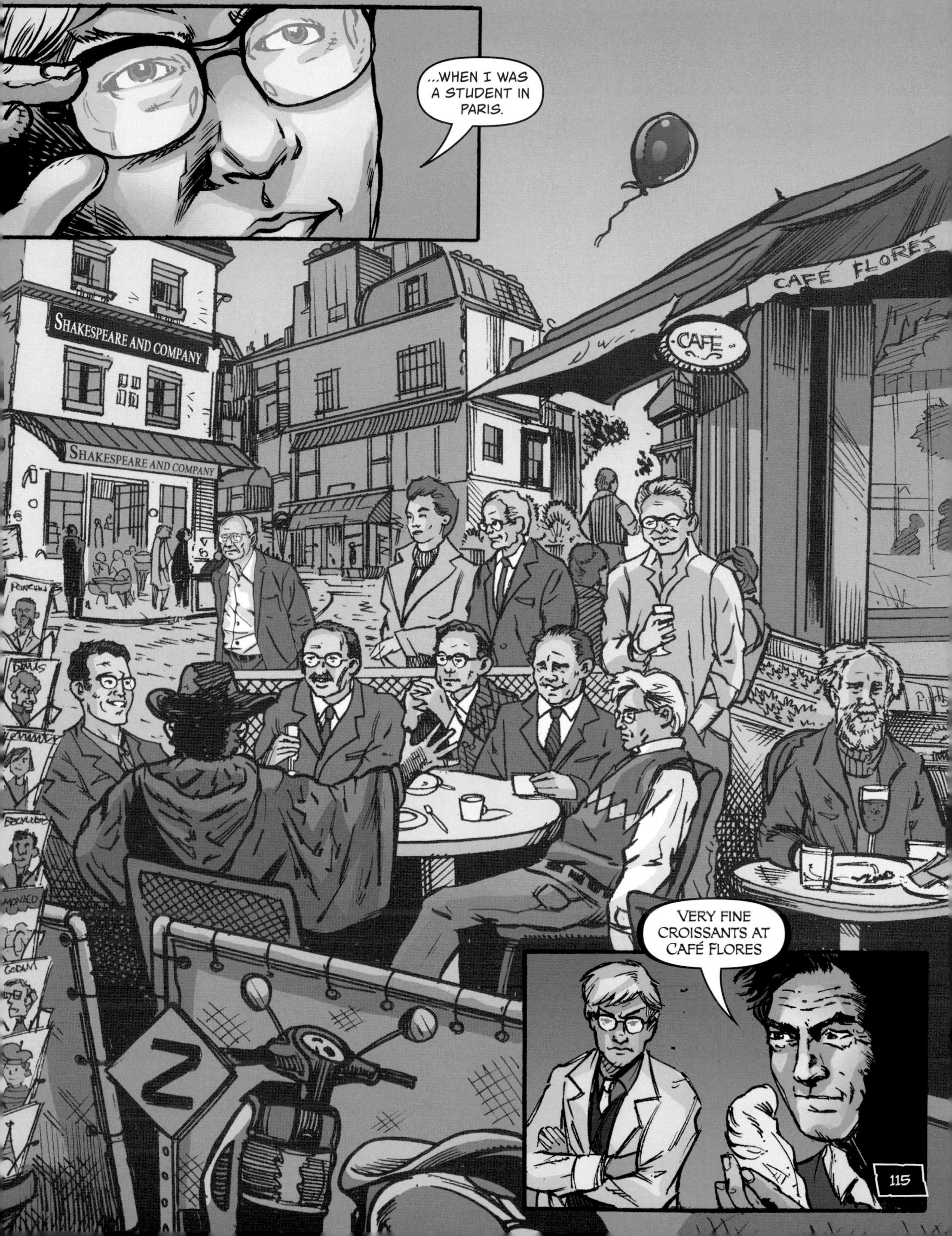

...WHEN I WAS A STUDENT IN PARIS.
CAFÉ FLORES
SHAKESPEARE AND COMPANY
SHAKESPEARE AND COMPANY
CAFÉ
VERY FINE CROISSANTS AT CAFÉ FLORES

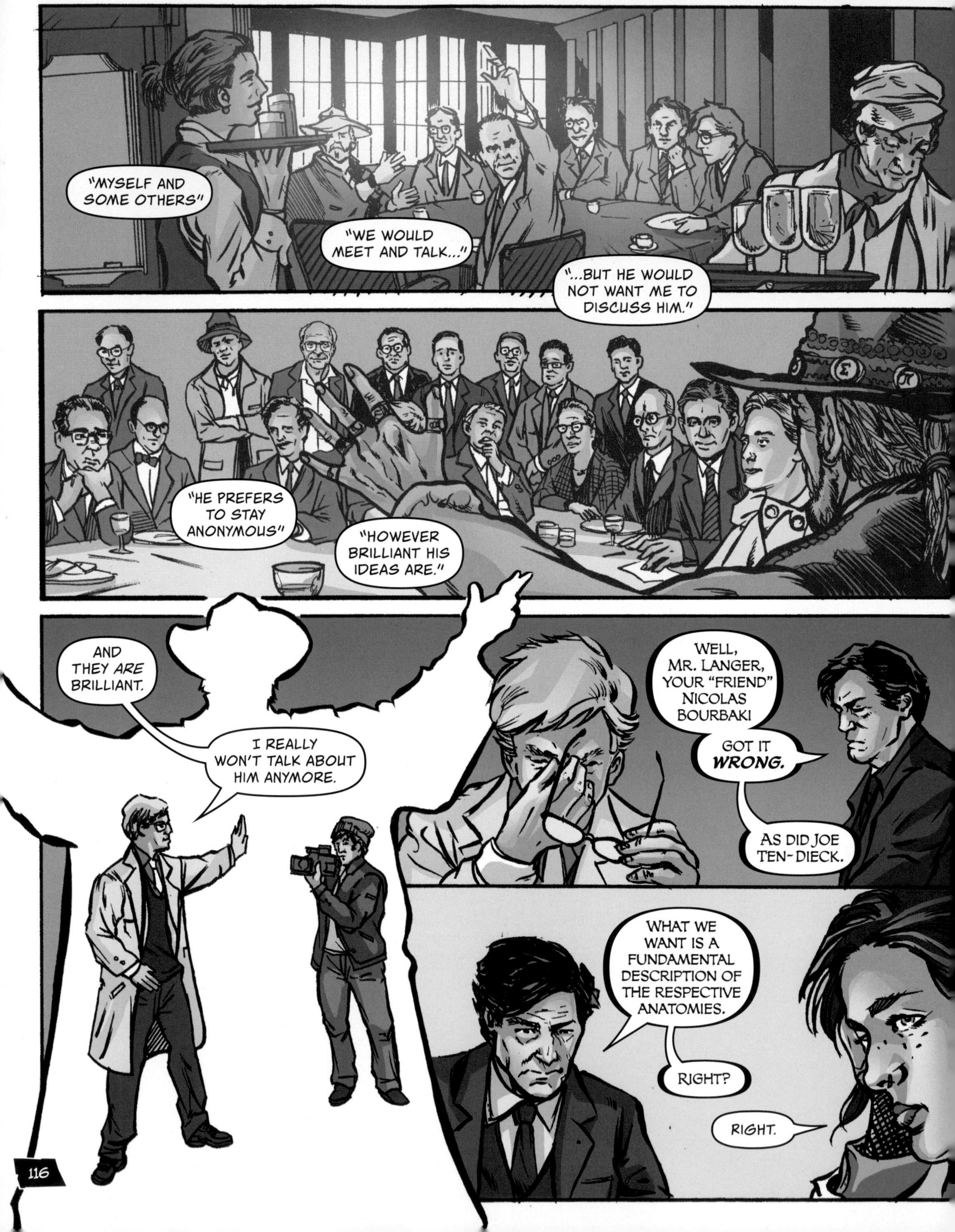

"MYSELF AND SOME OTHERS"
"WE WOULD MEET AND TALK..."
"...BUT HE WOULD NOT WANT ME TO DISCUSS HIM."
"HE PREFERS TO STAY ANONYMOUS"
"HOWEVER BRILLIANT HIS IDEAS ARE."
AND THEY *ARE* BRILLIANT.
I REALLY WON'T TALK ABOUT HIM ANYMORE.
WELL, MR. LANGER, YOUR "FRIEND" NICOLAS BOURBAKI
GOT IT **WRONG.**
AS DID JOE TEN-DIECK.
WHAT WE WANT IS A FUNDAMENTAL DESCRIPTION OF THE RESPECTIVE ANATOMIES.
RIGHT?
RIGHT.

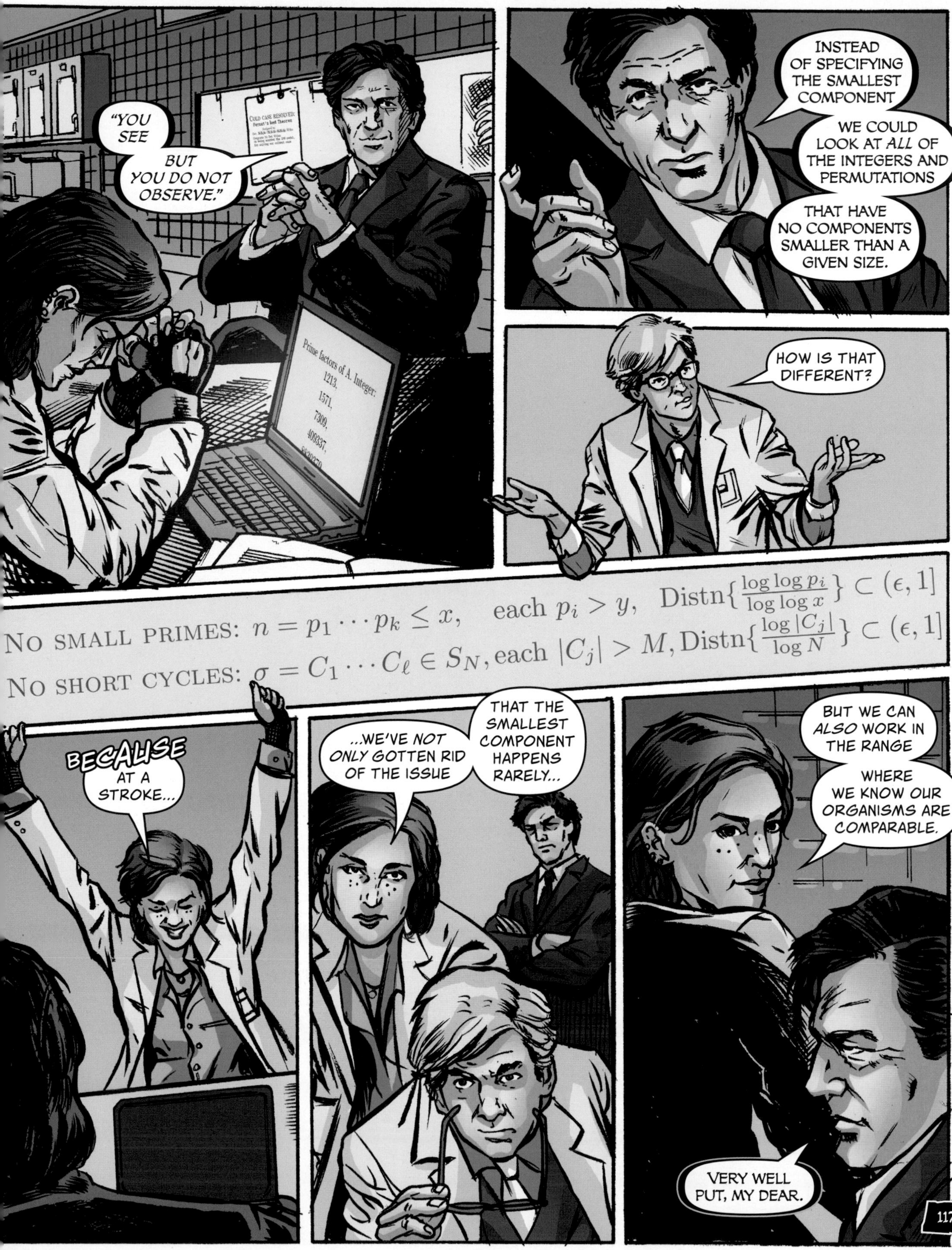
"YOU SEE
BUT YOU DO NOT OBSERVE."
COLD CASE RESOLVED: Fermat's Last Theorem
Prime factors of A. Integer: 1213, 1571, 7309, 409337,
INSTEAD OF SPECIFYING THE SMALLEST COMPONENT
WE COULD LOOK AT *ALL* OF THE INTEGERS AND PERMUTATIONS
THAT HAVE NO COMPONENTS SMALLER THAN A GIVEN SIZE.
HOW IS THAT DIFFERENT?
No SMALL PRIMES: $n = p_1 \cdots p_k \leq x, \quad \text{each } p_i > y, \quad \text{Distn}\{\frac{\log\log p_i}{\log\log x}\} \subset (\epsilon, 1]$
No SHORT CYCLES: $\sigma = C_1 \cdots C_\ell \in S_N, \text{each } |C_j| > M, \text{Distn}\{\frac{\log|C_j|}{\log N}\} \subset (\epsilon, 1]$
BECAUSE AT A STROKE...
...WE'VE *NOT* ONLY GOTTEN RID OF THE ISSUE
THAT THE SMALLEST COMPONENT HAPPENS RARELY...
BUT WE CAN *ALSO* WORK IN THE RANGE
WHERE WE KNOW OUR ORGANISMS ARE COMPARABLE.
VERY WELL PUT, MY DEAR.

THAT'S *GENIUS*, PROFESSOR!

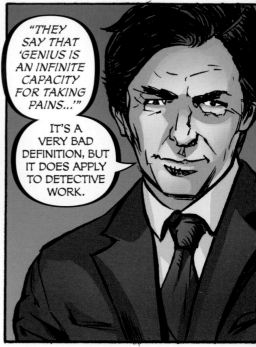

"THEY SAY THAT 'GENIUS IS AN INFINITE CAPACITY FOR TAKING PAINS...'"

IT'S A VERY BAD DEFINITION, BUT IT DOES APPLY TO DETECTIVE WORK.

...THE VALLEY OF FEAR...

HOUND OF THE BASKERVILLES...?

?????

.....

THE SIGN OF FOUR ???

A STUDY IN SCARLET!!!!

YES, MR. LANGER, YES.

NOW LET'S HOPE YOU CAN APPLY WHAT YOU HAVE LEARNED

FROM BOTH ME...

AND FROM MR. SHERLOCK HOLMES.

"SIXTEEN DAYS AFTER THE BODY WAS FOUND"

"VON NEUMANN FLIES HOME..."

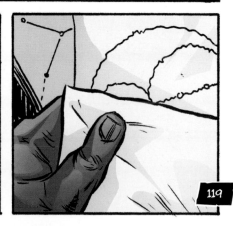

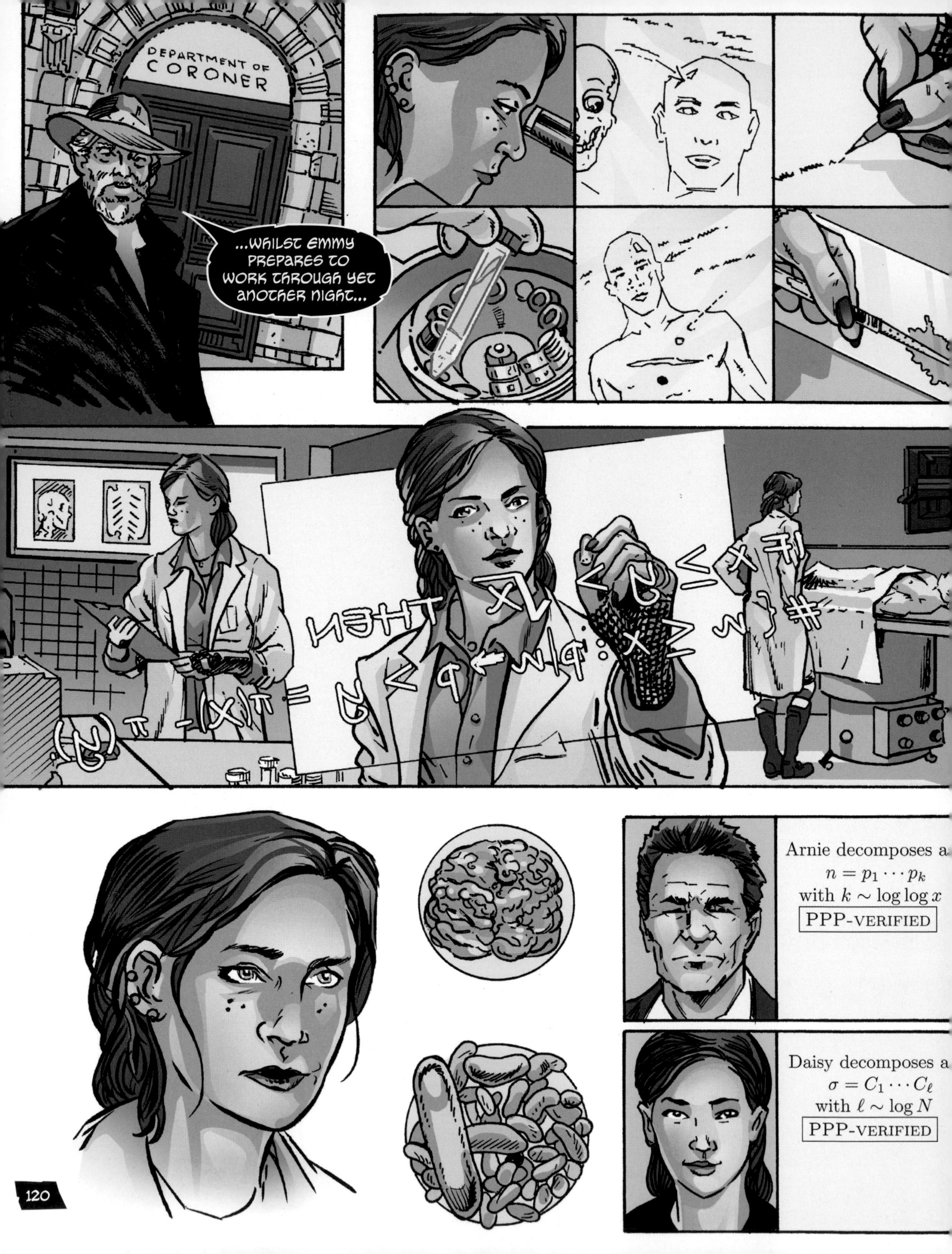
DEPARTMENT OF CORONER
...WHILST EMMY PREPARES TO WORK THROUGH YET ANOTHER NIGHT...
Arnie decomposes a $n = p_1 \cdots p_k$ with $k \sim \log \log x$
PPP-VERIFIED
Daisy decomposes a $\sigma = C_1 \cdots C_\ell$ with $\ell \sim \log N$
PPP-VERIFIED

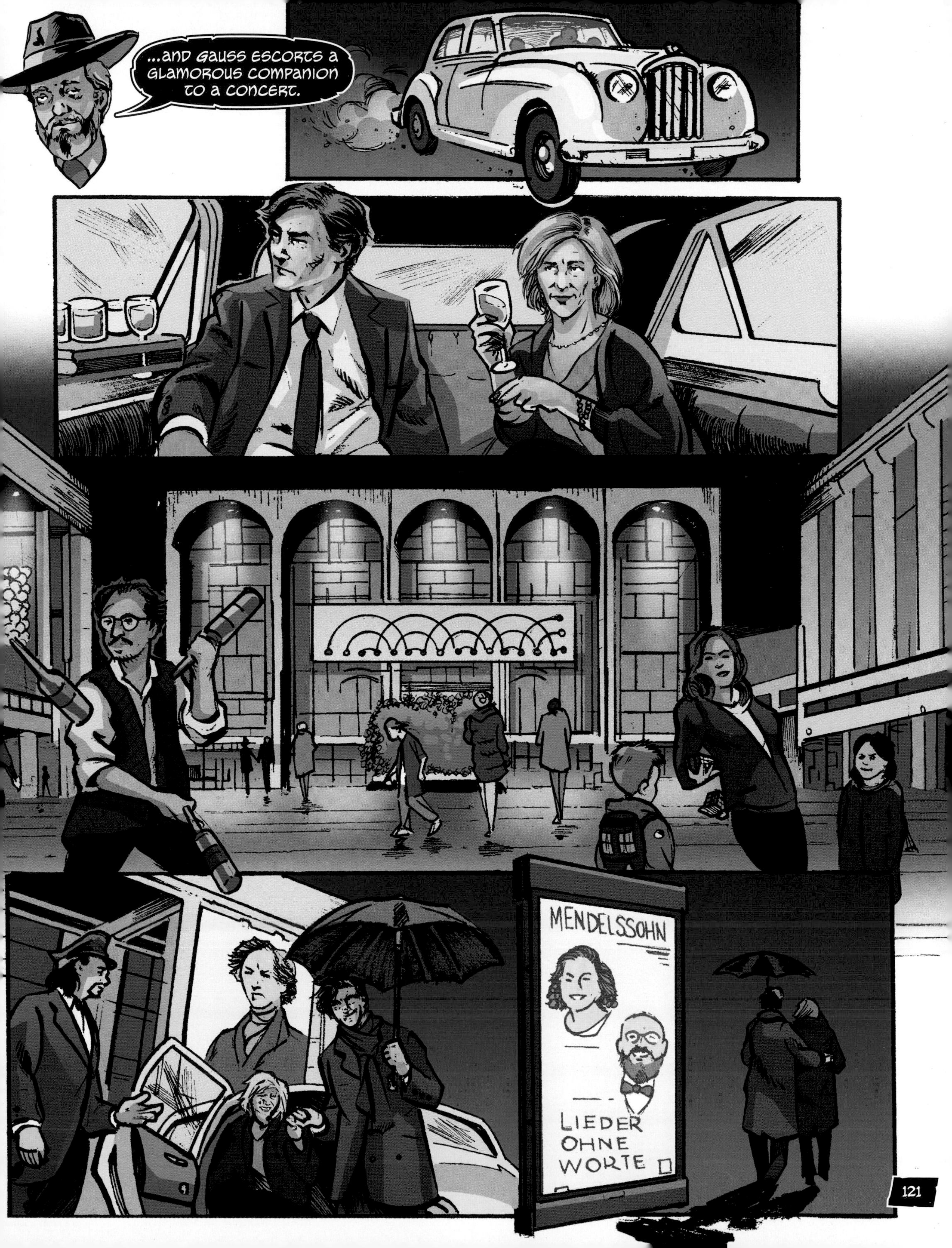
...and Gauss escorts a glamorous companion to a concert.
MENDELSSOHN
LIEDER OHNE WORTE

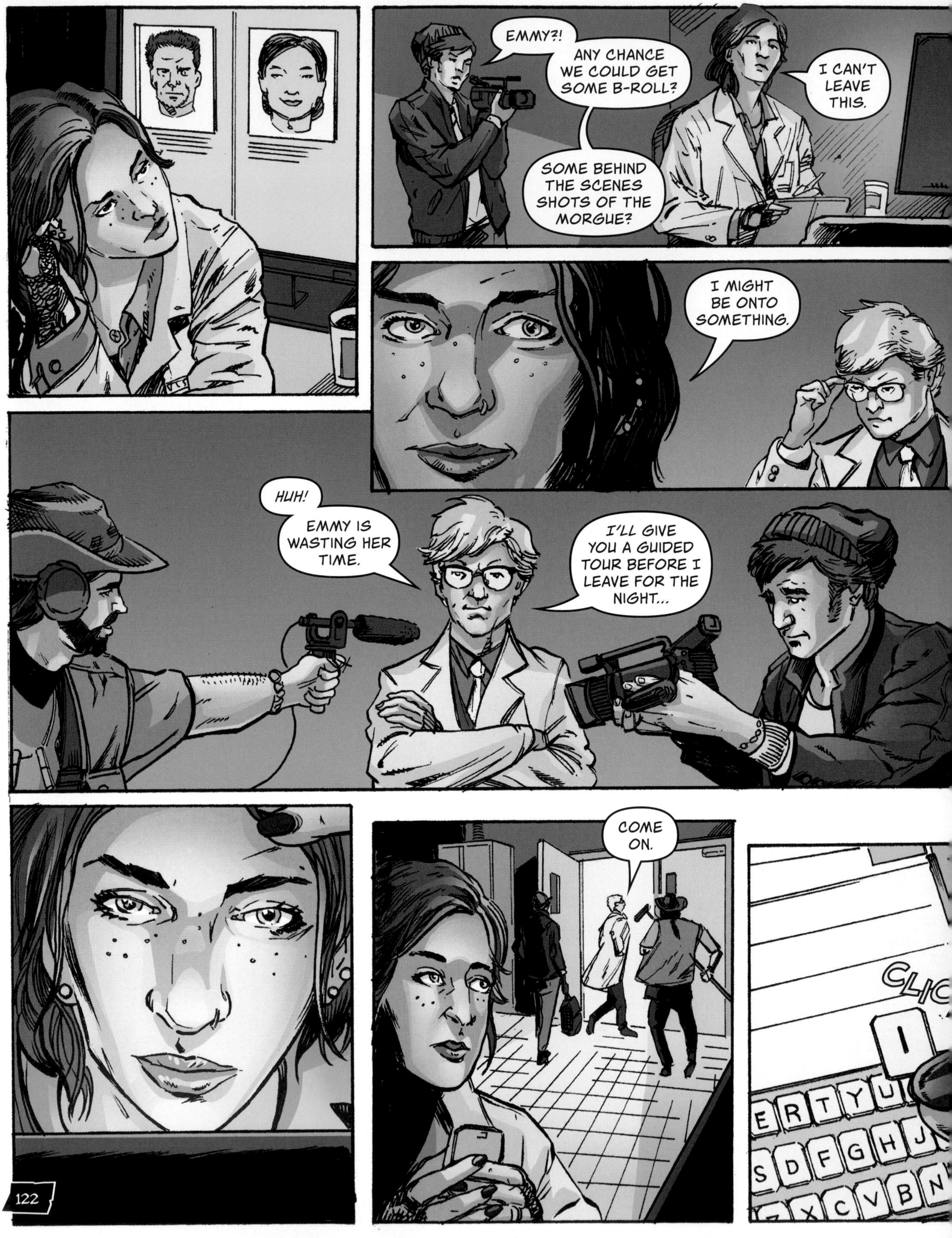
EMMY?!
ANY CHANCE WE COULD GET SOME B-ROLL?
SOME BEHIND THE SCENES SHOTS OF THE MORGUE?
I CAN'T LEAVE THIS.
I MIGHT BE ONTO SOMETHING.
HUH!
EMMY IS WASTING HER TIME.
I'LL GIVE YOU A GUIDED TOUR BEFORE I LEAVE FOR THE NIGHT...
COME ON.
CLIC
I
ERTYU
SDFGHJ
XCVBN

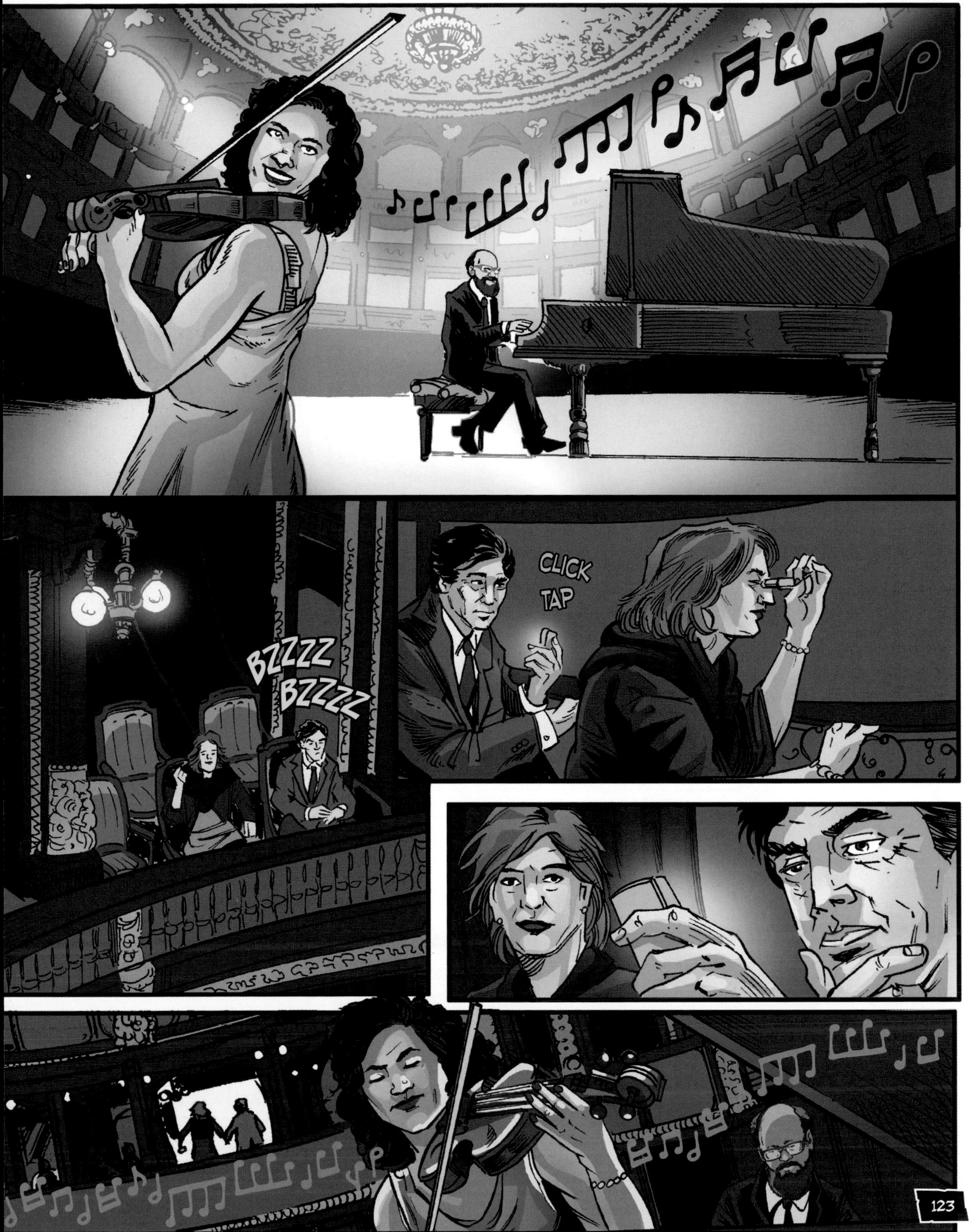
BZZZZ
BZZZZ
CLICK
TAP

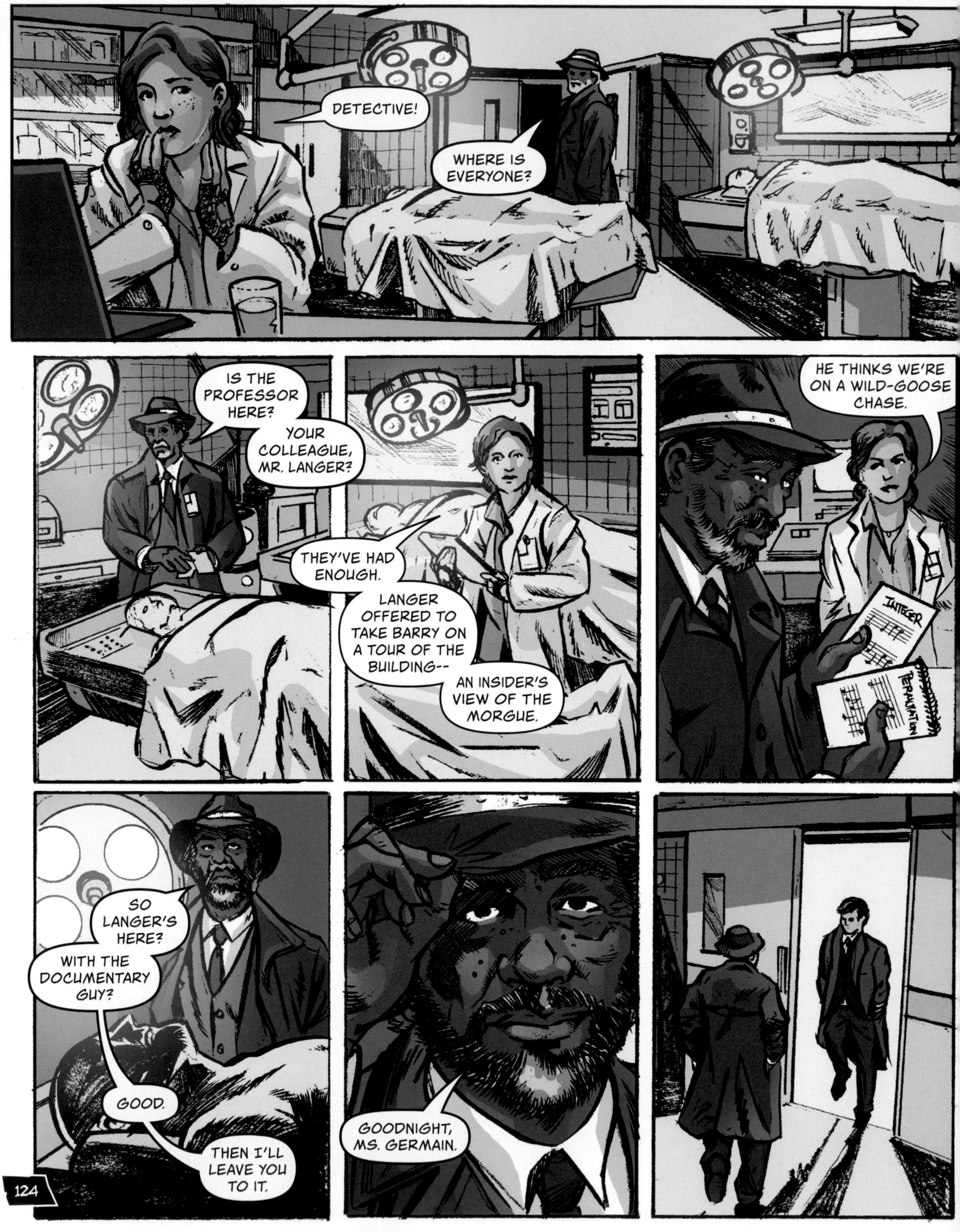
DETECTIVE!
WHERE IS EVERYONE?
IS THE PROFESSOR HERE?
YOUR COLLEAGUE, MR. LANGER?
THEY'VE HAD ENOUGH.
LANGER OFFERED TO TAKE BARRY ON A TOUR OF THE BUILDING--
AN INSIDER'S VIEW OF THE MORGUE.
HE THINKS WE'RE ON A WILD-GOOSE CHASE.
INTEGER
PERMUTATION
SO LANGER'S HERE?
WITH THE DOCUMENTARY GUY?
GOOD.
THEN I'LL LEAVE YOU TO IT.
GOODNIGHT, MS. GERMAIN.

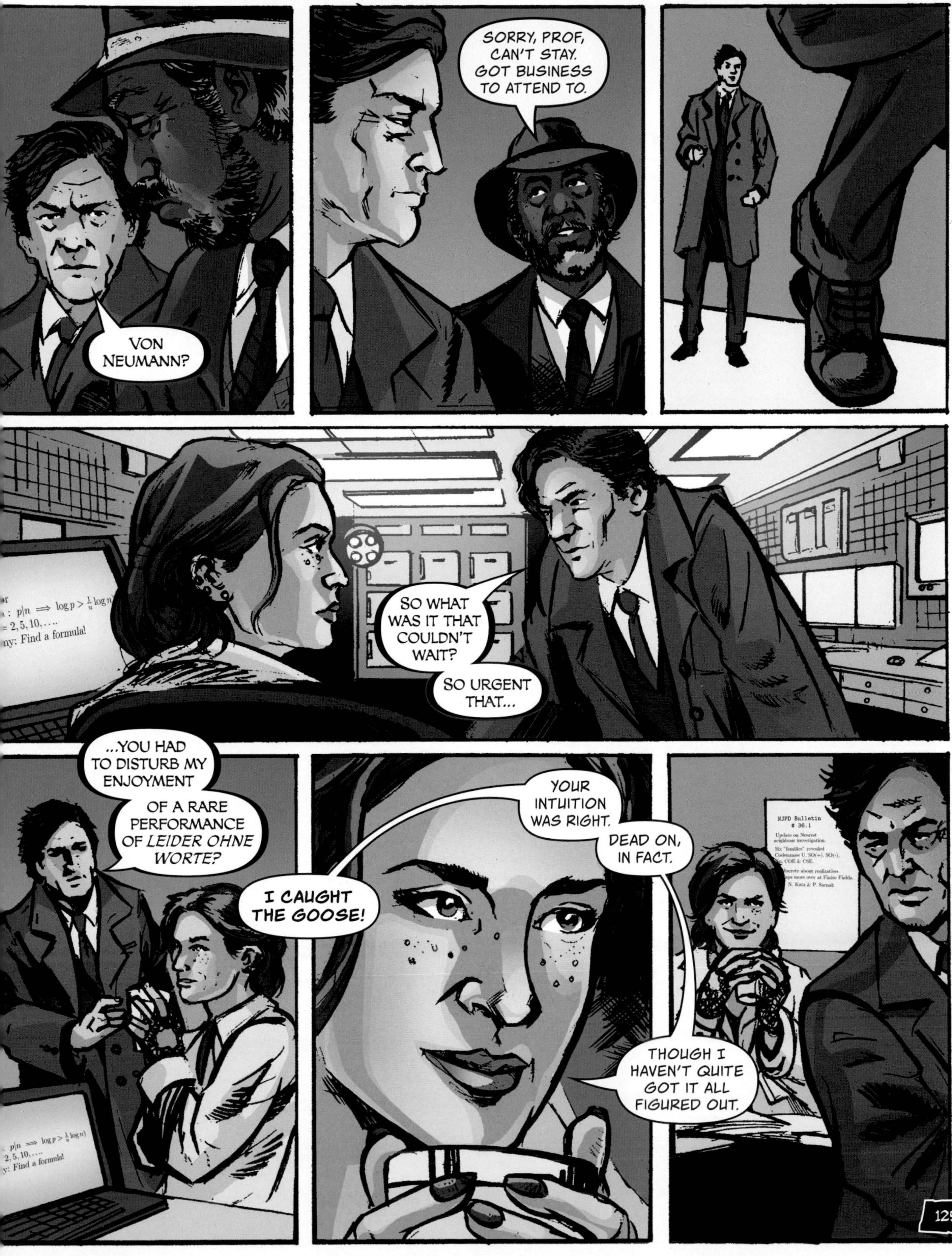

VON NEUMANN?
SORRY, PROF, CAN'T STAY. GOT BUSINESS TO ATTEND TO.
SO WHAT WAS IT THAT COULDN'T WAIT?
SO URGENT THAT...
...YOU HAD TO DISTURB MY ENJOYMENT
OF A RARE PERFORMANCE OF *LEIDER OHNE WORTE*?
I CAUGHT THE GOOSE!
YOUR INTUITION WAS RIGHT.
DEAD ON, IN FACT.
THOUGH I HAVEN'T QUITE GOT IT ALL FIGURED OUT.
NJPD Bulletin
# 36.1
Update on Nearest neighbour investigation.
Six "families" revealed
Codenames U, SO(+), SO(-), COE & CSE.
N. Katz & P. Sarnak

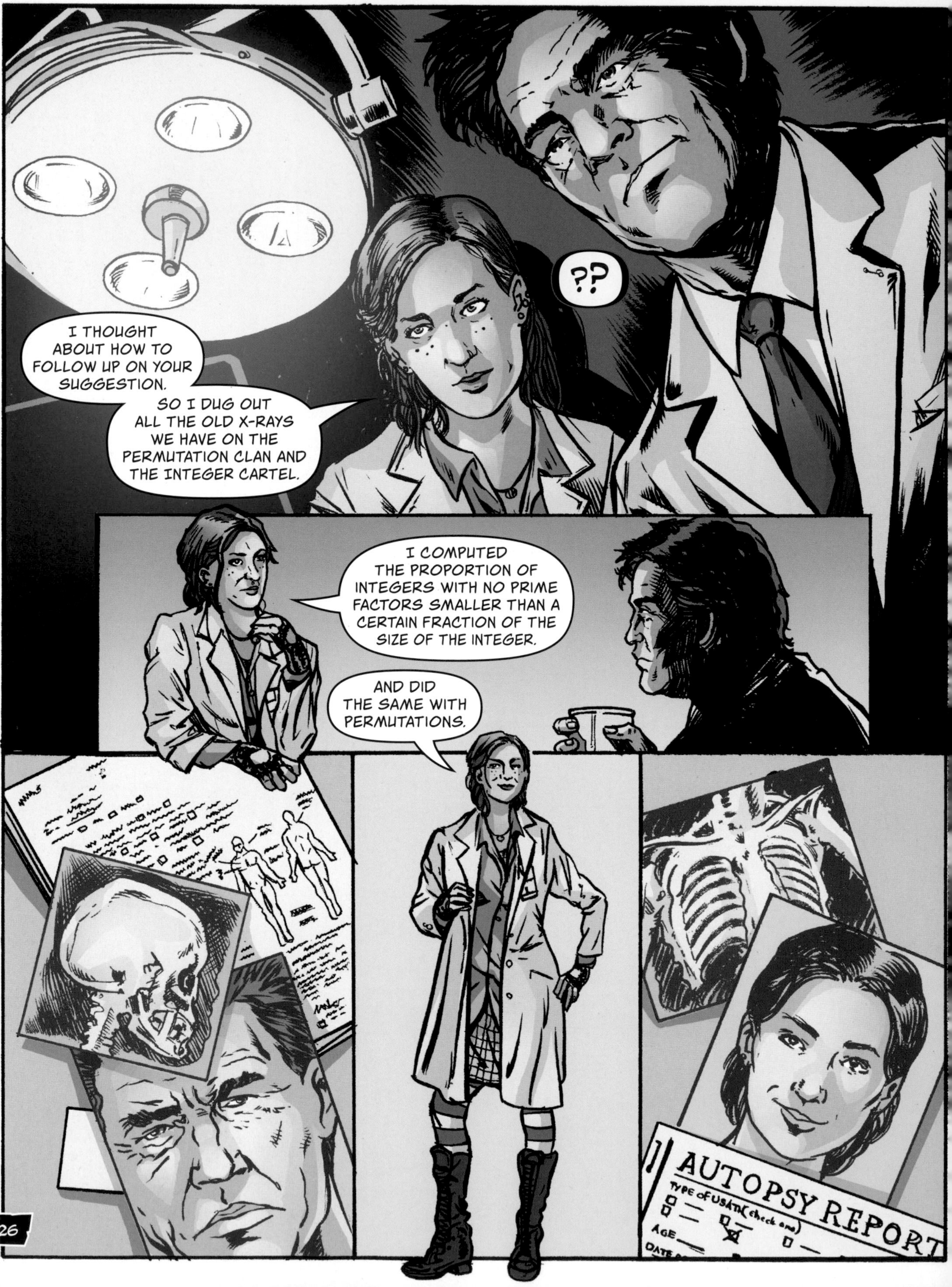
I THOUGHT ABOUT HOW TO FOLLOW UP ON YOUR SUGGESTION.
??
SO I DUG OUT ALL THE OLD X-RAYS WE HAVE ON THE PERMUTATION CLAN AND THE INTEGER CARTEL.
I COMPUTED THE PROPORTION OF INTEGERS WITH NO PRIME FACTORS SMALLER THAN A CERTAIN FRACTION OF THE SIZE OF THE INTEGER.
AND DID THE SAME WITH PERMUTATIONS.
AUTOPSY REPORT

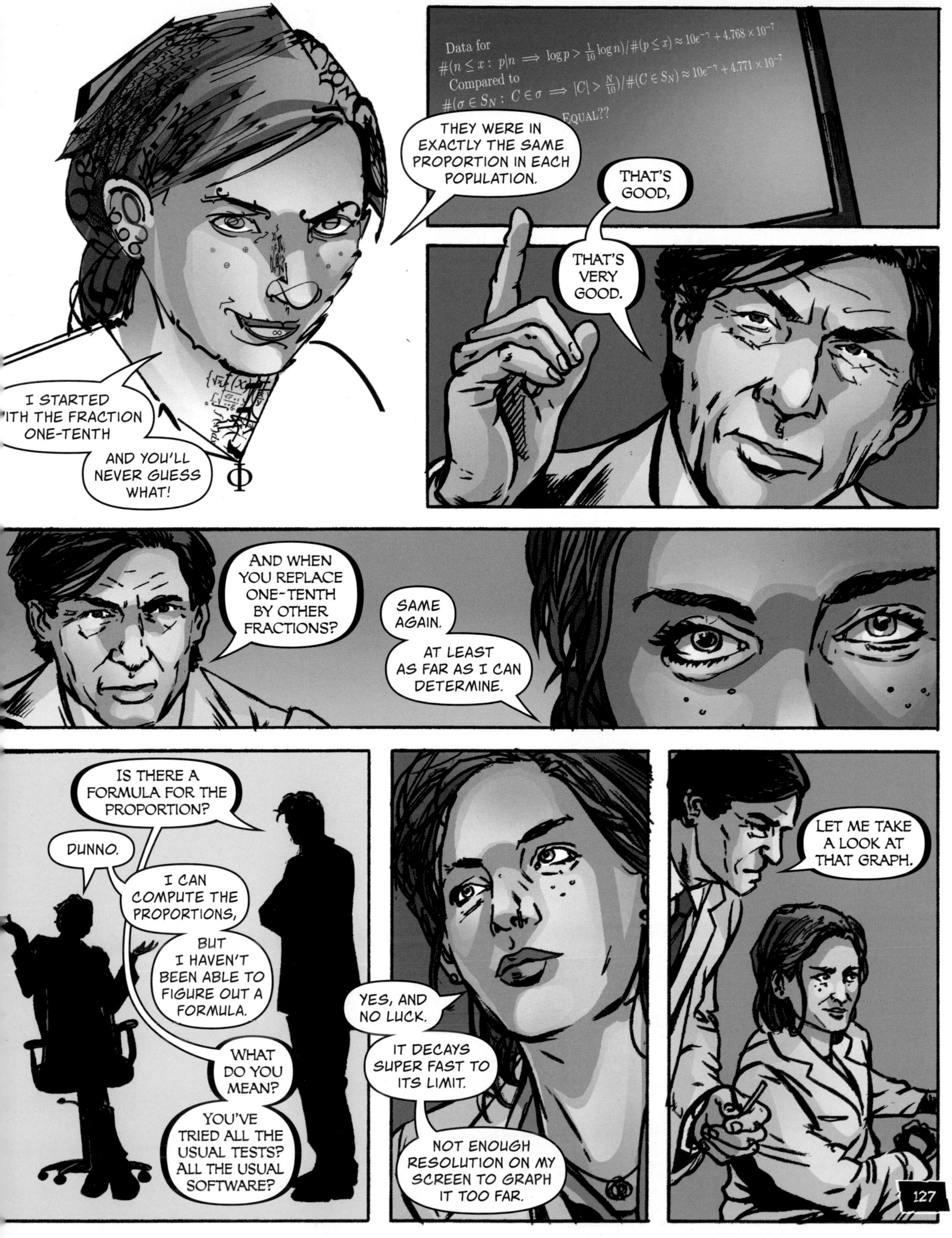

I STARTED ITH THE FRACTION ONE-TENTH
AND YOU'LL NEVER GUESS WHAT!
Data for
$\#(n \le x:\ p|n \implies \log p > \frac{1}{10}\log n)/\#(p \le x) \approx 10e^{-7} + 4.768 \times 10^{-7}$
Compared to
$\#(\sigma \in S_N:\ C \in \sigma \implies |C| > \frac{N}{10})/\#(C \in S_N) \approx 10e^{-7} + 4.771 \times 10^{-7}$
EQUAL??
THEY WERE IN EXACTLY THE SAME PROPORTION IN EACH POPULATION.
THAT'S GOOD,
THAT'S VERY GOOD.
AND WHEN YOU REPLACE ONE-TENTH BY OTHER FRACTIONS?
SAME AGAIN.
AT LEAST AS FAR AS I CAN DETERMINE.
IS THERE A FORMULA FOR THE PROPORTION?
DUNNO.
I CAN COMPUTE THE PROPORTIONS,
BUT I HAVEN'T BEEN ABLE TO FIGURE OUT A FORMULA.
WHAT DO YOU MEAN?
YOU'VE TRIED ALL THE USUAL TESTS? ALL THE USUAL SOFTWARE?
YES, AND NO LUCK.
IT DECAYS SUPER FAST TO ITS LIMIT.
NOT ENOUGH RESOLUTION ON MY SCREEN TO GRAPH IT TOO FAR.
LET ME TAKE A LOOK AT THAT GRAPH.

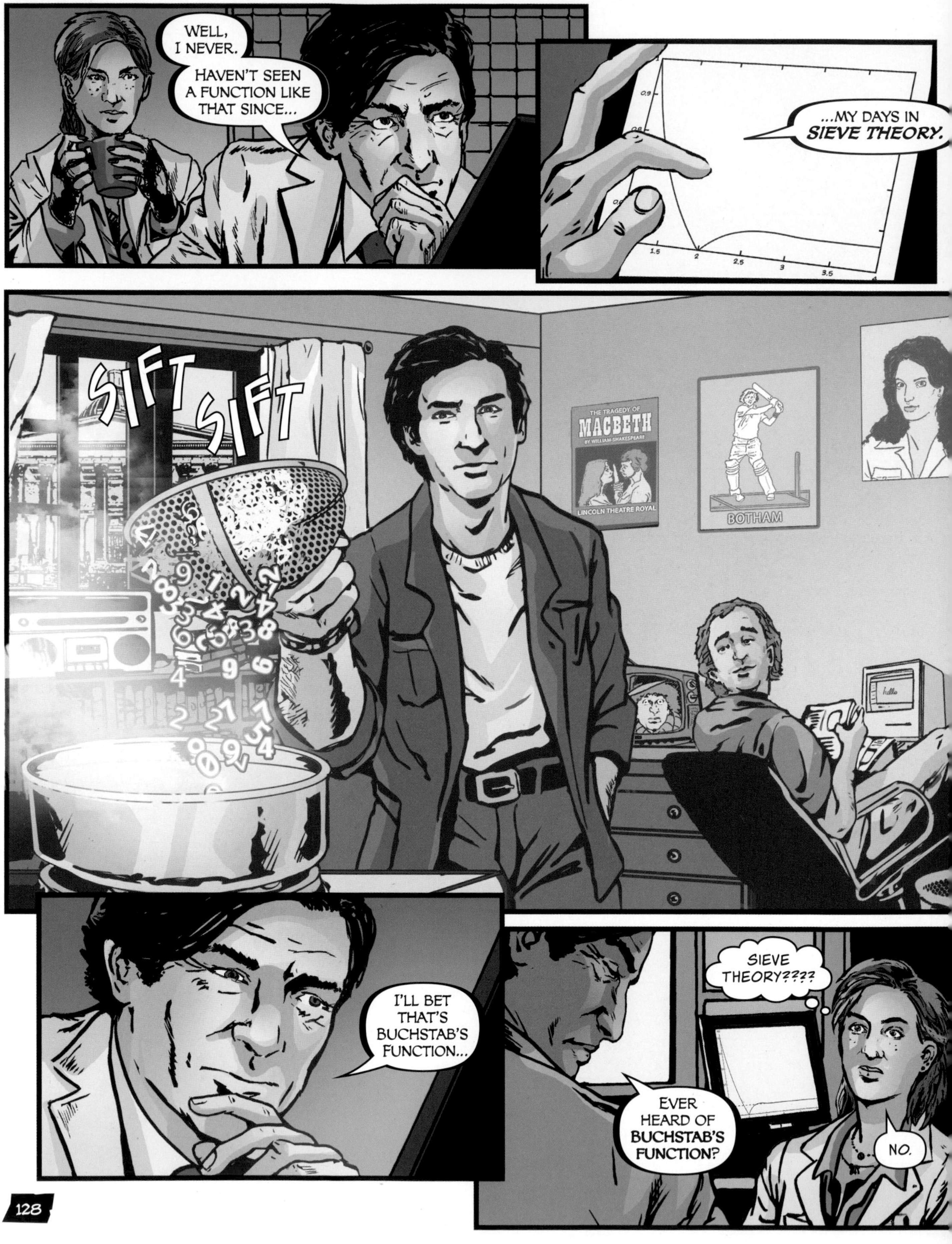
WELL, I NEVER.
HAVEN'T SEEN A FUNCTION LIKE THAT SINCE...
...MY DAYS IN SIEVE THEORY.
SIFT SIFT
THE TRAGEDY OF MACBETH BY WILLIAM SHAKESPEARE
LINCOLN THEATRE ROYAL
BOTHAM
I'LL BET THAT'S BUCHSTAB'S FUNCTION...
SIEVE THEORY????
EVER HEARD OF BUCHSTAB'S FUNCTION?
NO.

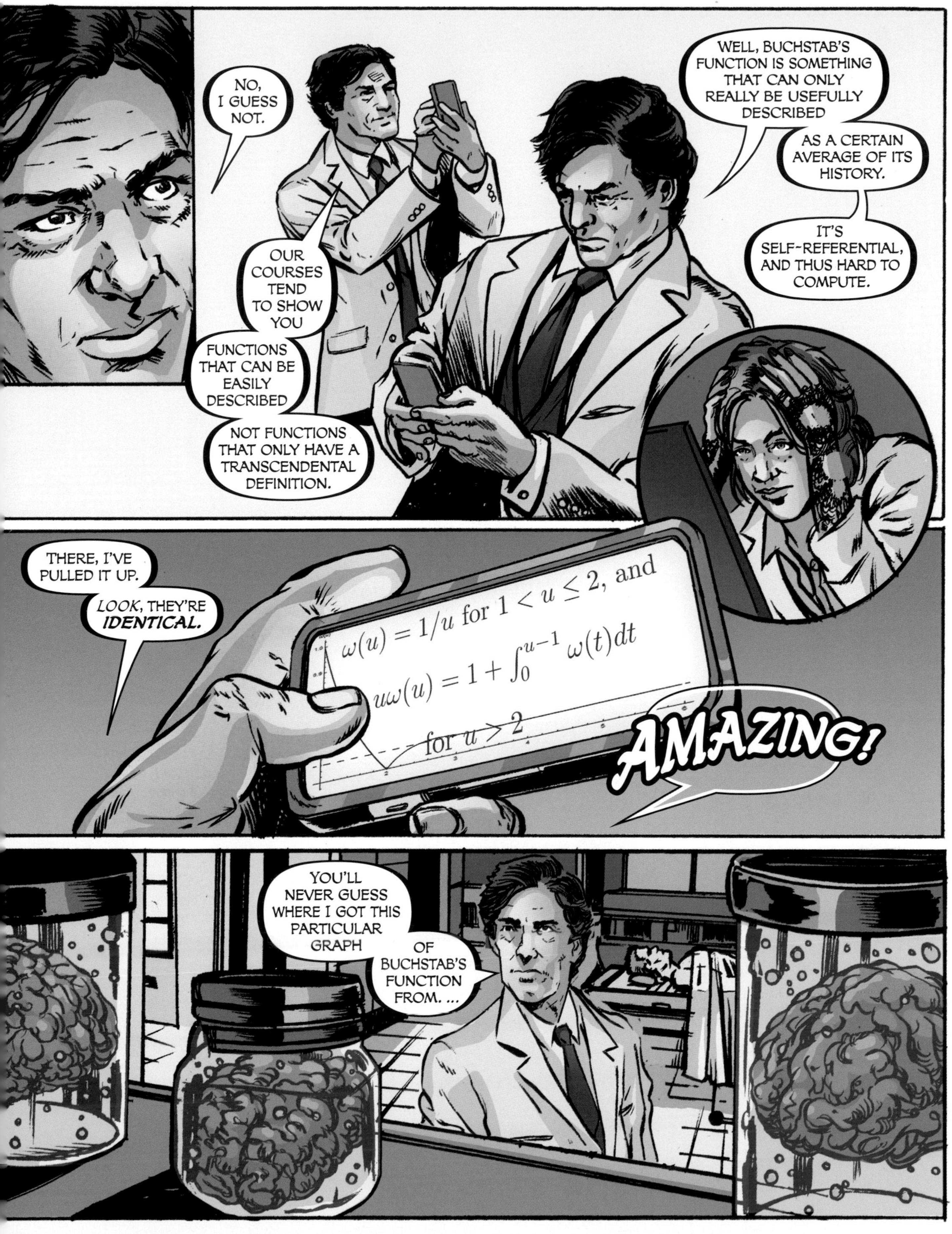
NO, I GUESS NOT.
WELL, BUCHSTAB'S FUNCTION IS SOMETHING THAT CAN ONLY REALLY BE USEFULLY DESCRIBED
AS A CERTAIN AVERAGE OF ITS HISTORY.
IT'S SELF-REFERENTIAL, AND THUS HARD TO COMPUTE.
OUR COURSES TEND TO SHOW YOU
FUNCTIONS THAT CAN BE EASILY DESCRIBED
NOT FUNCTIONS THAT ONLY HAVE A TRANSCENDENTAL DEFINITION.
THERE, I'VE PULLED IT UP.
LOOK, THEY'RE IDENTICAL.
$\omega(u) = 1/u$ for $1 < u \leq 2$, and
$u\omega(u) = 1 + \int_0^{u-1} \omega(t)dt$
for $u > 2$
AMAZING!
YOU'LL NEVER GUESS WHERE I GOT THIS PARTICULAR GRAPH
OF BUCHSTAB'S FUNCTION FROM. ...

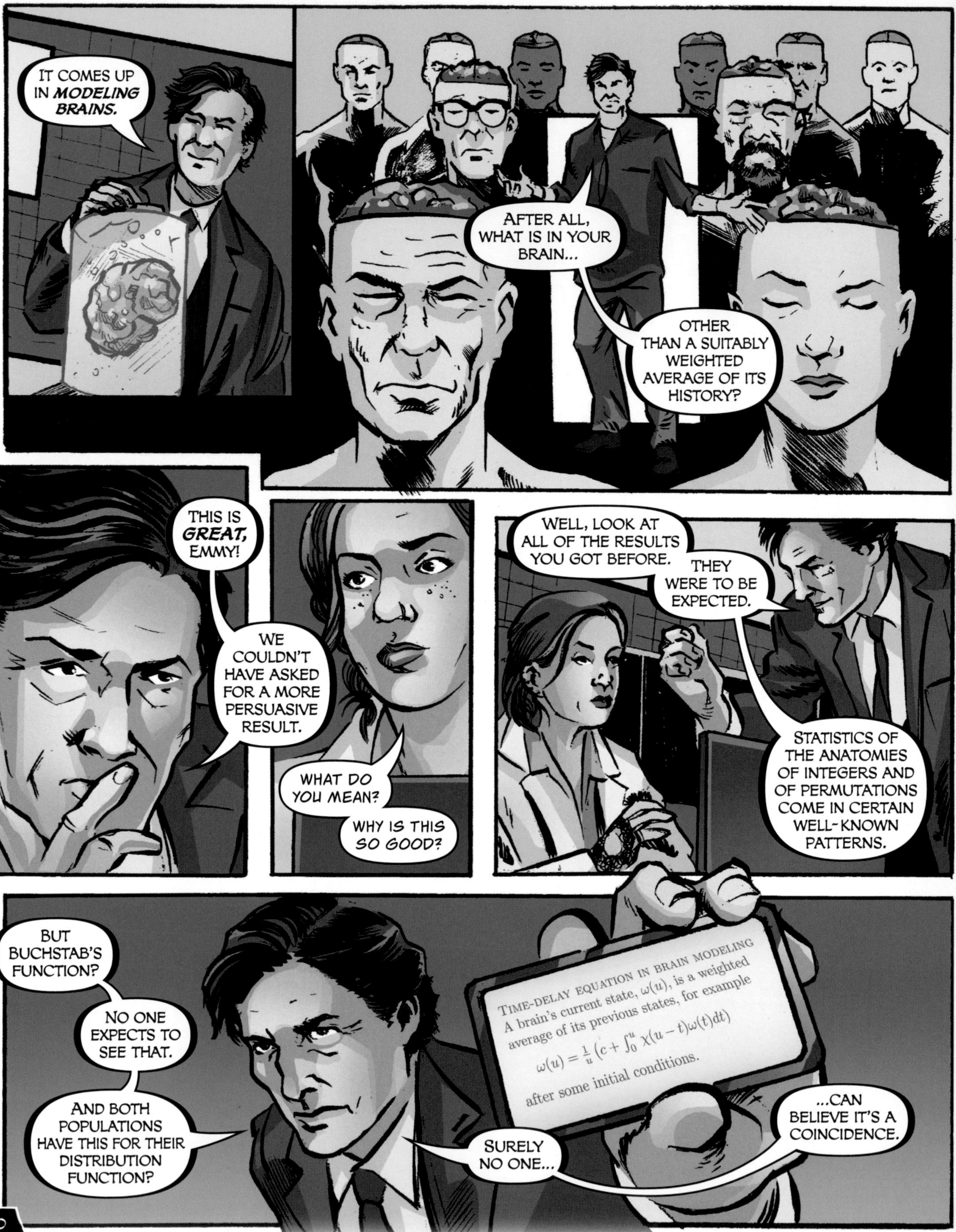

IT COMES UP IN *MODELING BRAINS.*
AFTER ALL, WHAT IS IN YOUR BRAIN...
OTHER THAN A SUITABLY WEIGHTED AVERAGE OF ITS HISTORY?
THIS IS ***GREAT,*** EMMY!
WE COULDN'T HAVE ASKED FOR A MORE PERSUASIVE RESULT.
WHAT DO YOU MEAN?
WHY IS THIS SO GOOD?
WELL, LOOK AT ALL OF THE RESULTS YOU GOT BEFORE.
THEY WERE TO BE EXPECTED.
STATISTICS OF THE ANATOMIES OF INTEGERS AND OF PERMUTATIONS COME IN CERTAIN WELL-KNOWN PATTERNS.
BUT BUCHSTAB'S FUNCTION?
NO ONE EXPECTS TO SEE THAT.
AND BOTH POPULATIONS HAVE THIS FOR THEIR DISTRIBUTION FUNCTION?
TIME-DELAY EQUATION IN BRAIN MODELING
A brain's current state, $\omega(u)$, is a weighted average of its previous states, for example
$\omega(u) = \frac{1}{u}\left(c + \int_0^u \chi(u-t)\omega(t)dt\right)$
after some initial conditions.
SURELY NO ONE...
...CAN BELIEVE IT'S A COINCIDENCE.

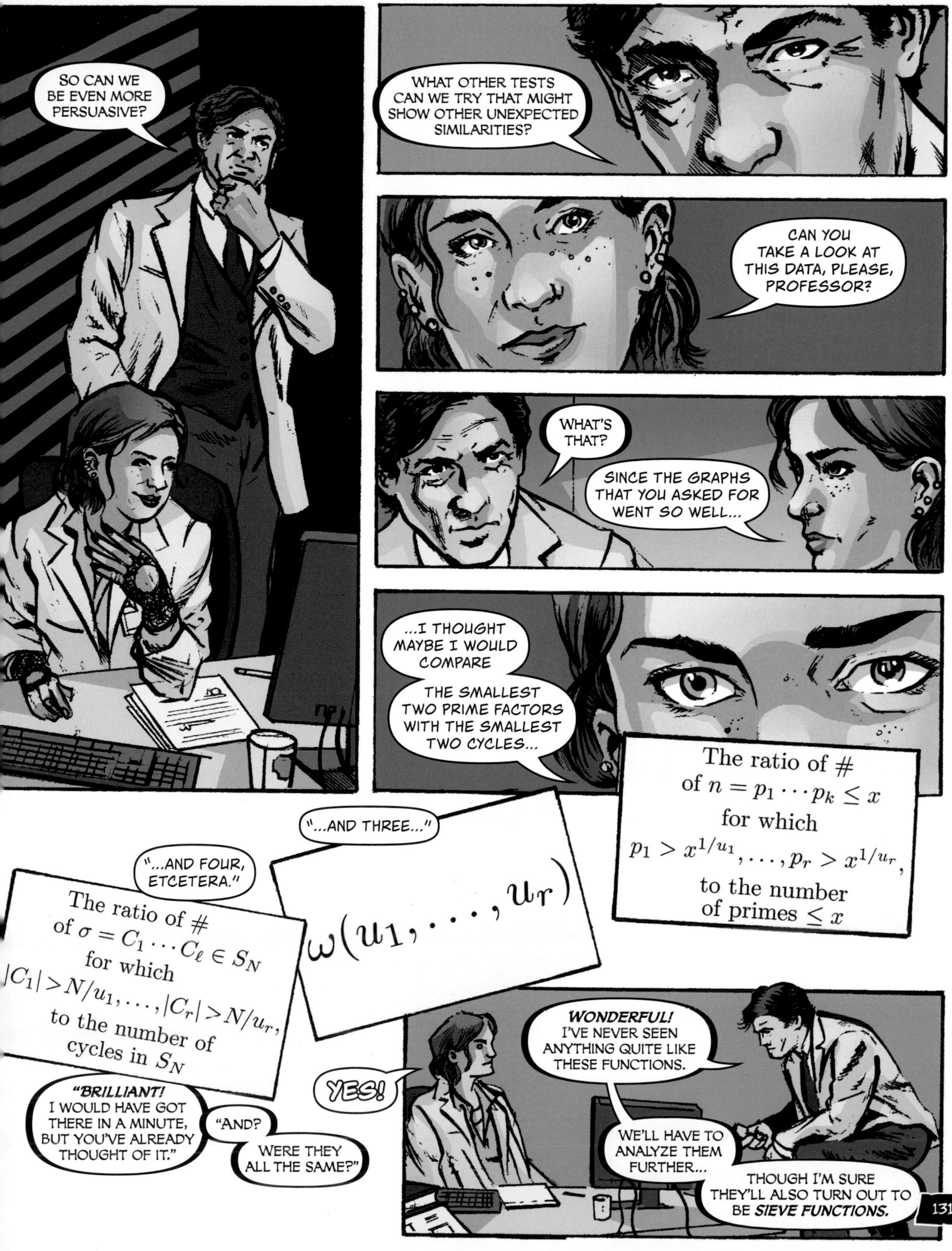

SO CAN WE BE EVEN MORE PERSUASIVE?
WHAT OTHER TESTS CAN WE TRY THAT MIGHT SHOW OTHER UNEXPECTED SIMILARITIES?
CAN YOU TAKE A LOOK AT THIS DATA, PLEASE, PROFESSOR?
WHAT'S THAT?
SINCE THE GRAPHS THAT YOU ASKED FOR WENT SO WELL...
...I THOUGHT MAYBE I WOULD COMPARE
THE SMALLEST TWO PRIME FACTORS WITH THE SMALLEST TWO CYCLES...
The ratio of # of $n = p_1 \cdots p_k \le x$ for which $p_1 > x^{1/u_1}, \ldots, p_r > x^{1/u_r}$, to the number of primes $\le x$
"...AND THREE..."
$\omega(u_1, \ldots, u_r)$
"...AND FOUR, ETCETERA."
The ratio of # of $\sigma = C_1 \cdots C_\ell \in S_N$ for which $|C_1| > N/u_1, \ldots, |C_r| > N/u_r$, to the number of cycles in $S_N$
"BRILLIANT! I WOULD HAVE GOT THERE IN A MINUTE, BUT YOU'VE ALREADY THOUGHT OF IT."
"AND?
WERE THEY ALL THE SAME?"
YES!
WONDERFUL! I'VE NEVER SEEN ANYTHING QUITE LIKE THESE FUNCTIONS.
WE'LL HAVE TO ANALYZE THEM FURTHER...
THOUGH I'M SURE THEY'LL ALSO TURN OUT TO BE SIEVE FUNCTIONS.

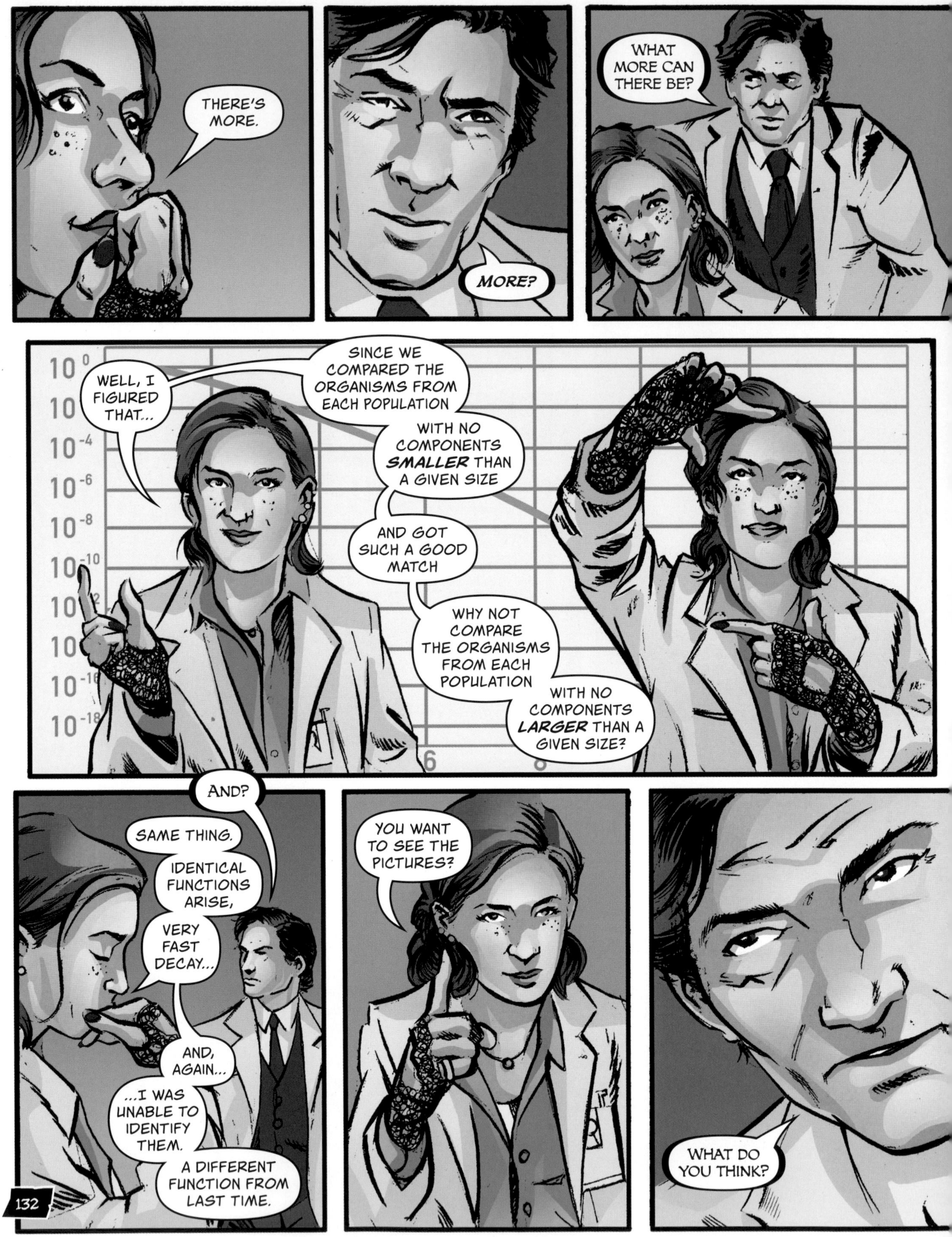

THERE'S MORE.
MORE?
WHAT MORE CAN THERE BE?
WELL, I FIGURED THAT...
SINCE WE COMPARED THE ORGANISMS FROM EACH POPULATION
WITH NO COMPONENTS SMALLER THAN A GIVEN SIZE
AND GOT SUCH A GOOD MATCH
WHY NOT COMPARE THE ORGANISMS FROM EACH POPULATION
WITH NO COMPONENTS LARGER THAN A GIVEN SIZE?
AND?
SAME THING.
IDENTICAL FUNCTIONS ARISE,
VERY FAST DECAY...
AND, AGAIN...
...I WAS UNABLE TO IDENTIFY THEM.
A DIFFERENT FUNCTION FROM LAST TIME.
YOU WANT TO SEE THE PICTURES?
WHAT DO YOU THINK?

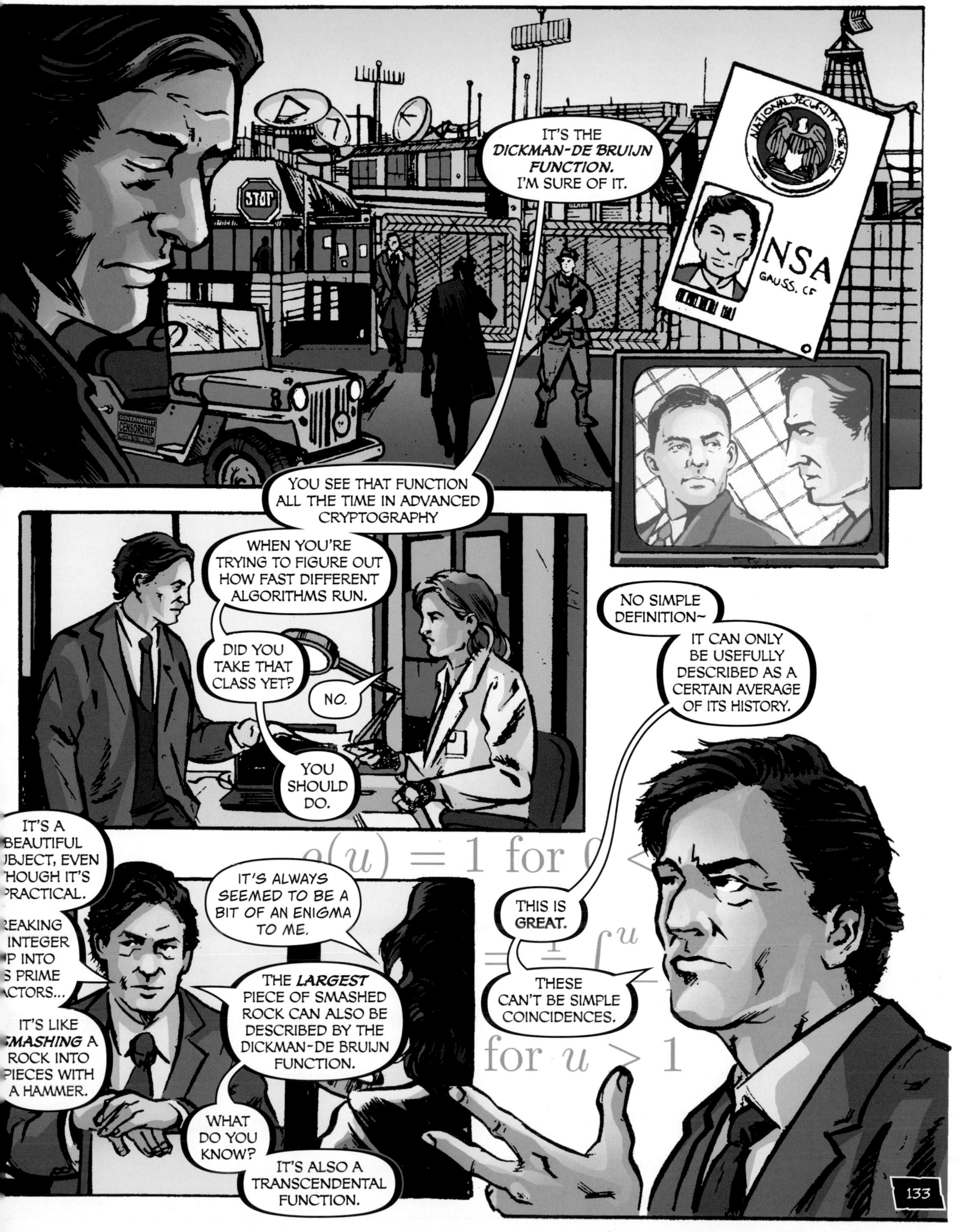
IT'S THE DICKMAN-DE BRUIJN FUNCTION. I'M SURE OF IT.
STOP
GOVERNMENT CENSORSHIP
NATIONAL SECURITY AGENCY
NSA
GAUSS, CF
YOU SEE THAT FUNCTION ALL THE TIME IN ADVANCED CRYPTOGRAPHY
WHEN YOU'RE TRYING TO FIGURE OUT HOW FAST DIFFERENT ALGORITHMS RUN.
DID YOU TAKE THAT CLASS YET?
NO.
YOU SHOULD DO.
IT'S A BEAUTIFUL BJECT, EVEN HOUGH IT'S PRACTICAL.
REAKING INTEGER P INTO S PRIME ACTORS...
IT'S LIKE SMASHING A ROCK INTO PIECES WITH A HAMMER.
IT'S ALWAYS SEEMED TO BE A BIT OF AN ENIGMA TO ME.
THE LARGEST PIECE OF SMASHED ROCK CAN ALSO BE DESCRIBED BY THE DICKMAN-DE BRUIJN FUNCTION.
WHAT DO YOU KNOW?
IT'S ALSO A TRANSCENDENTAL FUNCTION.
NO SIMPLE DEFINITION~
IT CAN ONLY BE USEFULLY DESCRIBED AS A CERTAIN AVERAGE OF ITS HISTORY.
THIS IS GREAT.
THESE CAN'T BE SIMPLE COINCIDENCES.
for u > 1

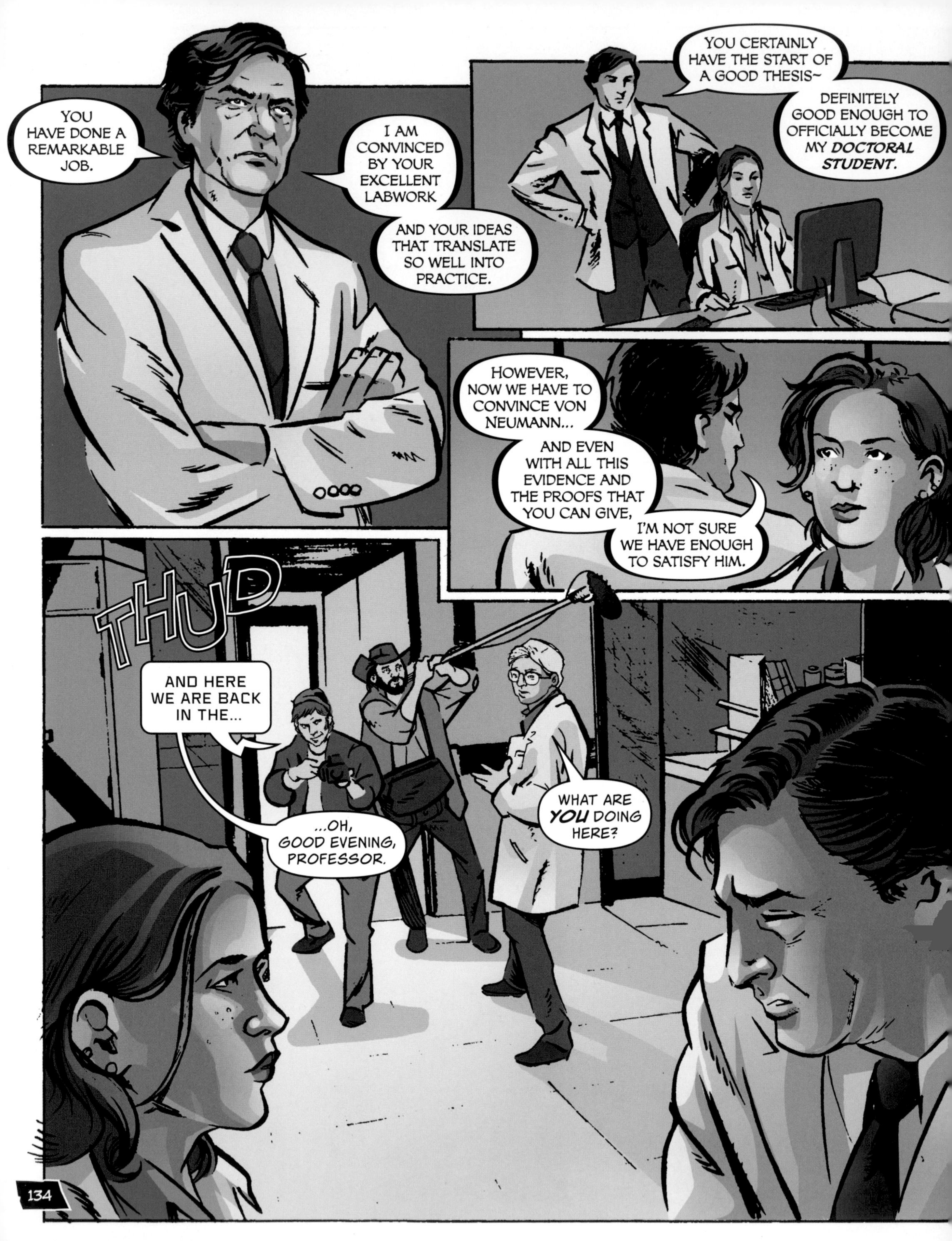
YOU HAVE DONE A REMARKABLE JOB.
I AM CONVINCED BY YOUR EXCELLENT LABWORK
AND YOUR IDEAS THAT TRANSLATE SO WELL INTO PRACTICE.
YOU CERTAINLY HAVE THE START OF A GOOD THESIS~
DEFINITELY GOOD ENOUGH TO OFFICIALLY BECOME MY ***DOCTORAL STUDENT***.
HOWEVER, NOW WE HAVE TO CONVINCE VON NEUMANN...
AND EVEN WITH ALL THIS EVIDENCE AND THE PROOFS THAT YOU CAN GIVE,
I'M NOT SURE WE HAVE ENOUGH TO SATISFY HIM.
THUD
AND HERE WE ARE BACK IN THE...
...OH, GOOD EVENING, PROFESSOR.
WHAT ARE ***YOU*** DOING HERE?

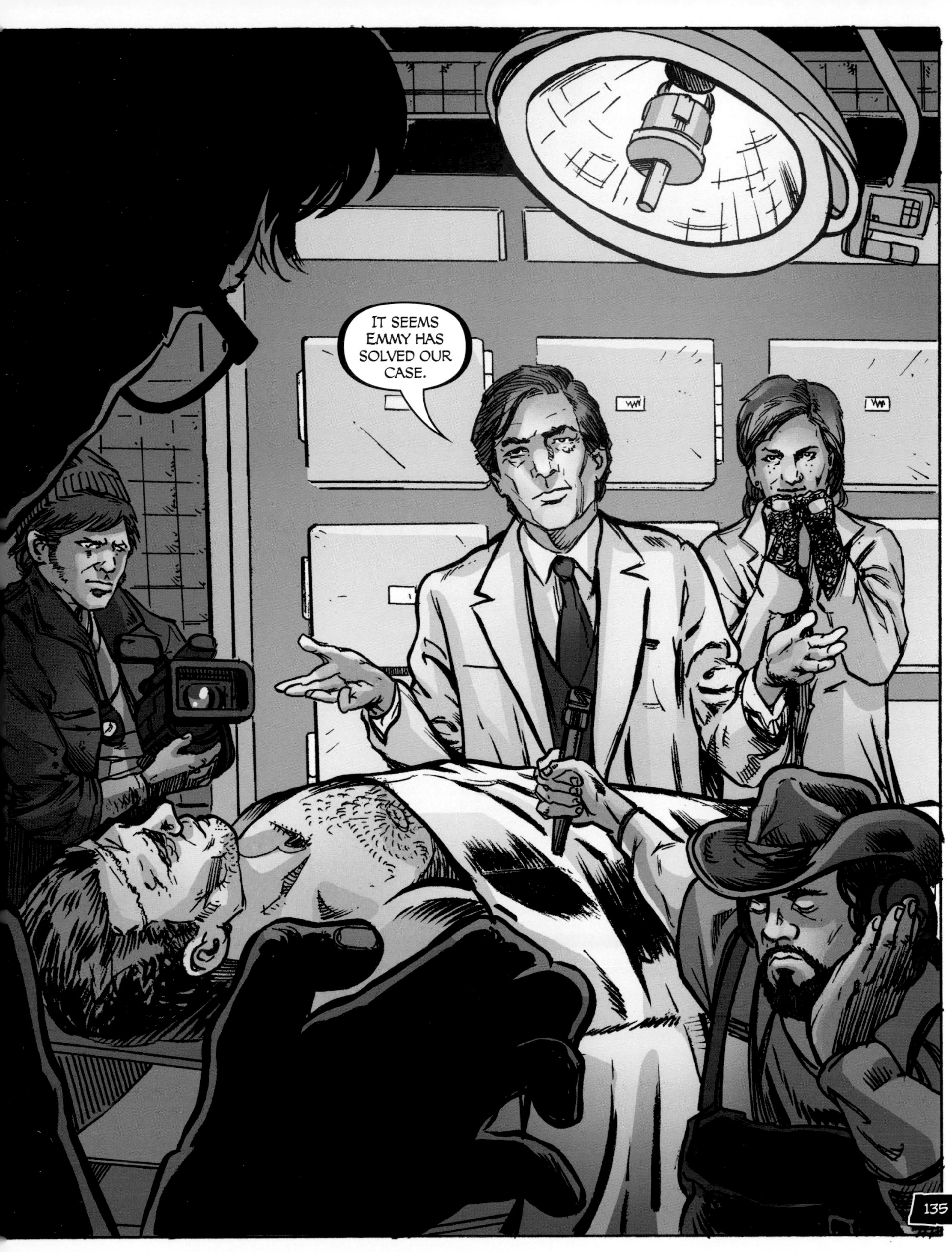
IT SEEMS EMMY HAS SOLVED OUR CASE.

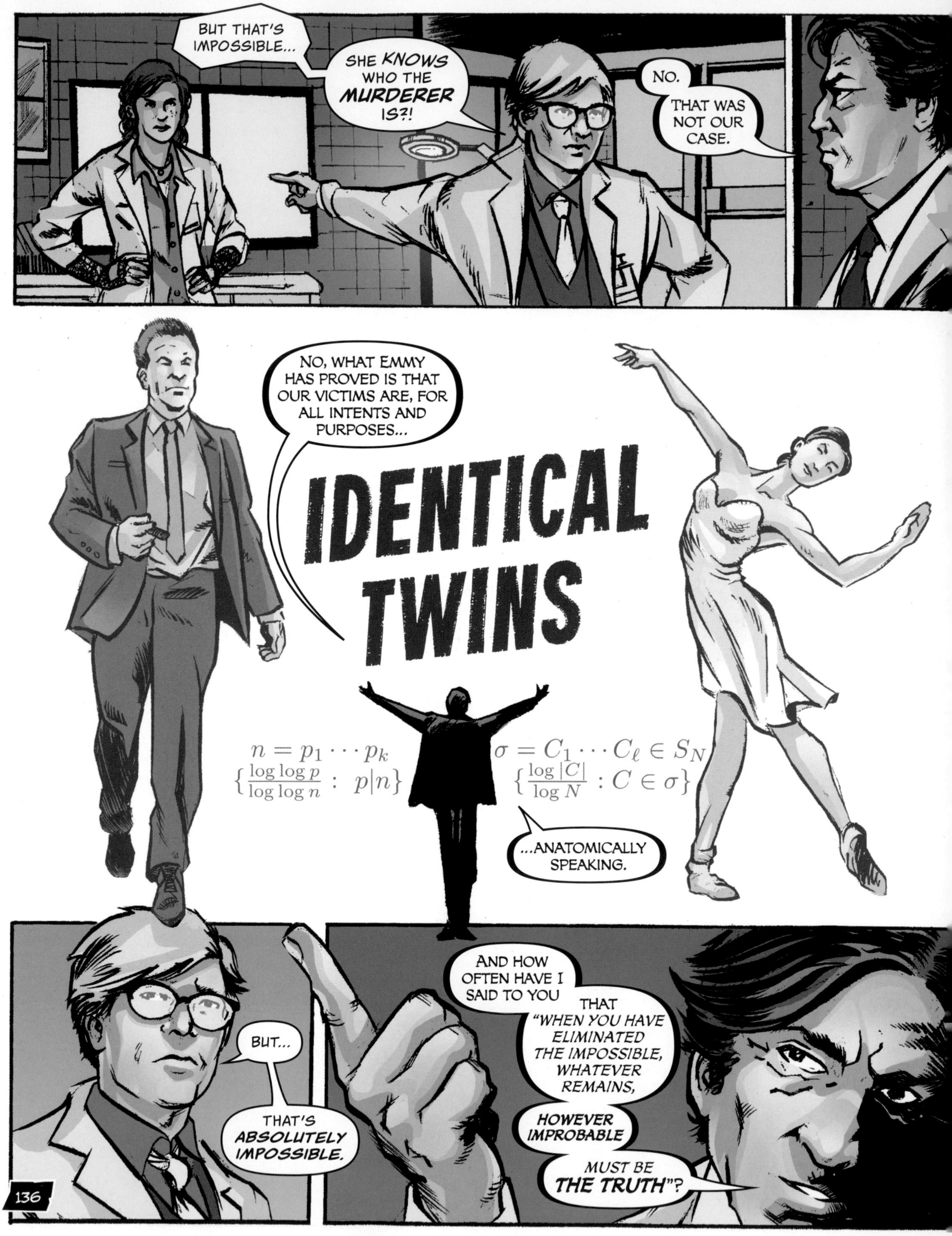

BUT THAT'S IMPOSSIBLE...
SHE KNOWS WHO THE MURDERER IS?!
NO.
THAT WAS NOT OUR CASE.
NO, WHAT EMMY HAS PROVED IS THAT OUR VICTIMS ARE, FOR ALL INTENTS AND PURPOSES...
IDENTICAL TWINS
$n = p_1 \cdots p_k$
$\left\{\frac{\log\log p}{\log\log n} : p|n\right\}$
$\sigma = C_1 \cdots C_\ell \in S_N$
$\left\{\frac{\log|C|}{\log N} : C \in \sigma\right\}$
...ANATOMICALLY SPEAKING.
BUT...
THAT'S ABSOLUTELY IMPOSSIBLE.
AND HOW OFTEN HAVE I SAID TO YOU
THAT "WHEN YOU HAVE ELIMINATED THE IMPOSSIBLE, WHATEVER REMAINS,
HOWEVER IMPROBABLE
MUST BE THE TRUTH"?

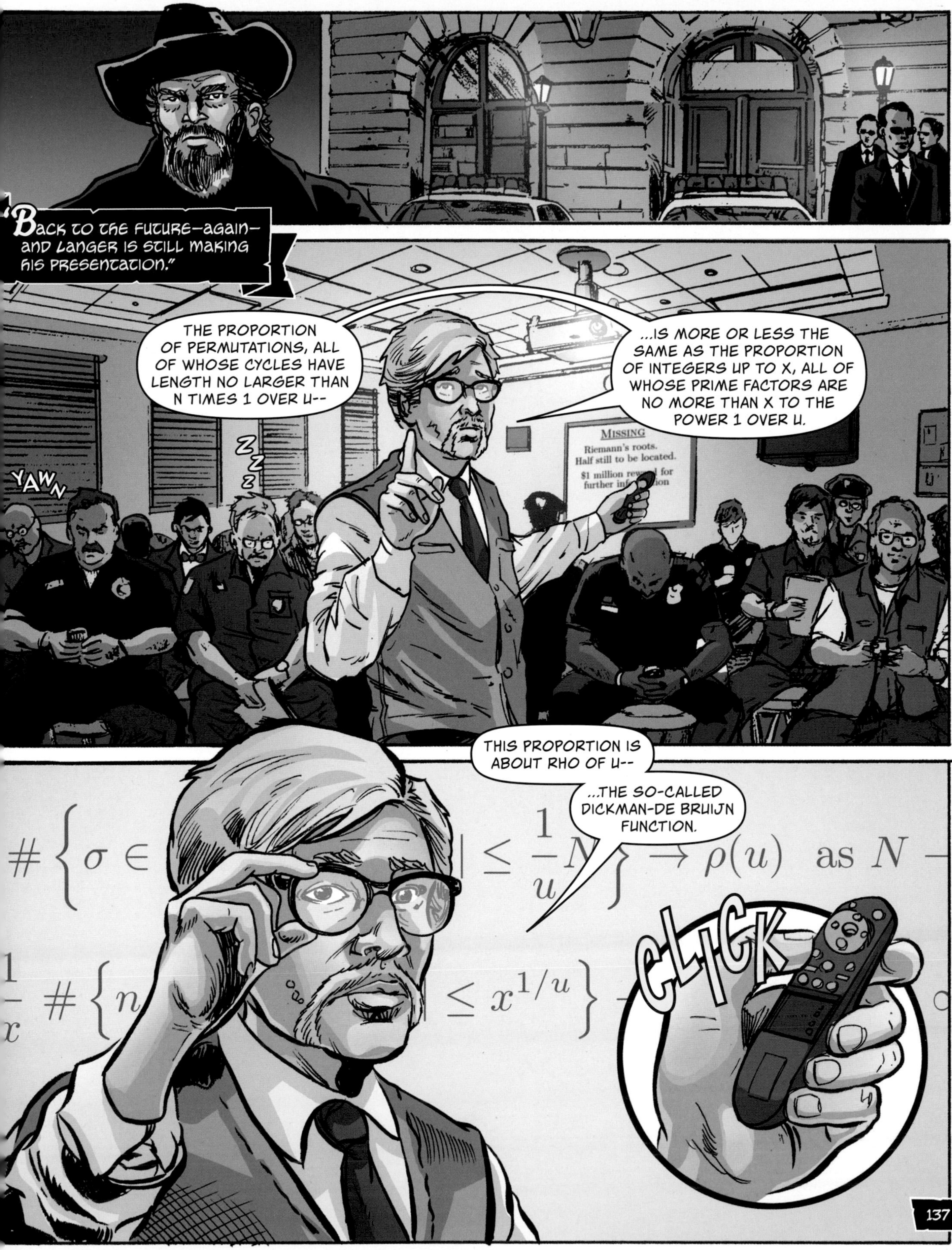
"Back to the future—again—and Langer is still making his presentation."
THE PROPORTION OF PERMUTATIONS, ALL OF WHOSE CYCLES HAVE LENGTH NO LARGER THAN N TIMES 1 OVER U--
...IS MORE OR LESS THE SAME AS THE PROPORTION OF INTEGERS UP TO X, ALL OF WHOSE PRIME FACTORS ARE NO MORE THAN X TO THE POWER 1 OVER U.
MISSING
Riemann's roots.
Half still to be located.
ZZZ
YAWN
THIS PROPORTION IS ABOUT RHO OF U--
...THE SO-CALLED DICKMAN-DE BRUIJN FUNCTION.
CLICK

ckman de Bruijn function ρ(u): = 1 for 0 < u ≤ 1, and
dt for u > 1.
simpler definition.
WHICH YOU CAN SEE ON THIS SLIDE.
NOTICE THAT RHO U IS GIVEN AS AN AVERAGE OF ITS RECENT HISTORY, AND THERE DOESN'T SEEM TO BE A SIMPLER WAY TO DESCRIBE IT.
o have
THE "JOINT DISTRIBUTION" OF THE K LARGEST CYCLES OF A PERMUTATION, AND OF THE K LARGEST PRIME FACTORS OF AN INTEGER, ARE ALSO IDENTICAL, VERY COMPLICATED FUNCTIONS.

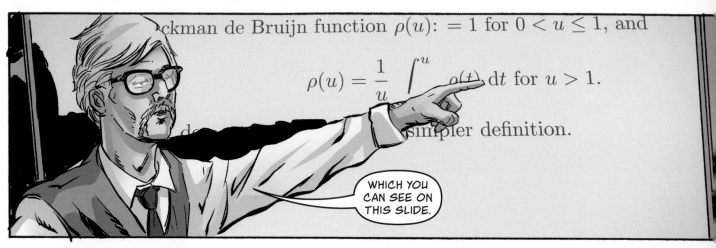

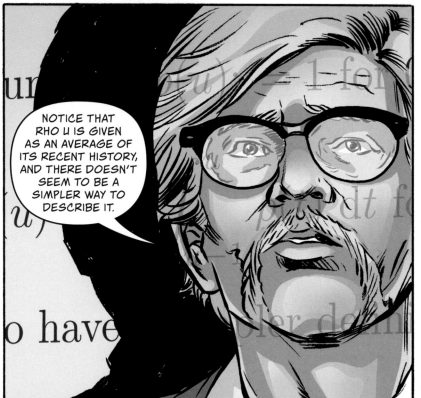

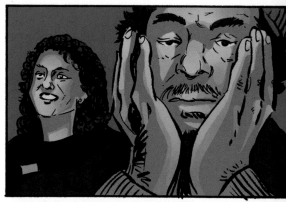

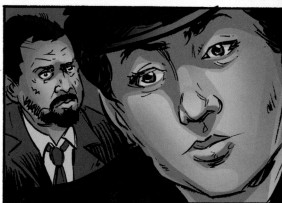

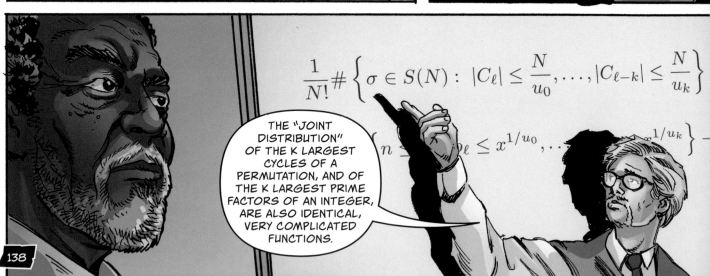

$$\left\{ \sigma \in S(N) : \right. \quad \left. \frac{1}{u} N \right\} \to u\omega(u)$$

SIMILAR REMARKS APPLY WHEN COMPARING THE PROPORTION OF PERMUTATIONS WHOSE CYCLES HAVE LENGTH

NO SMALLER THAN *N* TIMES *1* OVER *U*, TO THE PROPORTION OF INTEGERS UP TO *X*...

...ALL OF WHOSE PRIME FACTORS ARE NO LESS THAN *X* TO THE POWER *1* OVER *U*.

$$\# \left\{ n \leq x : \text{ All } \quad x^{1/u} \right\} \to u\omega(u)$$

$$u) \cdot \#\{C \in S(N)\}$$

$$u) \cdot \#\{p \leq x\}$$

THAT OUR TWO POPULATIONS SHOULD DISPLAY THESE FEATURES IN COMMON,

*BOTH* BEING DESCRIBED...

...BY THESE ENORMOUSLY COMPLICATED FUNCTIONS, IS VERY SURPRISING,

AND SHOULD MAKE US SUSPICIOUS THAT THIS IS SOMETHING MORE...

POLICE

GULP
GULP
GULP

GULP

GULP
GULP
GULP

GULP

AHHHH

...THAN MERE COINCIDENCE.

$$= \frac{1}{u} \int_0^{u-}$$

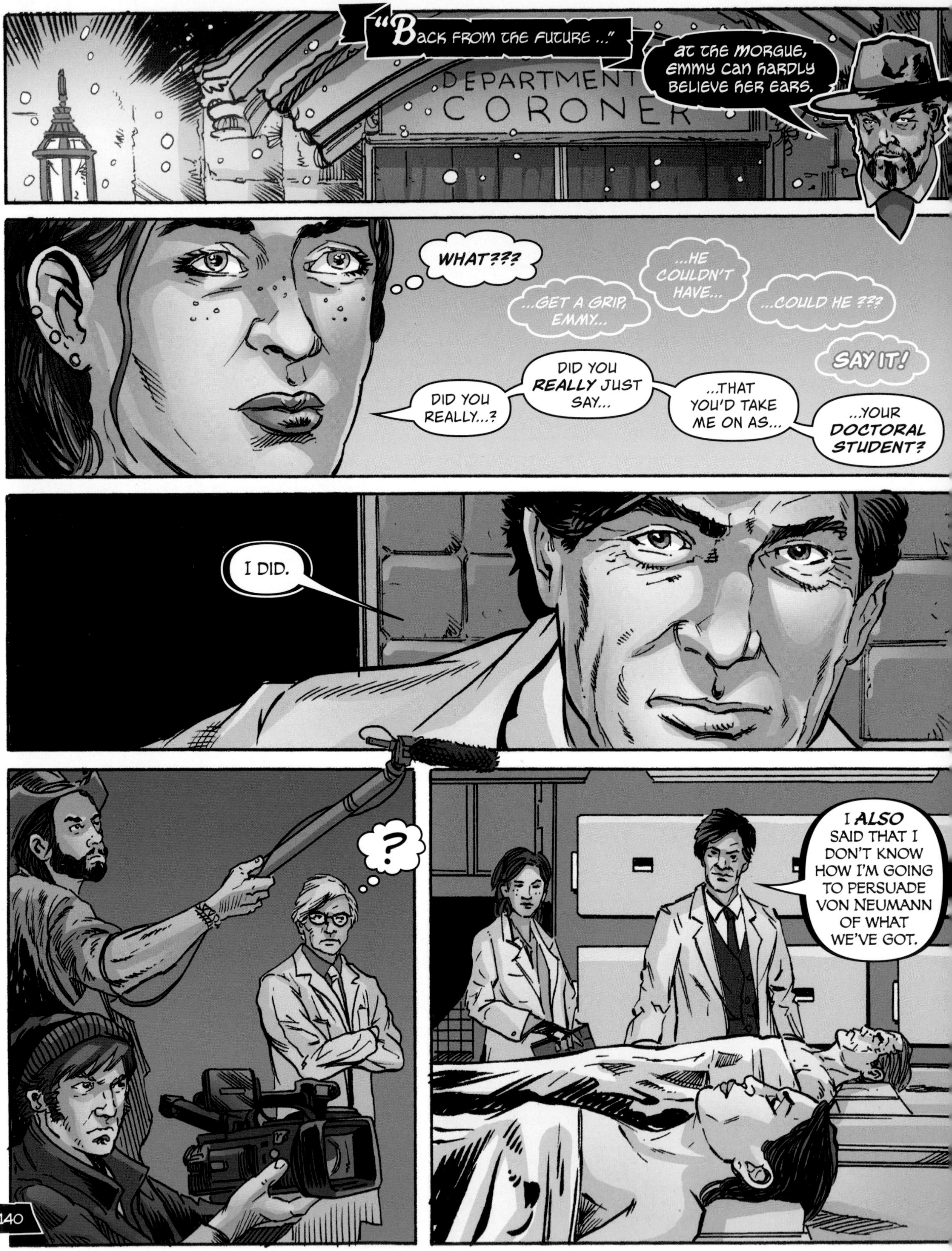
"Back from the future ..."
DEPARTMENT CORONER
at the morgue, Emmy can hardly believe her ears.
WHAT???
...GET A GRIP, EMMY...
...HE COULDN'T HAVE...
...COULD HE ???
SAY IT!
DID YOU REALLY...?
DID YOU REALLY JUST SAY...
...THAT YOU'D TAKE ME ON AS...
...YOUR DOCTORAL STUDENT?
I DID.
?
I ALSO SAID THAT I DON'T KNOW HOW I'M GOING TO PERSUADE VON NEUMANN OF WHAT WE'VE GOT.

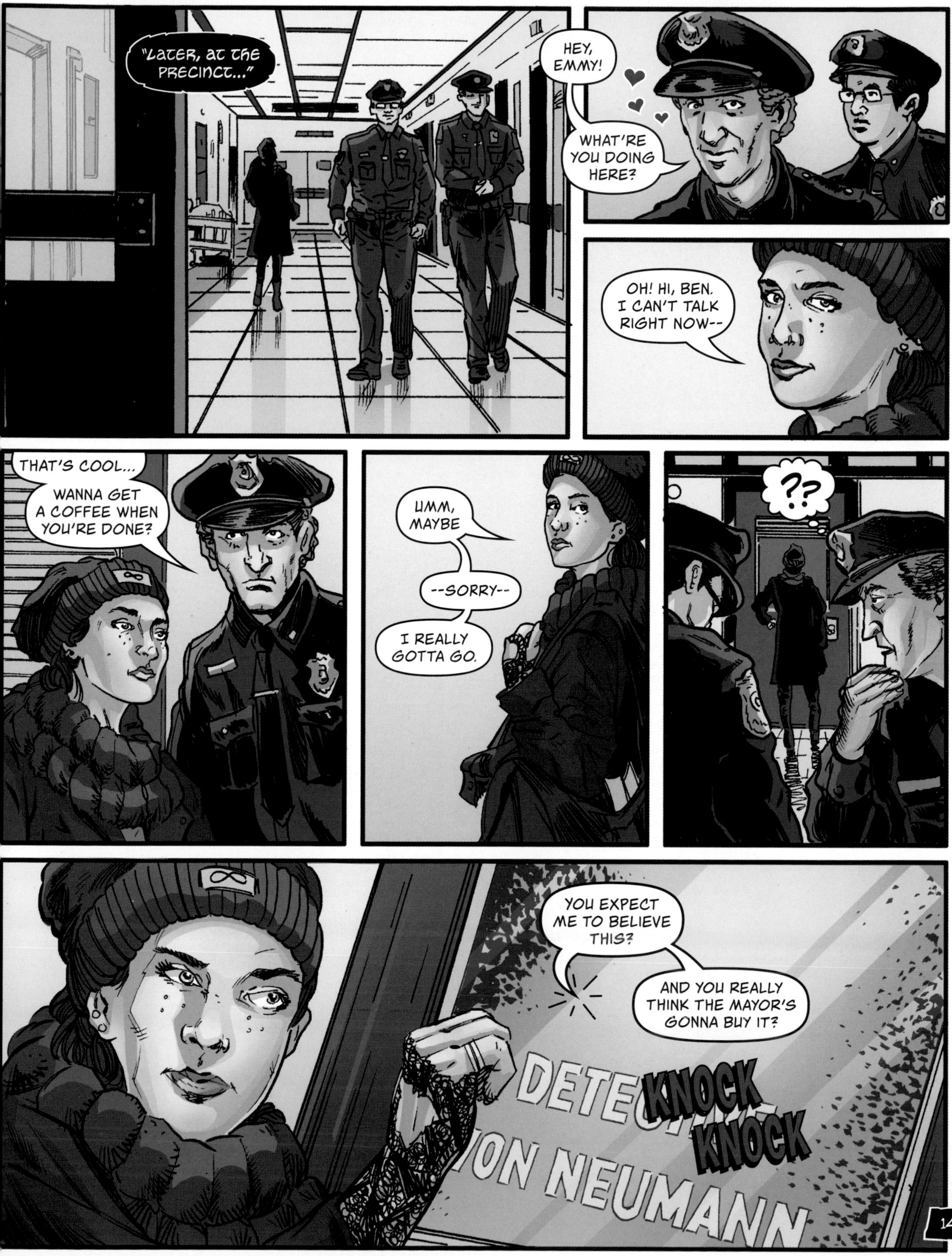
"Later, at the precinct..."
HEY, EMMY!
WHAT'RE YOU DOING HERE?
OH! HI, BEN. I CAN'T TALK RIGHT NOW--
THAT'S COOL...
WANNA GET A COFFEE WHEN YOU'RE DONE?
UMM, MAYBE
--SORRY--
I REALLY GOTTA GO.
??
YOU EXPECT ME TO BELIEVE THIS?
AND YOU REALLY THINK THE MAYOR'S GONNA BUY IT?
DETECTIVE VON NEUMANN
KNOCK
KNOCK

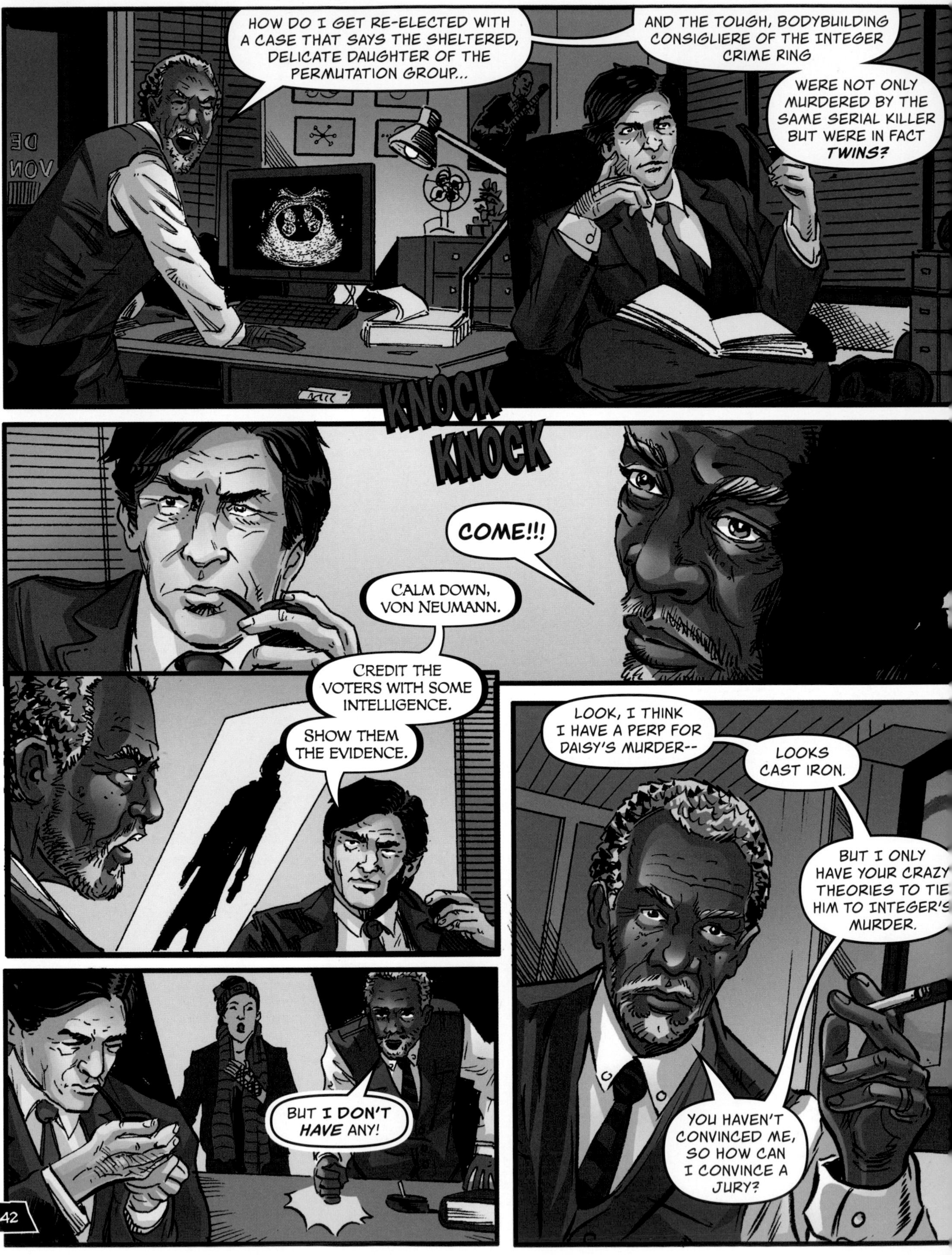
HOW DO I GET RE-ELECTED WITH A CASE THAT SAYS THE SHELTERED, DELICATE DAUGHTER OF THE PERMUTATION GROUP...
AND THE TOUGH, BODYBUILDING CONSIGLIERE OF THE INTEGER CRIME RING
WERE NOT ONLY MURDERED BY THE SAME SERIAL KILLER BUT WERE IN FACT TWINS?
KNOCK KNOCK
COME!!!
CALM DOWN, VON NEUMANN.
CREDIT THE VOTERS WITH SOME INTELLIGENCE.
SHOW THEM THE EVIDENCE.
BUT I DON'T HAVE ANY!
LOOK, I THINK I HAVE A PERP FOR DAISY'S MURDER--
LOOKS CAST IRON.
BUT I ONLY HAVE YOUR CRAZY THEORIES TO TIE HIM TO INTEGER'S MURDER.
YOU HAVEN'T CONVINCED ME, SO HOW CAN I CONVINCE A JURY?

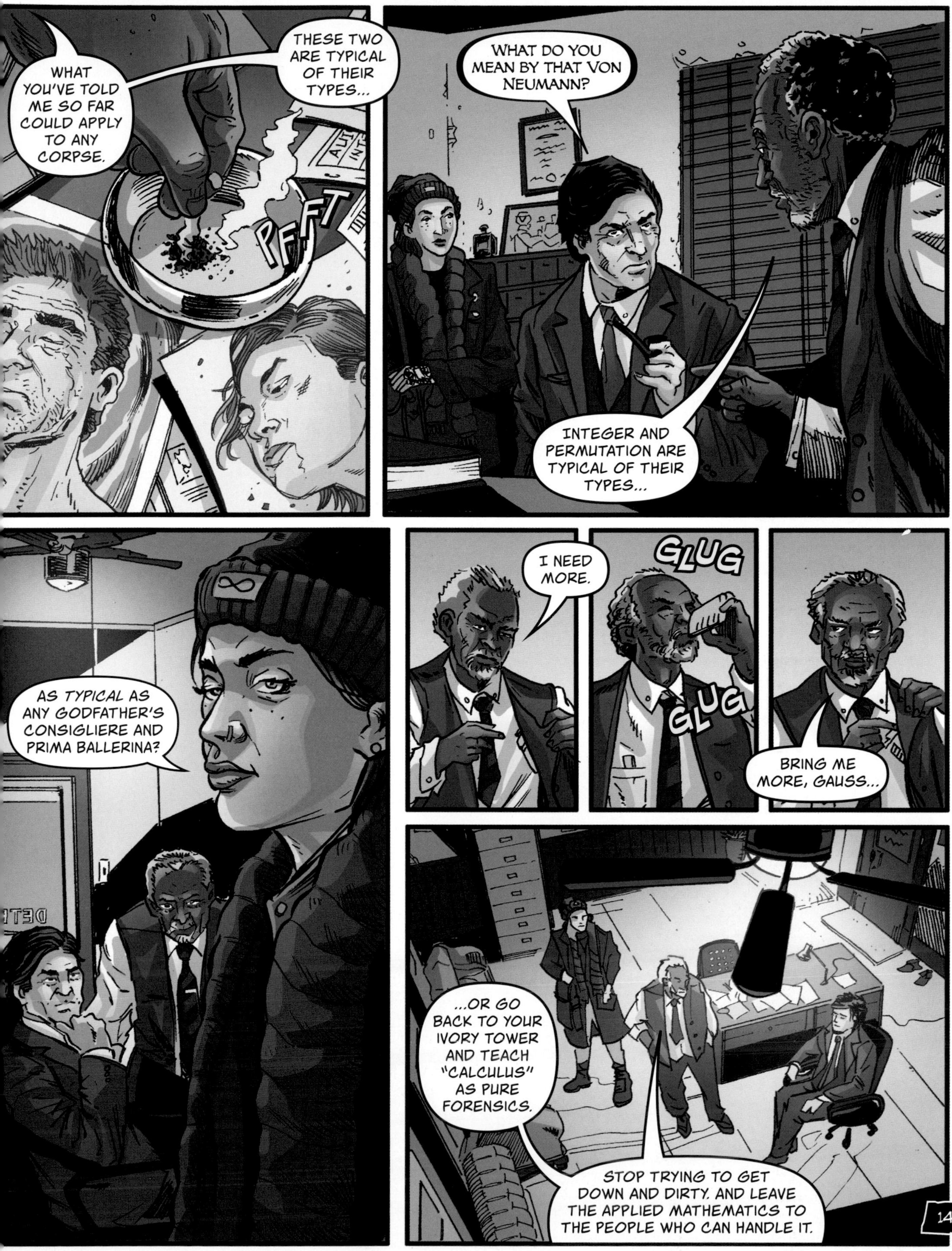
WHAT YOU'VE TOLD ME SO FAR COULD APPLY TO ANY CORPSE.
THESE TWO ARE TYPICAL OF THEIR TYPES...
PFFT
WHAT DO YOU MEAN BY THAT VON NEUMANN?
INTEGER AND PERMUTATION ARE TYPICAL OF THEIR TYPES...
AS *TYPICAL* AS ANY GODFATHER'S CONSIGLIERE AND PRIMA BALLERINA?
I NEED MORE.
GLUG
GLUG
BRING ME MORE, GAUSS...
...OR GO BACK TO YOUR IVORY TOWER AND TEACH "CALCULUS" AS PURE FORENSICS.
STOP TRYING TO GET DOWN AND DIRTY. AND LEAVE THE APPLIED MATHEMATICS TO THE PEOPLE WHO CAN HANDLE IT.

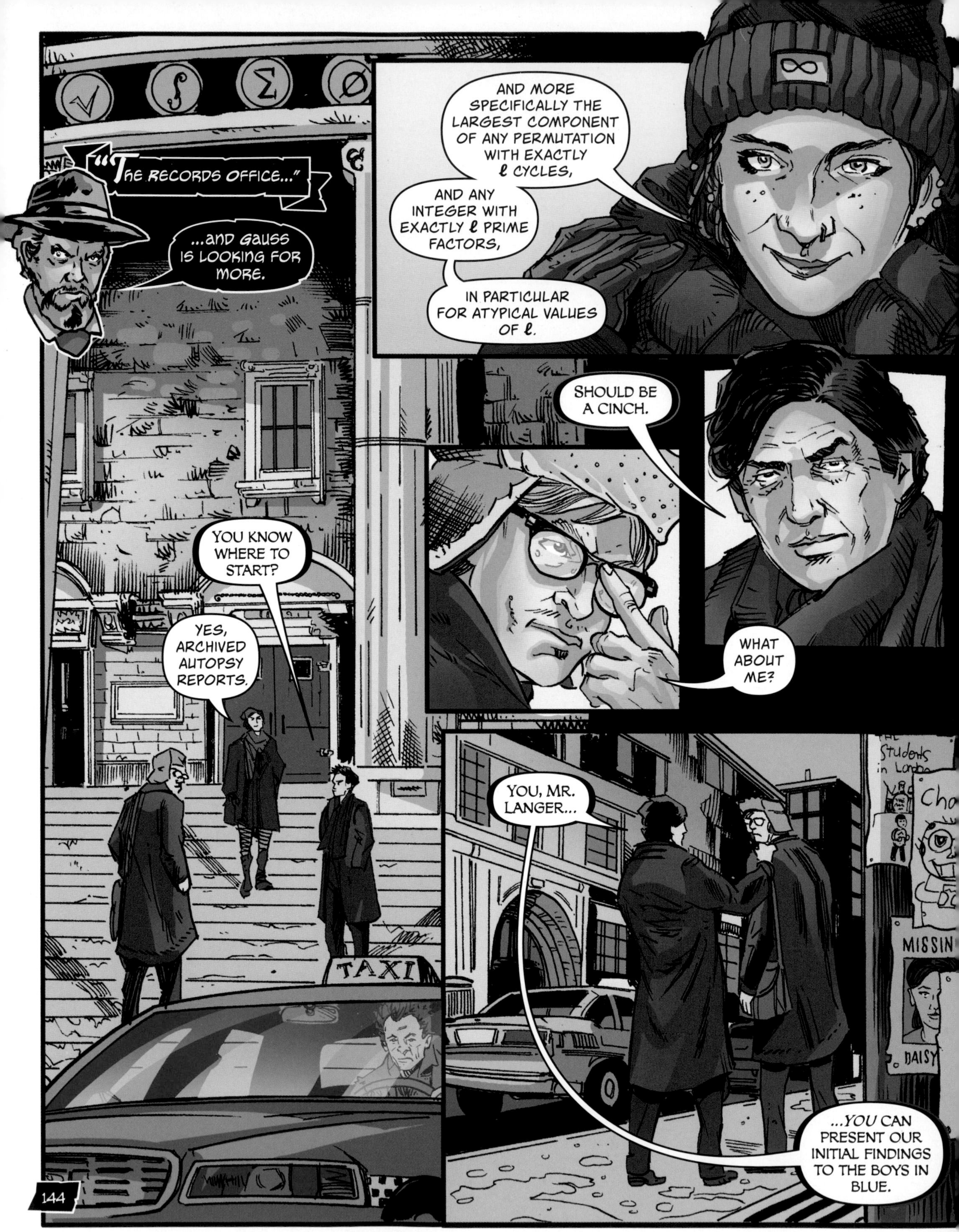
"THE RECORDS OFFICE..."
...AND GAUSS IS LOOKING FOR MORE.
AND MORE SPECIFICALLY THE LARGEST COMPONENT OF ANY PERMUTATION WITH EXACTLY ℓ CYCLES,
AND ANY INTEGER WITH EXACTLY ℓ PRIME FACTORS,
IN PARTICULAR FOR ATYPICAL VALUES OF ℓ.
YOU KNOW WHERE TO START?
YES, ARCHIVED AUTOPSY REPORTS.
SHOULD BE A CINCH.
WHAT ABOUT ME?
YOU, MR. LANGER...
...YOU CAN PRESENT OUR INITIAL FINDINGS TO THE BOYS IN BLUE.
TAXI
MISSIN
DAISY

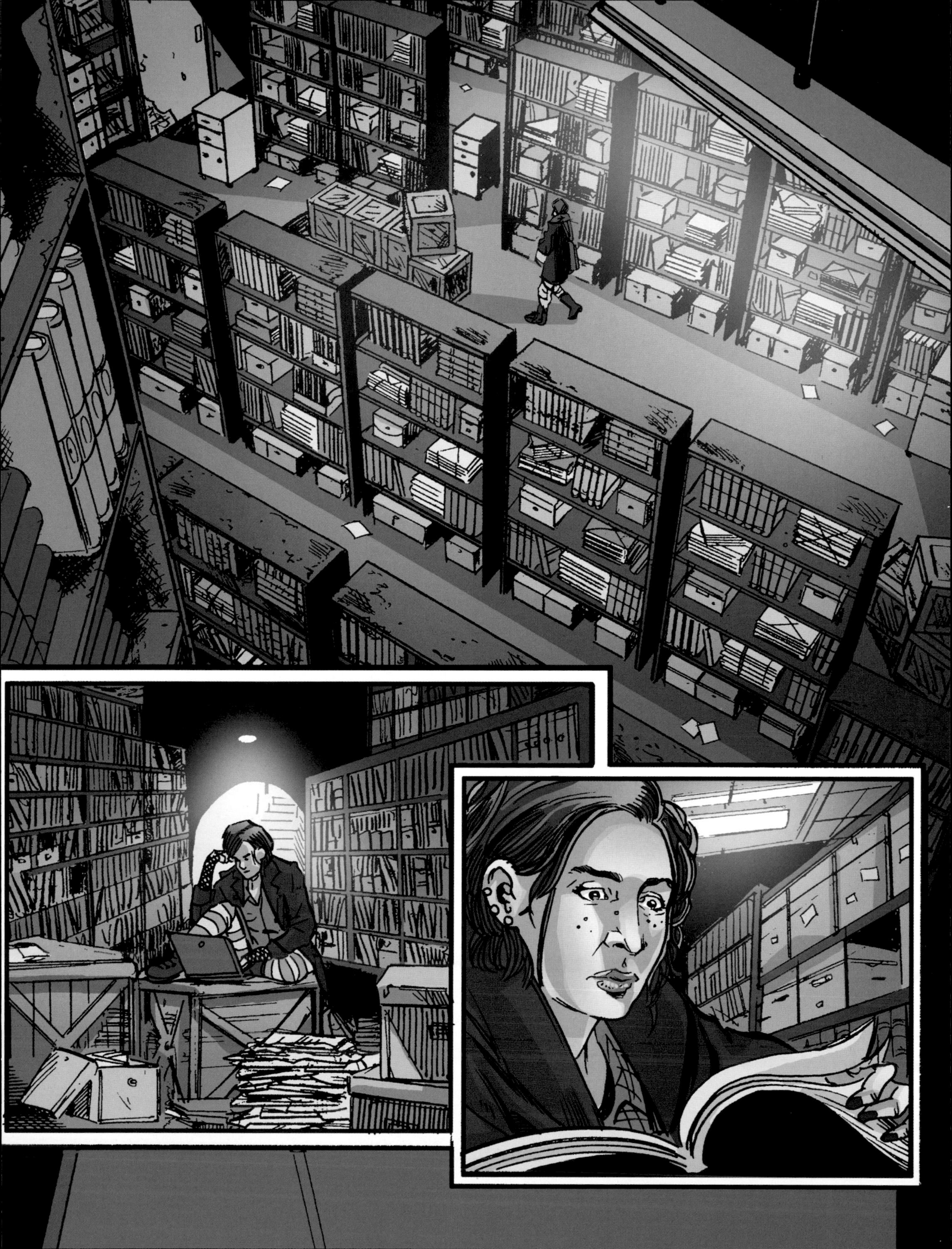

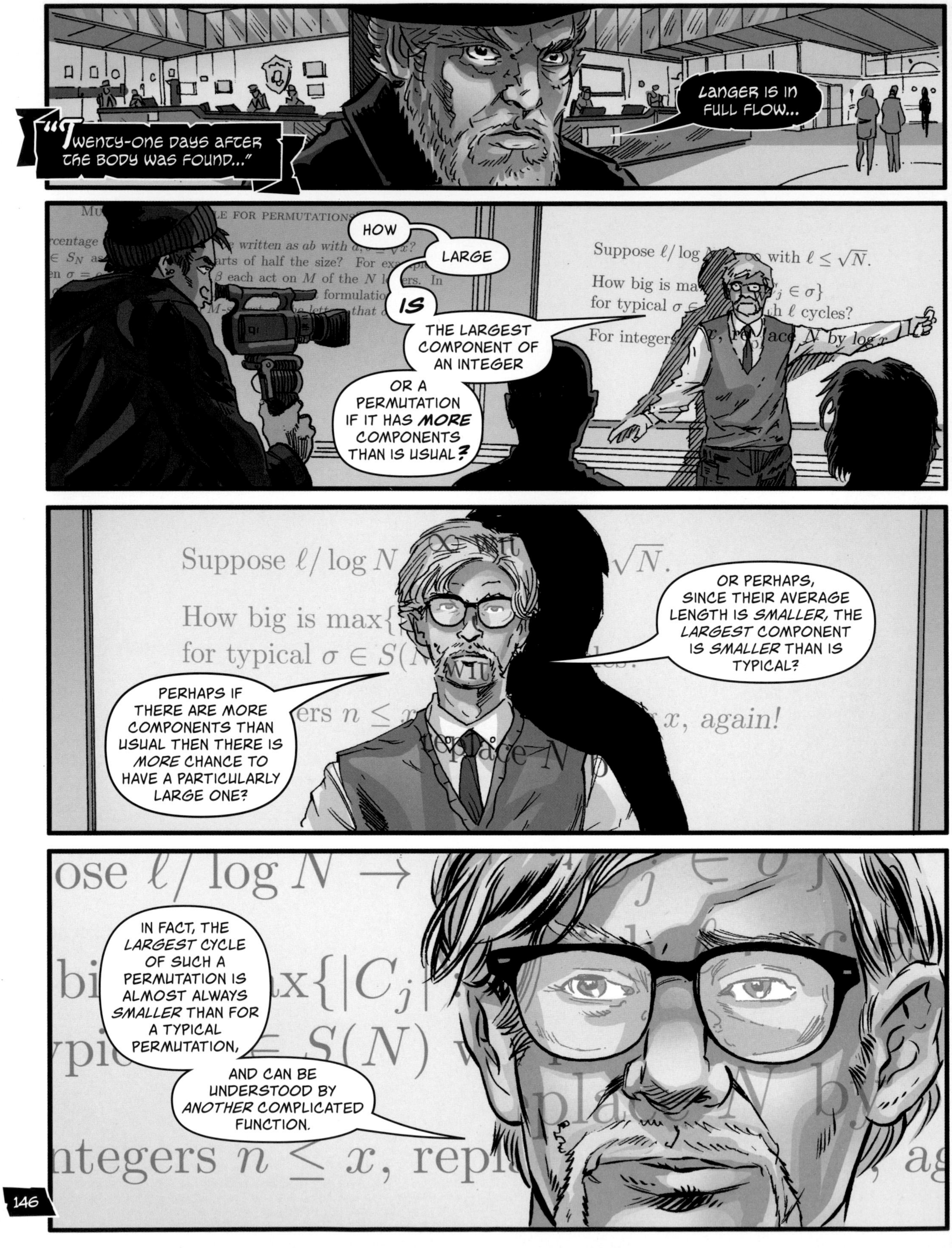
"Twenty-one days after the body was found..."
Langer is in full flow...
How large is the largest component of an integer
or a permutation if it has more components than is usual?
Perhaps if there are more components than usual then there is more chance to have a particularly large one?
Or perhaps, since their average length is smaller, the largest component is smaller than is typical?
In fact, the largest cycle of such a permutation is almost always smaller than for a typical permutation,
and can be understood by another complicated function.

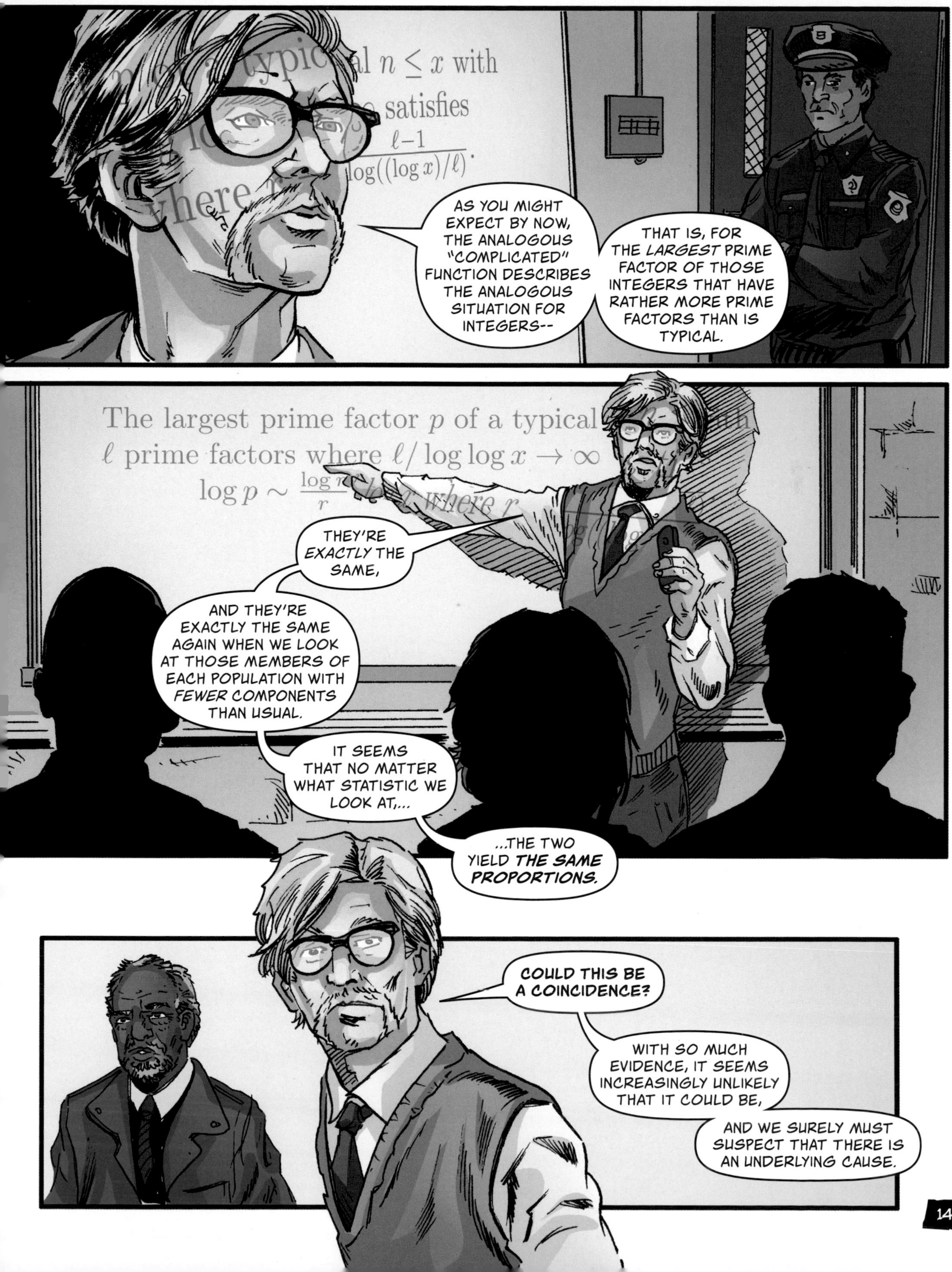
AS YOU MIGHT EXPECT BY NOW, THE ANALOGOUS "COMPLICATED" FUNCTION DESCRIBES THE ANALOGOUS SITUATION FOR INTEGERS--
THAT IS, FOR THE *LARGEST* PRIME FACTOR OF THOSE INTEGERS THAT HAVE RATHER MORE PRIME FACTORS THAN IS TYPICAL.
The largest prime factor $p$ of a typical
$\ell$ prime factors where $\ell/\log\log x \to \infty$
THEY'RE *EXACTLY* THE SAME,
AND THEY'RE EXACTLY THE SAME AGAIN WHEN WE LOOK AT THOSE MEMBERS OF EACH POPULATION WITH *FEWER* COMPONENTS THAN USUAL.
IT SEEMS THAT NO MATTER WHAT STATISTIC WE LOOK AT,...
...THE TWO YIELD ***THE SAME PROPORTIONS.***
**COULD THIS BE A COINCIDENCE?**
WITH SO MUCH EVIDENCE, IT SEEMS INCREASINGLY UNLIKELY THAT IT COULD BE,
AND WE SURELY MUST SUSPECT THAT THERE IS AN UNDERLYING CAUSE.

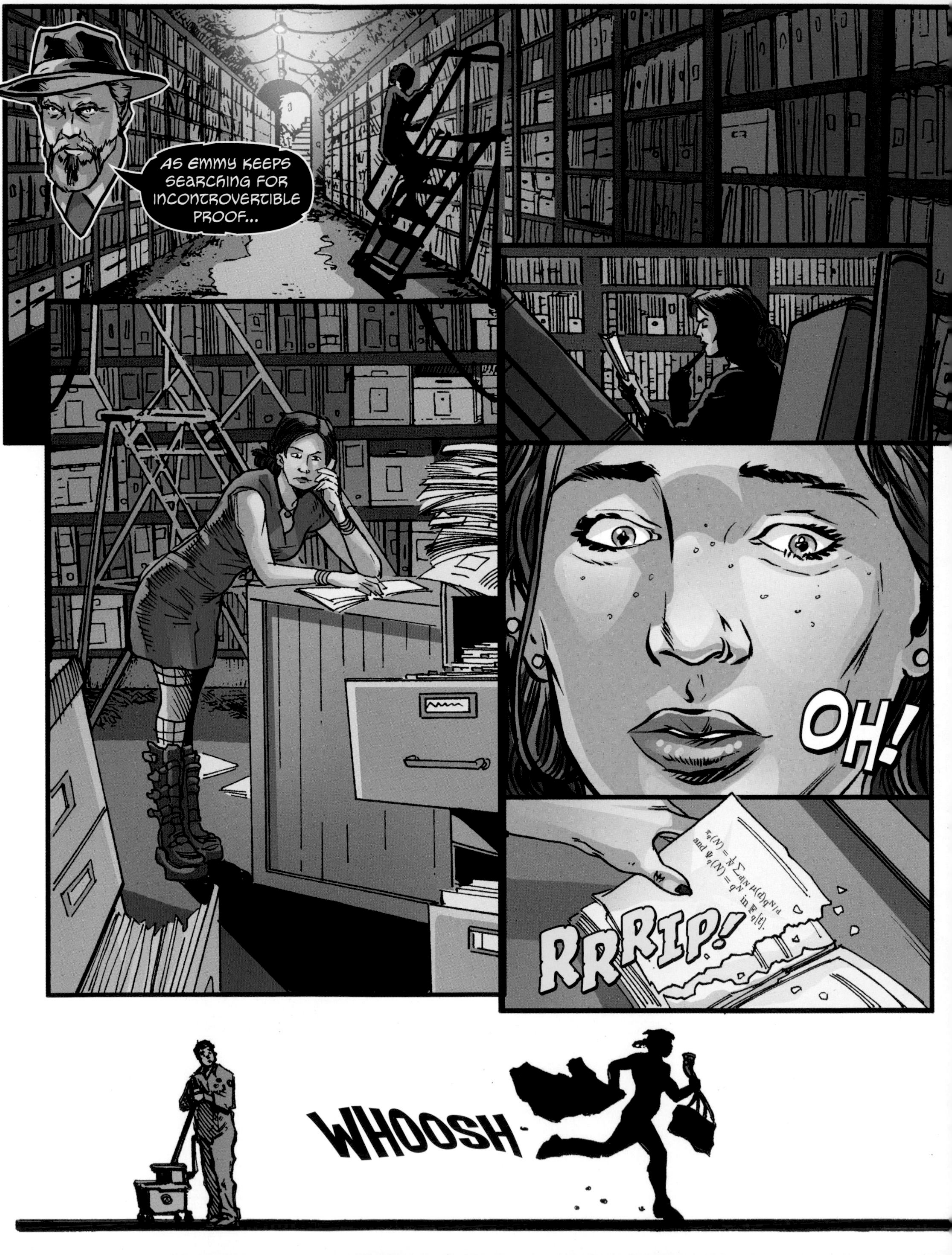
AS EMMY KEEPS SEARCHING FOR INCONTROVERTIBLE PROOF...
OH!
RRRIP!
WHOOSH

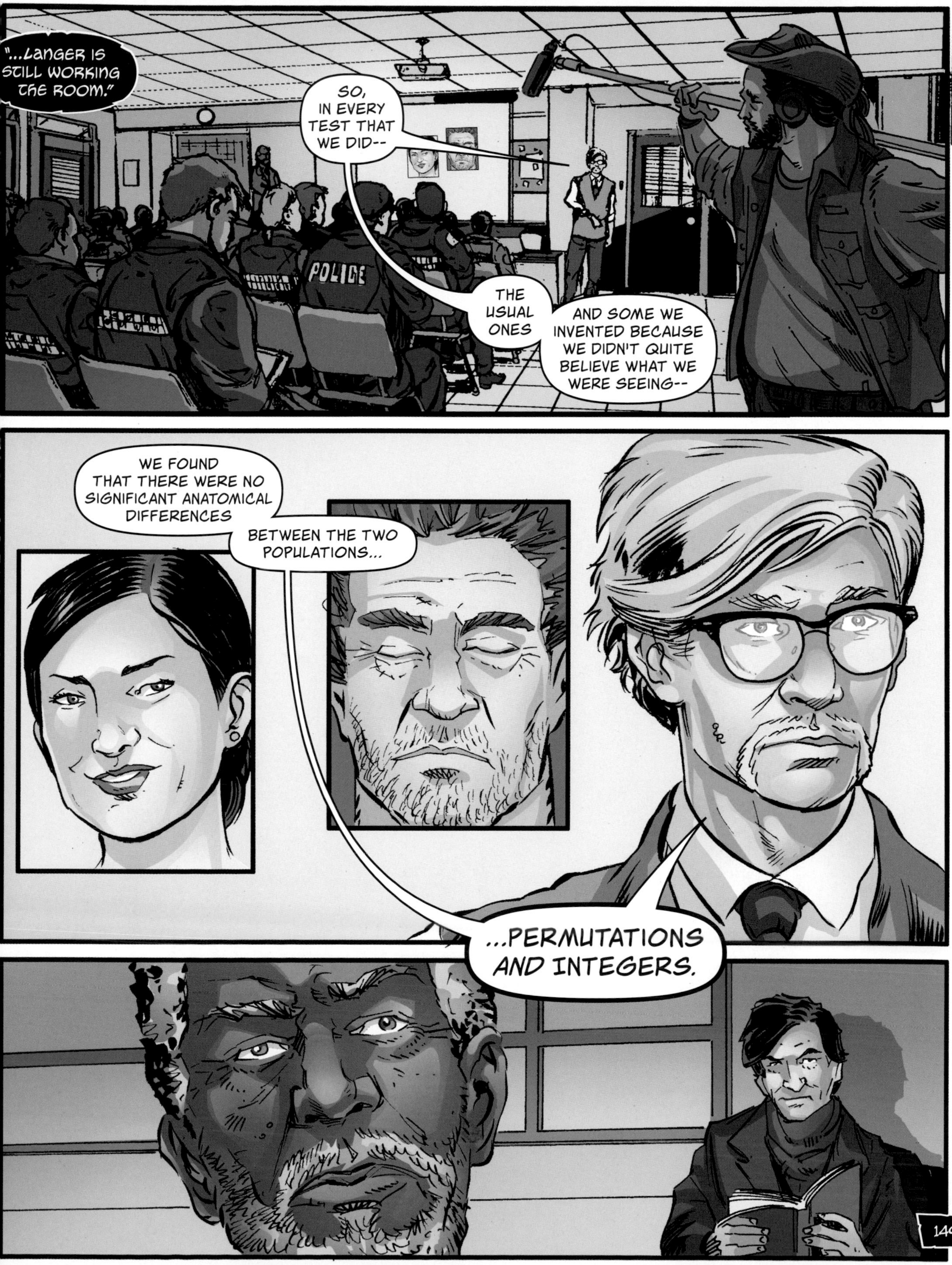

"...LANGER IS STILL WORKING THE ROOM."
SO, IN EVERY TEST THAT WE DID--
THE USUAL ONES
AND SOME WE INVENTED BECAUSE WE DIDN'T QUITE BELIEVE WHAT WE WERE SEEING--
POLICE
WE FOUND THAT THERE WERE NO SIGNIFICANT ANATOMICAL DIFFERENCES
BETWEEN THE TWO POPULATIONS...
...PERMUTATIONS AND INTEGERS.

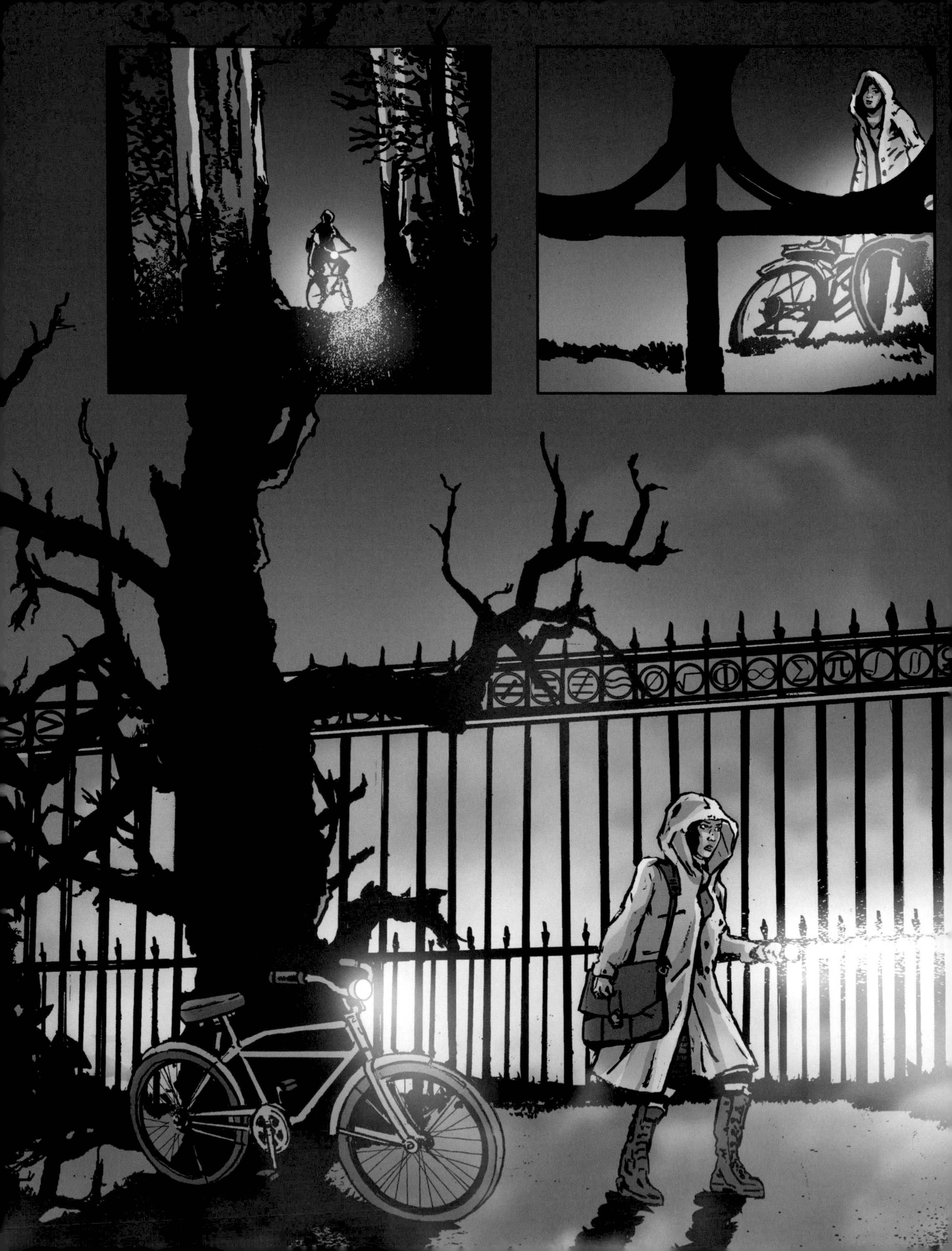

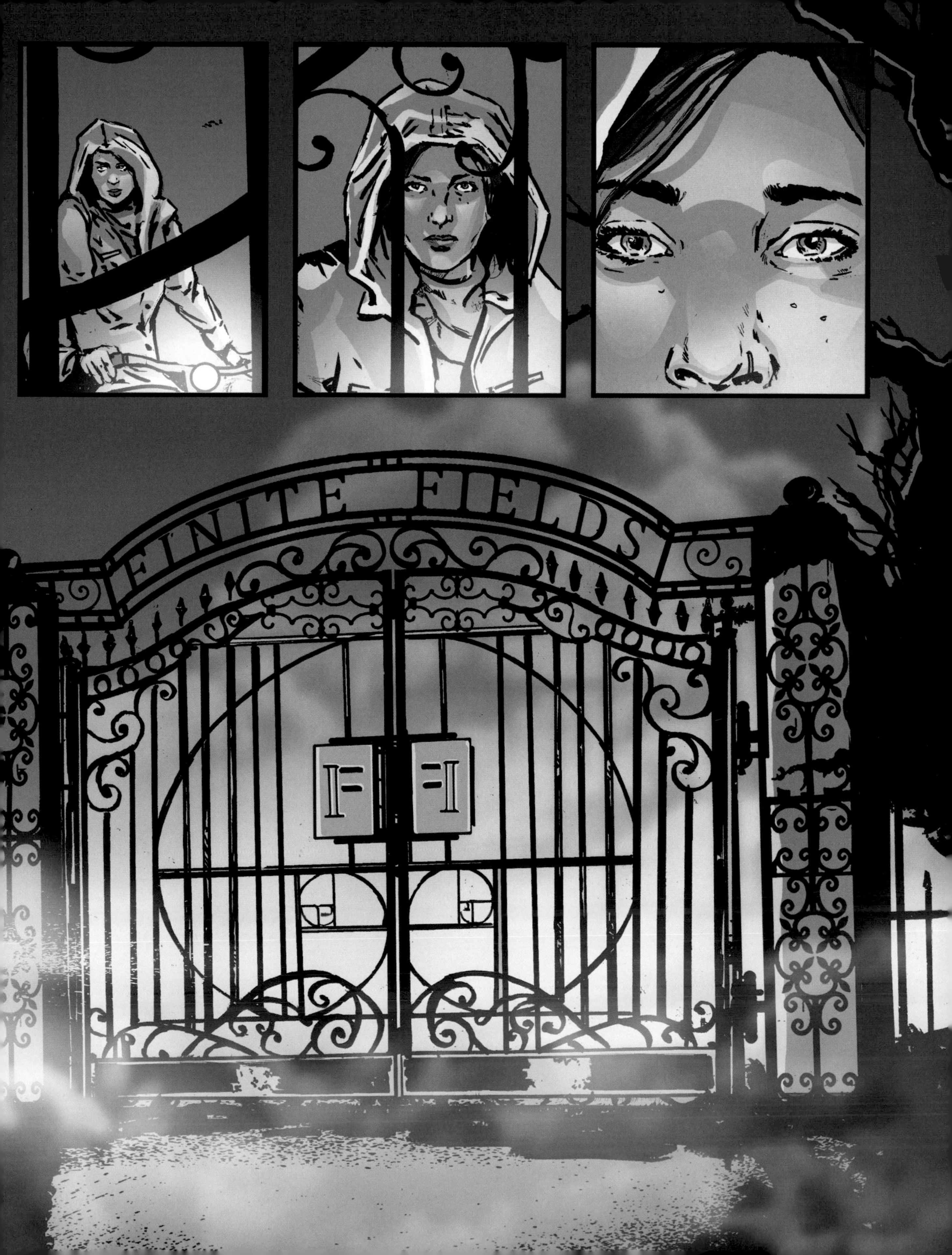
FINITE FIELDS

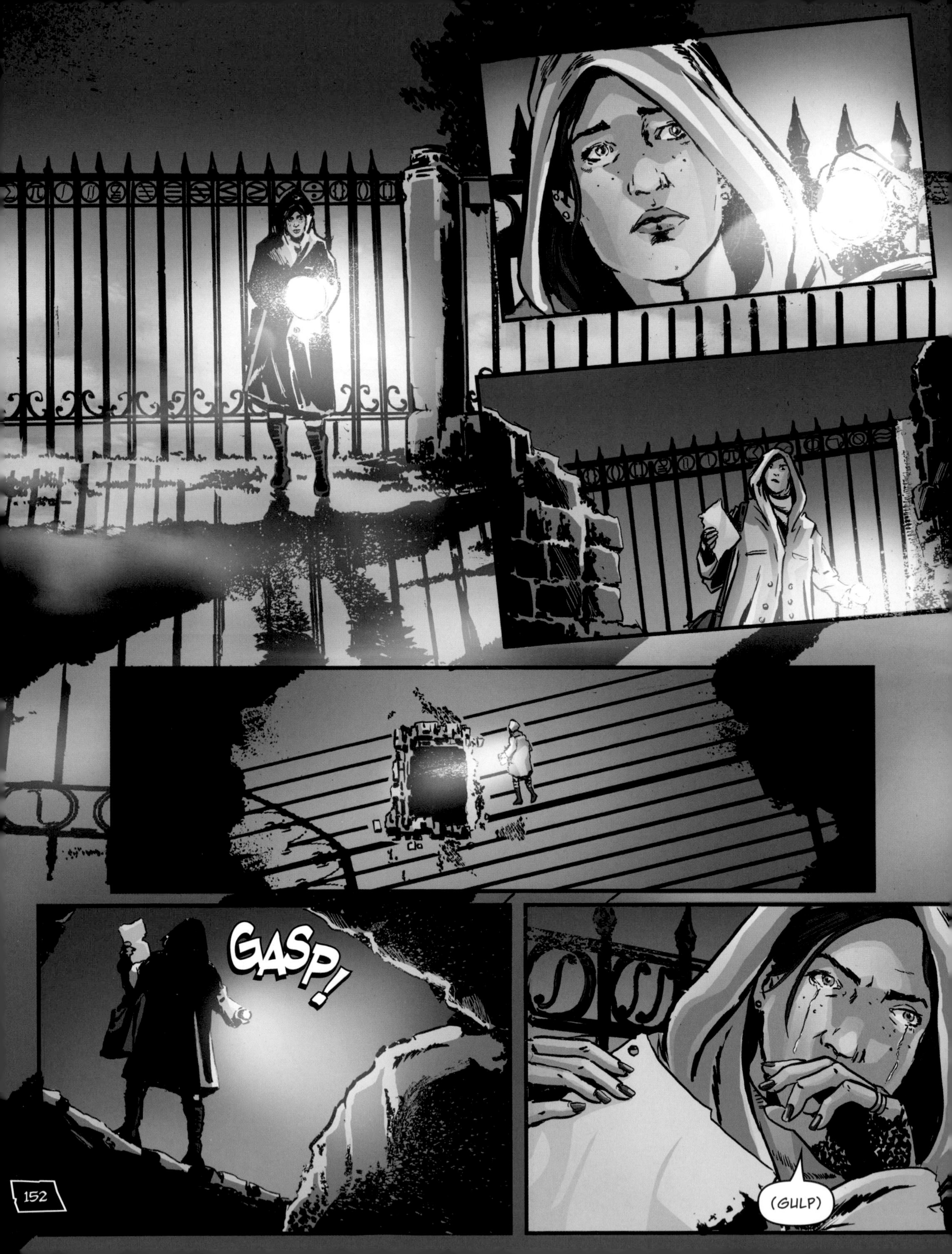
GASP!
(GULP)

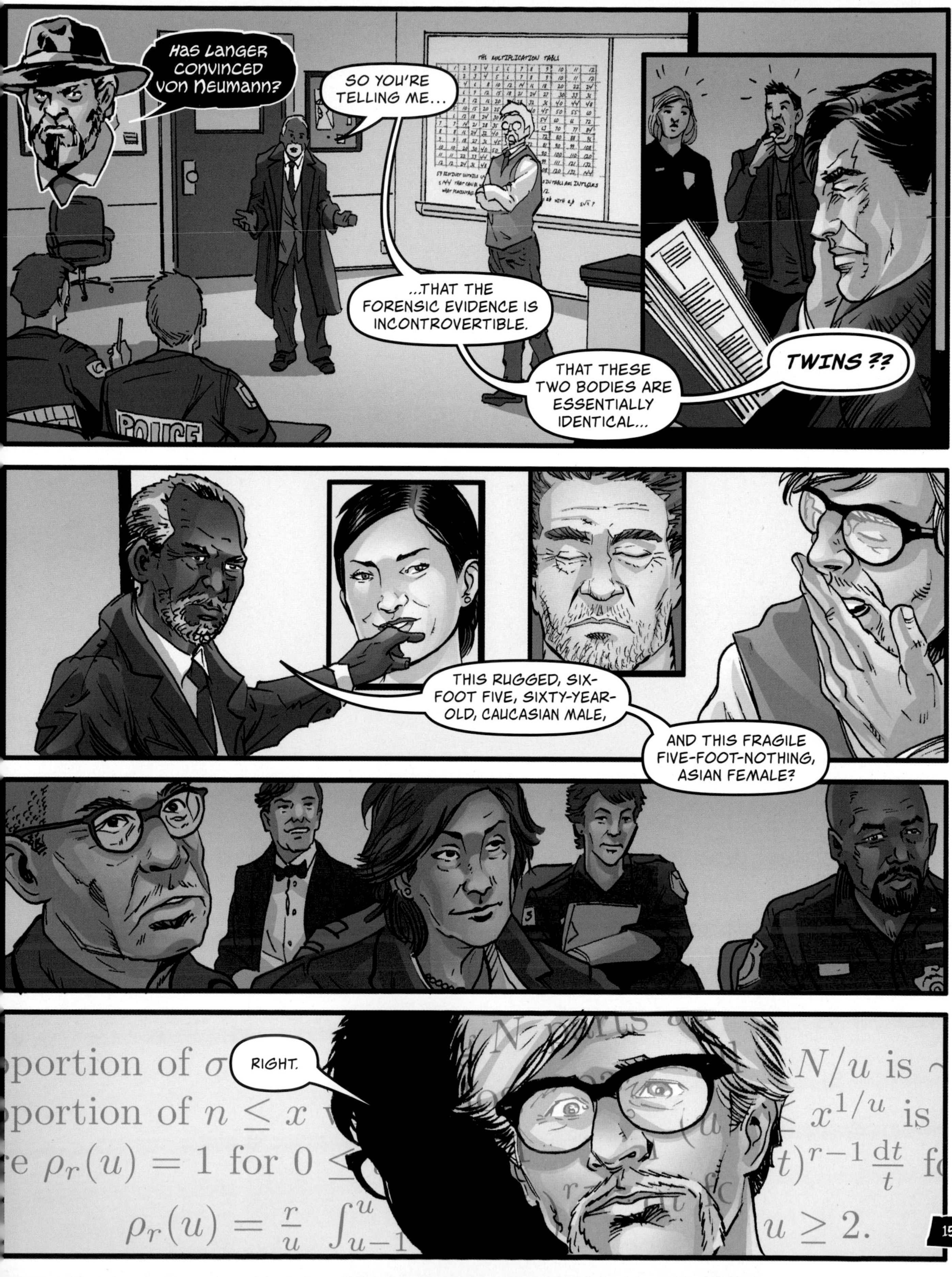
Has Langer convinced von Neumann?
SO YOU'RE TELLING ME...
...THAT THE FORENSIC EVIDENCE IS INCONTROVERTIBLE.
THAT THESE TWO BODIES ARE ESSENTIALLY IDENTICAL...
TWINS ??
THIS RUGGED, SIX-FOOT FIVE, SIXTY-YEAR-OLD, CAUCASIAN MALE,
AND THIS FRAGILE FIVE-FOOT-NOTHING, ASIAN FEMALE?
RIGHT.

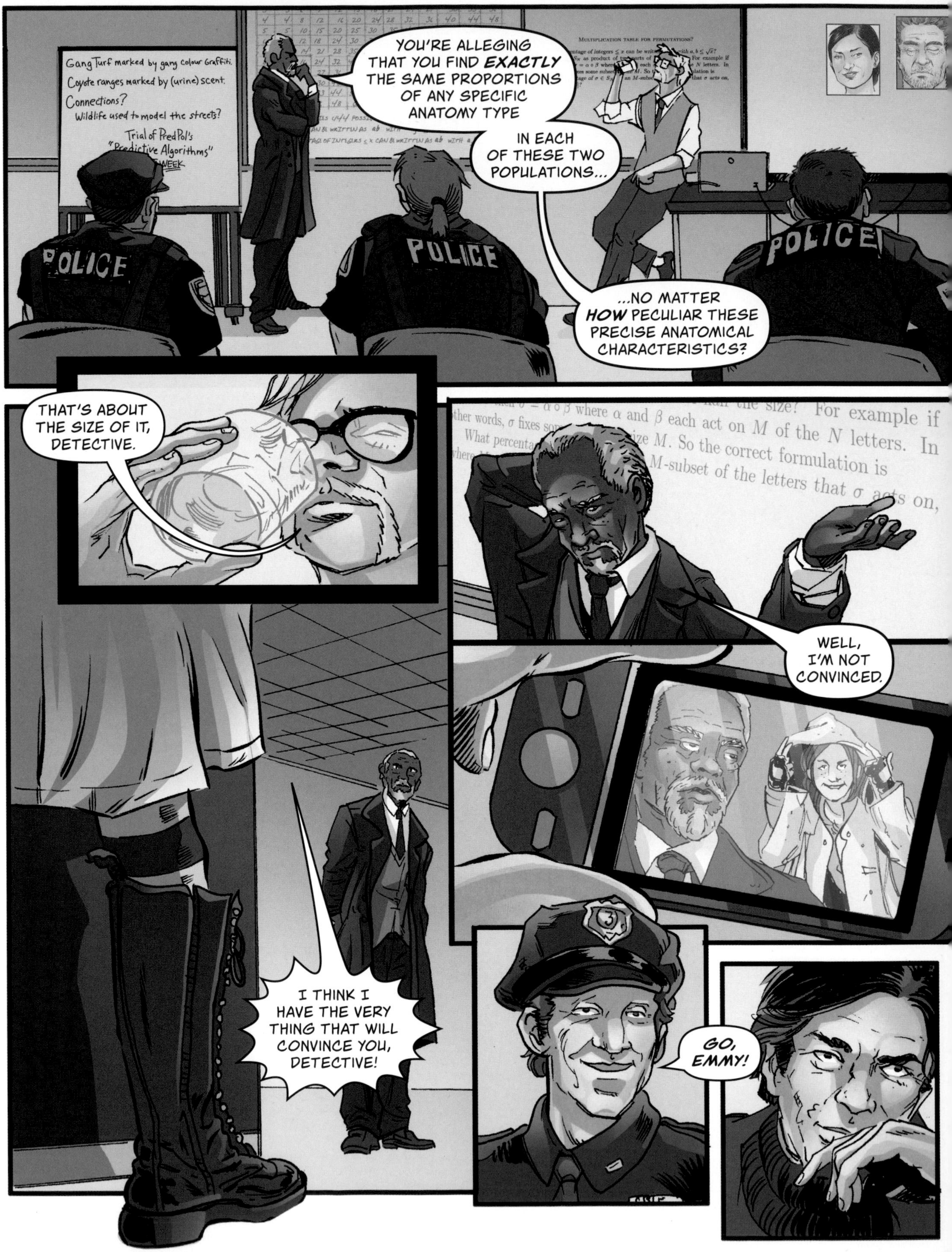
Gang Turf marked by gang colour Graffiti.
Coyote ranges marked by (urine) scent.
Connections?
Wildlife used to model the streets?
Trial of PredPol's "Predictive Algorithms"
POLICE
POLICE
YOU'RE ALLEGING THAT YOU FIND EXACTLY THE SAME PROPORTIONS OF ANY SPECIFIC ANATOMY TYPE
IN EACH OF THESE TWO POPULATIONS...
...NO MATTER HOW PECULIAR THESE PRECISE ANATOMICAL CHARACTERISTICS?
POLICE
THAT'S ABOUT THE SIZE OF IT, DETECTIVE.
WELL, I'M NOT CONVINCED.
I THINK I HAVE THE VERY THING THAT WILL CONVINCE YOU, DETECTIVE!
GO, EMMY!

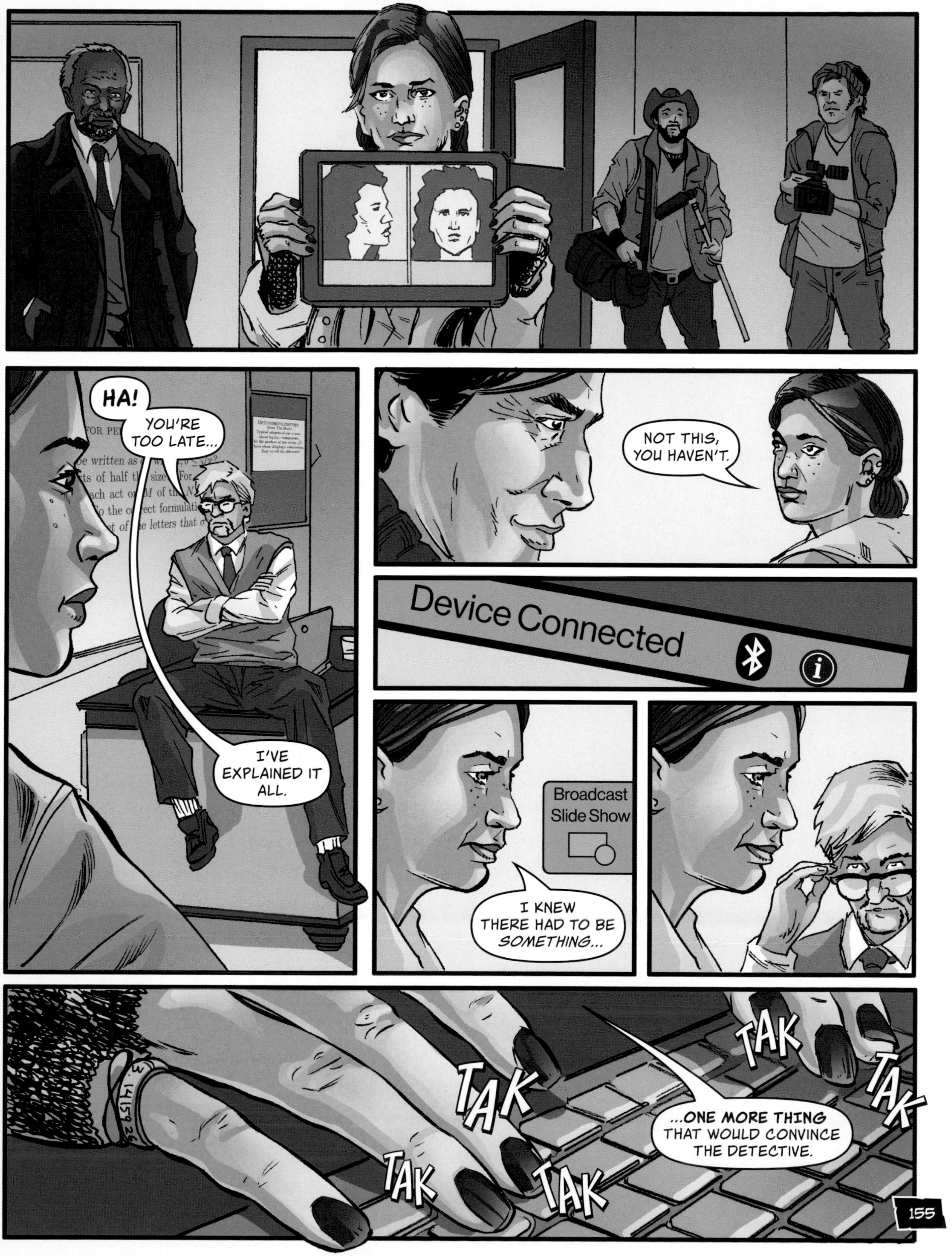
HA!
YOU'RE TOO LATE...
I'VE EXPLAINED IT ALL.
NOT THIS, YOU HAVEN'T.
Device Connected
Broadcast Slide Show
I KNEW THERE HAD TO BE *SOMETHING*...
TAK
TAK
TAK
TAK
TAK
...ONE MORE THING THAT WOULD CONVINCE THE DETECTIVE.

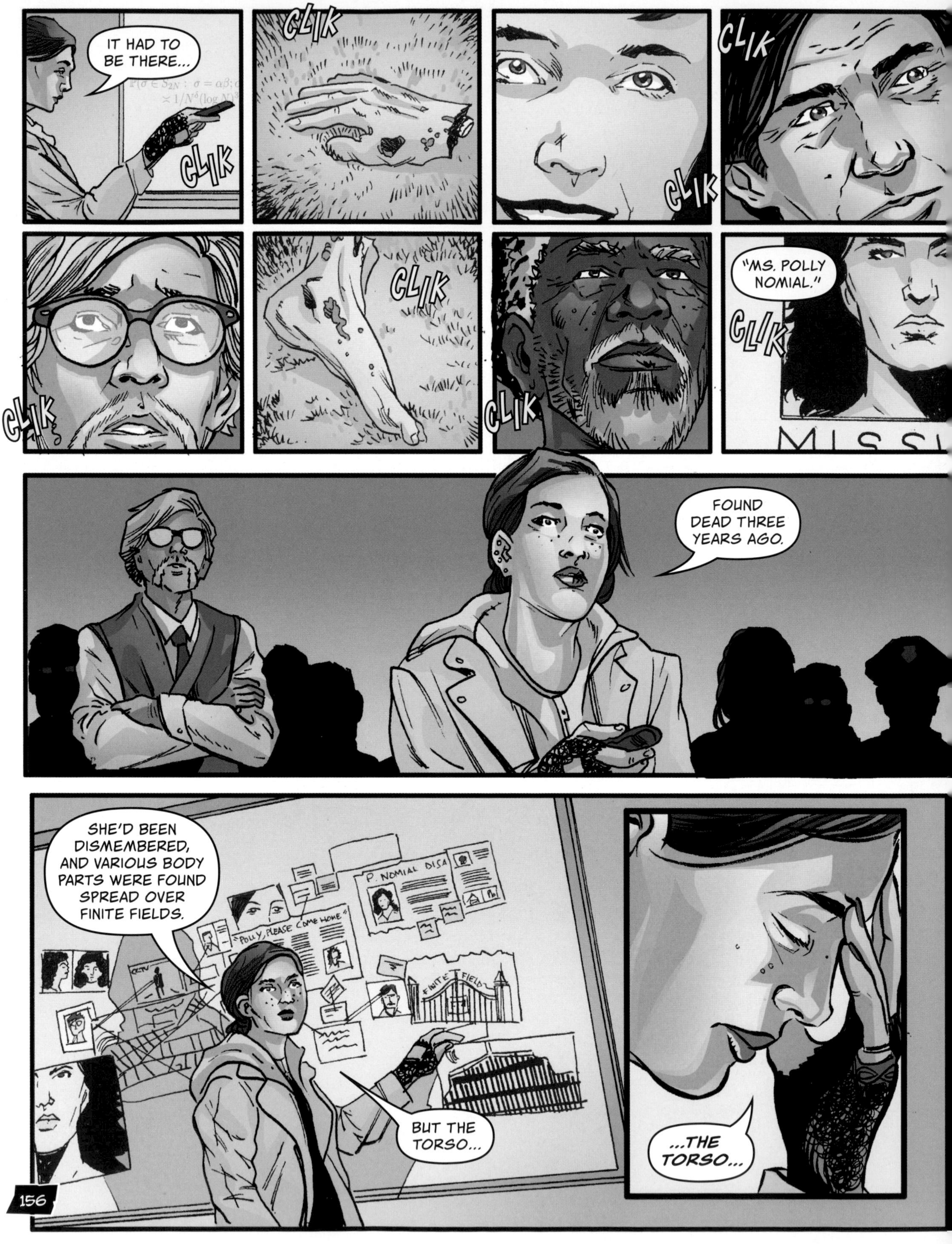
IT HAD TO BE THERE...
CLIK
CLIK
CLIK
CLIK
CLIK
CLIK
CLIK
"MS. POLLY NOMIAL."
CLIK
FOUND DEAD THREE YEARS AGO.
SHE'D BEEN DISMEMBERED, AND VARIOUS BODY PARTS WERE FOUND SPREAD OVER FINITE FIELDS.
P. NOMIAL DISA
"POLLY, PLEASE COME HOME"
FINITE FIELD
BUT THE TORSO...
...THE TORSO...

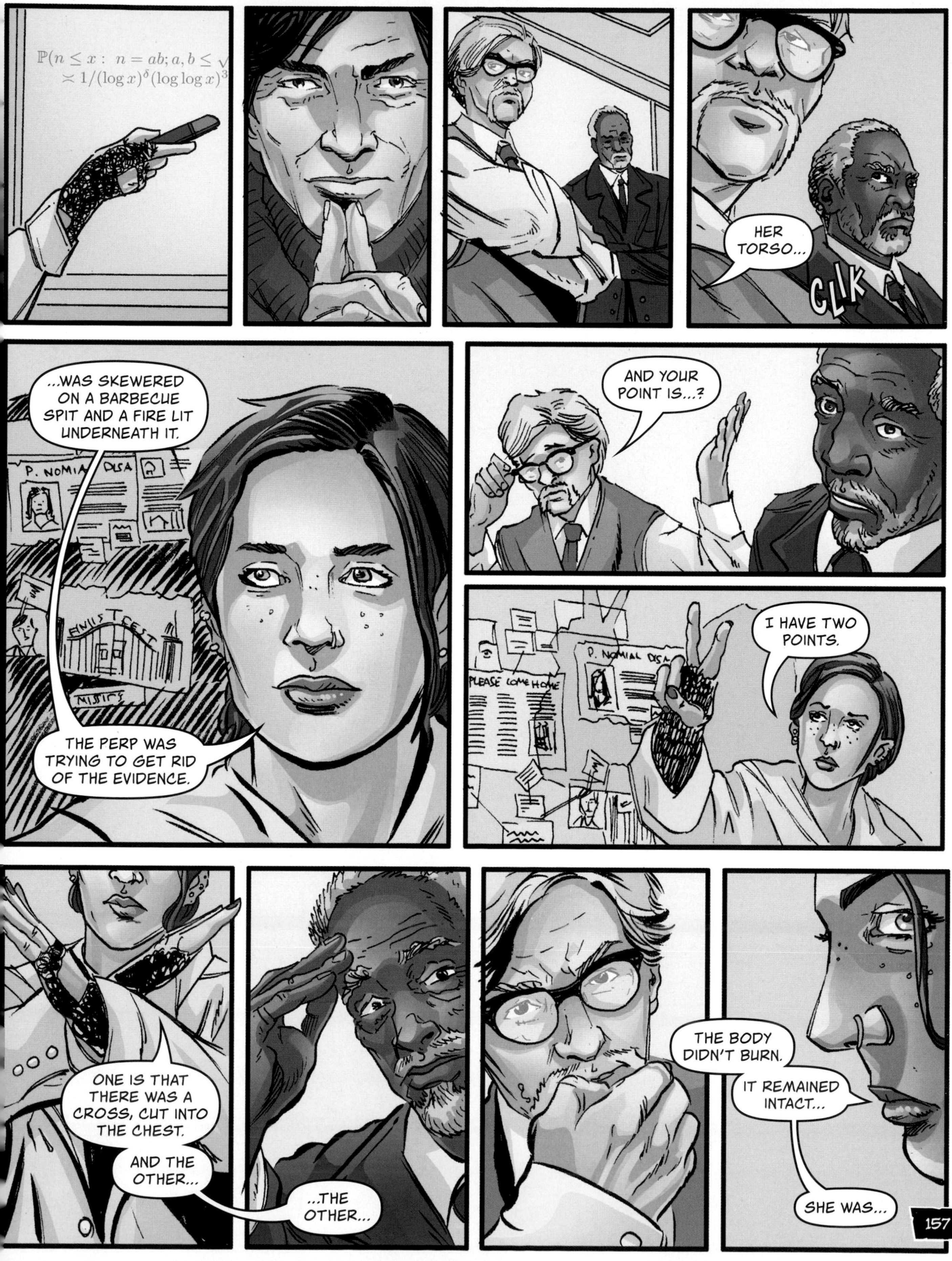
$\mathbb{P}(n \le x:\ n = ab; a, b \le \sqrt{}$
$\asymp 1/(\log x)^{\delta}(\log\log x)^{3}$
HER TORSO...
CLIK
...WAS SKEWERED ON A BARBECUE SPIT AND A FIRE LIT UNDERNEATH IT.
THE PERP WAS TRYING TO GET RID OF THE EVIDENCE.
AND YOUR POINT IS...?
I HAVE TWO POINTS.
PLEASE COME HOME
ONE IS THAT THERE WAS A CROSS, CUT INTO THE CHEST.
AND THE OTHER...
...THE OTHER...
THE BODY DIDN'T BURN.
IT REMAINED INTACT...
SHE WAS...

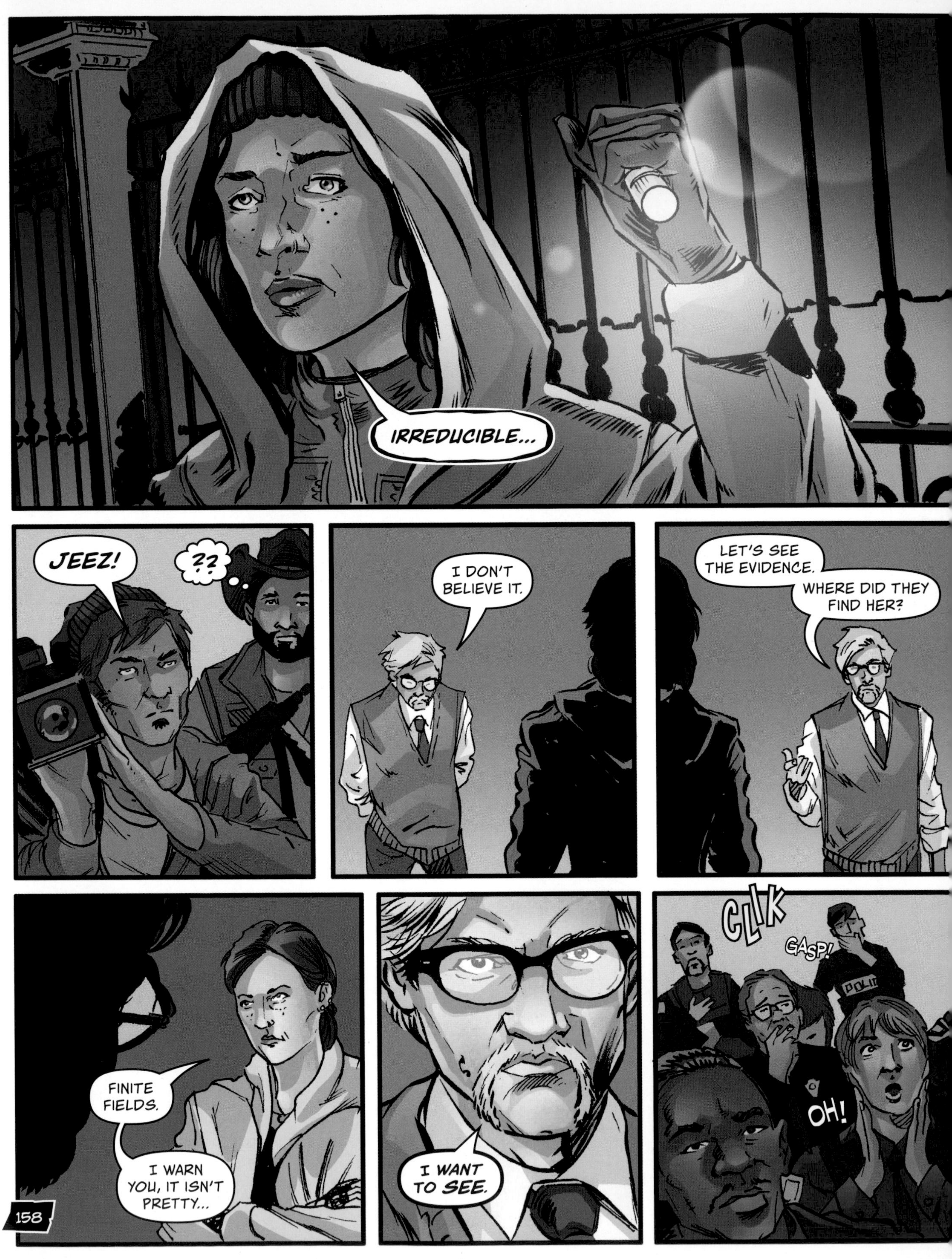
IRREDUCIBLE...
JEEZ!
??
I DON'T BELIEVE IT.
LET'S SEE THE EVIDENCE.
WHERE DID THEY FIND HER?
FINITE FIELDS.
I WARN YOU, IT ISN'T PRETTY...
I WANT TO SEE.
CLIK
GASP!
OH!

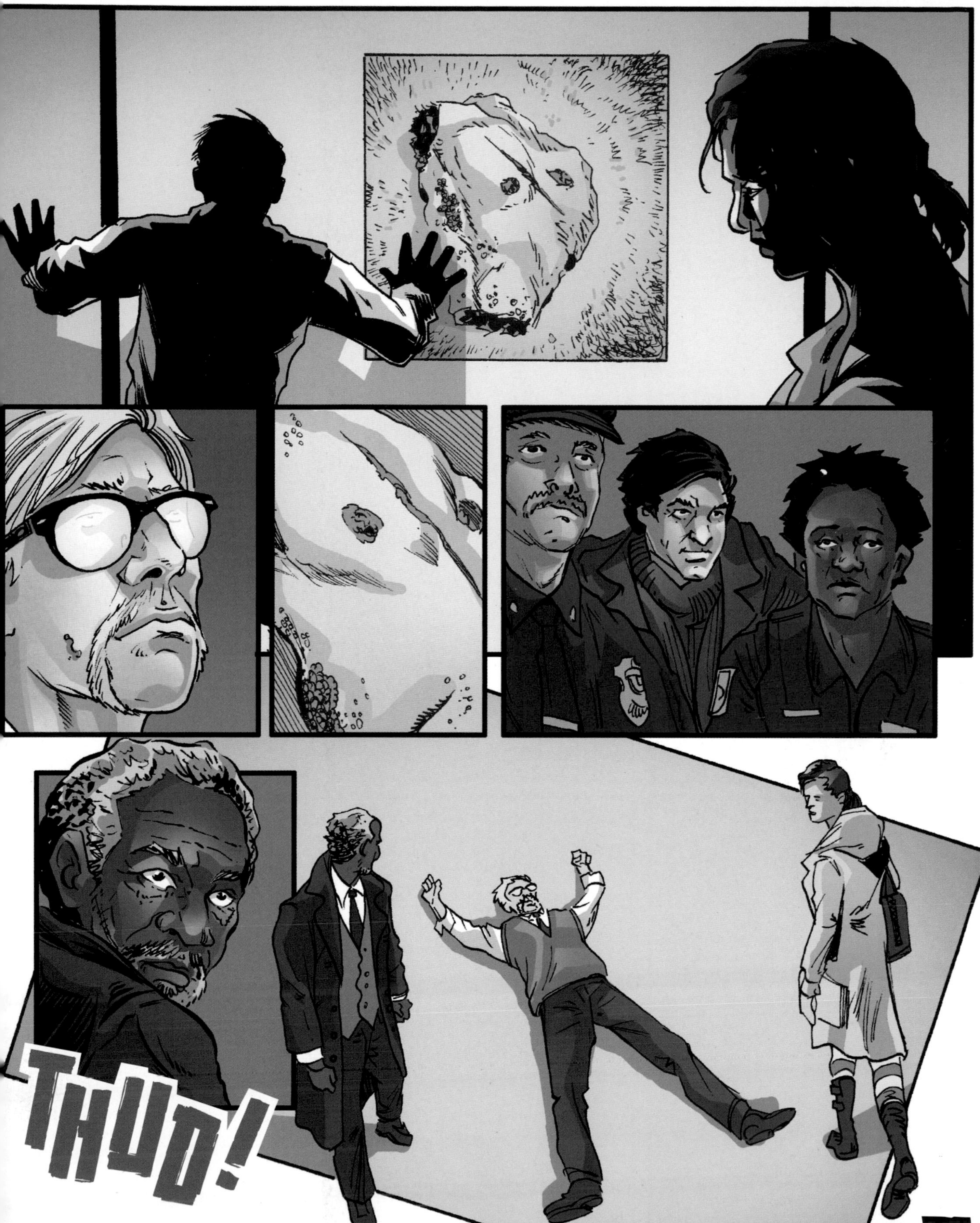
THUD!

CLAP CLAP

CLAP

CLAP

A CIRCUS SHOW.

A TRICK.

CLAP

CLAP

CLAP

IT'S NOT ENOUGH.

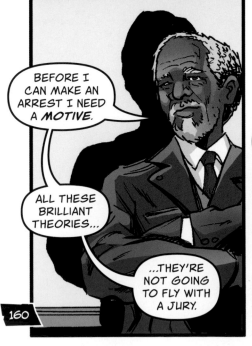

BEFORE I CAN MAKE AN ARREST I NEED A *MOTIVE.*

ALL THESE BRILLIANT THEORIES...

...THEY'RE NOT GOING TO FLY WITH A JURY.

$H_0 = H_1 + \frac{1}{\sqrt{1 - v^2/c^2}}$

$R_{ab} - \frac{1}{2} R g_{ab} = \frac{8\pi G}{c^4} T_{ab}$

SMACK

$E = mc^2$

IT'S IMPOSSIBLE TO CONCENTRATE IN HERE!

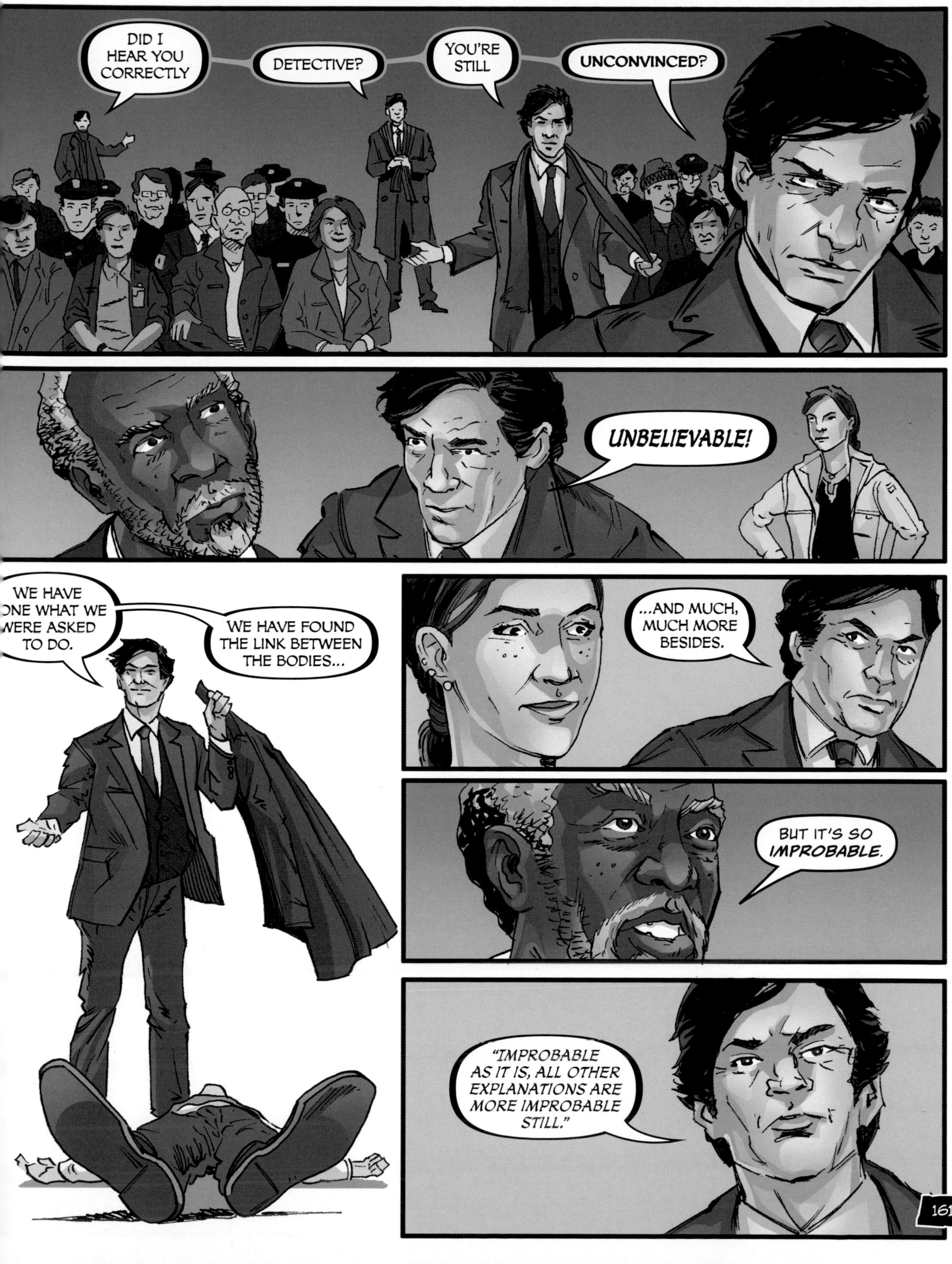

DID I HEAR YOU CORRECTLY
DETECTIVE?
YOU'RE STILL
UNCONVINCED?
UNBELIEVABLE!
WE HAVE DONE WHAT WE WERE ASKED TO DO.
WE HAVE FOUND THE LINK BETWEEN THE BODIES...
...AND MUCH, MUCH MORE BESIDES.
BUT IT'S SO *IMPROBABLE*.
"IMPROBABLE AS IT IS, ALL OTHER EXPLANATIONS ARE MORE IMPROBABLE STILL."

EVEN IF I GRANT YOU THAT, I HAVE TO UNDERSTAND *WHY.*

WHY ARE THEIR ANATOMIES SO SIMILAR?

DO THEY HAVE THE SAME DNA?

IS ONE MODELED ON THE OTHER?

YOU DON'T SEEM TO BE TALKING ABOUT GENETICS.

SO HOW CAN THE MATH TELL YOU ALL THIS WITH *NO BIOLOGICAL EXPLANATION?*

LANGER OUTLINED OUR RESPONSES IN HIS PRESENTATION, DETECTIVE.

WE HAVE PROPOSED TWO POSSIBLE EXPLANATIONS.

ONE FROM *PROBABILITY THEORY...*

...THE OTHER FROM *ANALYTIC COMBINATORICS.*

TURN TO YOUR HAND-OUTS, SECTION 5.1.

ADMITTEDLY NEITHER IS INCREDIBLY PERSUASIVE, BUT IT'S AS MUCH AS WE KNOW FOR NOW.

5.1.a) (Det. Kingman and the ABT swat
Define a Poisson-Dirichlet point process on
taining uniformity by conditioning on previously
Successfully used by Det Ewens for DNA Allele co
5.1.b) (Dets/ Panario-Richmond). Decompos

$a + b(1 - r_2)i($

ts using analys

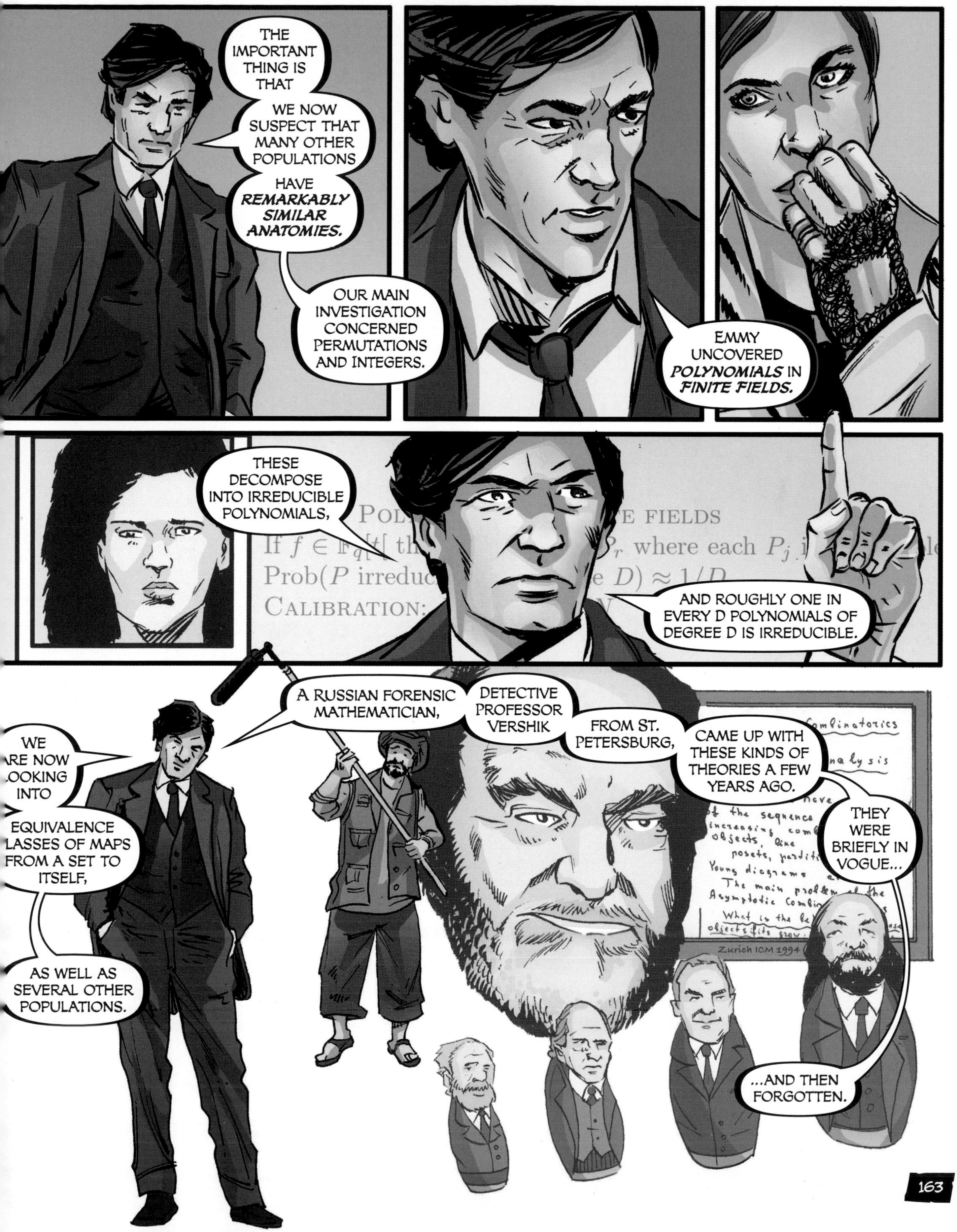
THE IMPORTANT THING IS THAT
WE NOW SUSPECT THAT MANY OTHER POPULATIONS
HAVE ***REMARKABLY SIMILAR ANATOMIES.***
OUR MAIN INVESTIGATION CONCERNED PERMUTATIONS AND INTEGERS.
EMMY UNCOVERED ***POLYNOMIALS*** IN ***FINITE FIELDS.***
THESE DECOMPOSE INTO IRREDUCIBLE POLYNOMIALS,
AND ROUGHLY ONE IN EVERY D POLYNOMIALS OF DEGREE D IS IRREDUCIBLE.
WE RE NOW OOKING INTO
EQUIVALENCE LASSES OF MAPS FROM A SET TO ITSELF,
AS WELL AS SEVERAL OTHER POPULATIONS.
A RUSSIAN FORENSIC MATHEMATICIAN,
DETECTIVE PROFESSOR VERSHIK
FROM ST. PETERSBURG,
CAME UP WITH THESE KINDS OF THEORIES A FEW YEARS AGO.
THEY WERE BRIEFLY IN VOGUE...
...AND THEN FORGOTTEN.
Zurich ICM 1994

BUT HE WAS RIGHT, VON NEUMANN.

HE WAS *RIGHT*.

EVEN IF HE DIDN'T HAVE ALL THE EVIDENCE THAT WE HAVE.

HE WAS RIGHT.

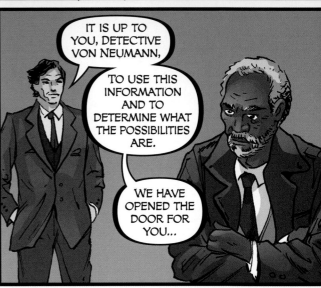

IT IS UP TO YOU, DETECTIVE VON NEUMANN,

TO USE THIS INFORMATION AND TO DETERMINE WHAT THE POSSIBILITIES ARE.

WE HAVE OPENED THE DOOR FOR YOU...

...IT IS UP TO YOU TO DECIDE WHETHER TO GO THROUGH IT.

WELL...

...WE STILL HAVEN'T FOUND THE MOTIVE

BUT WE DO HAVE THE *PERP!*

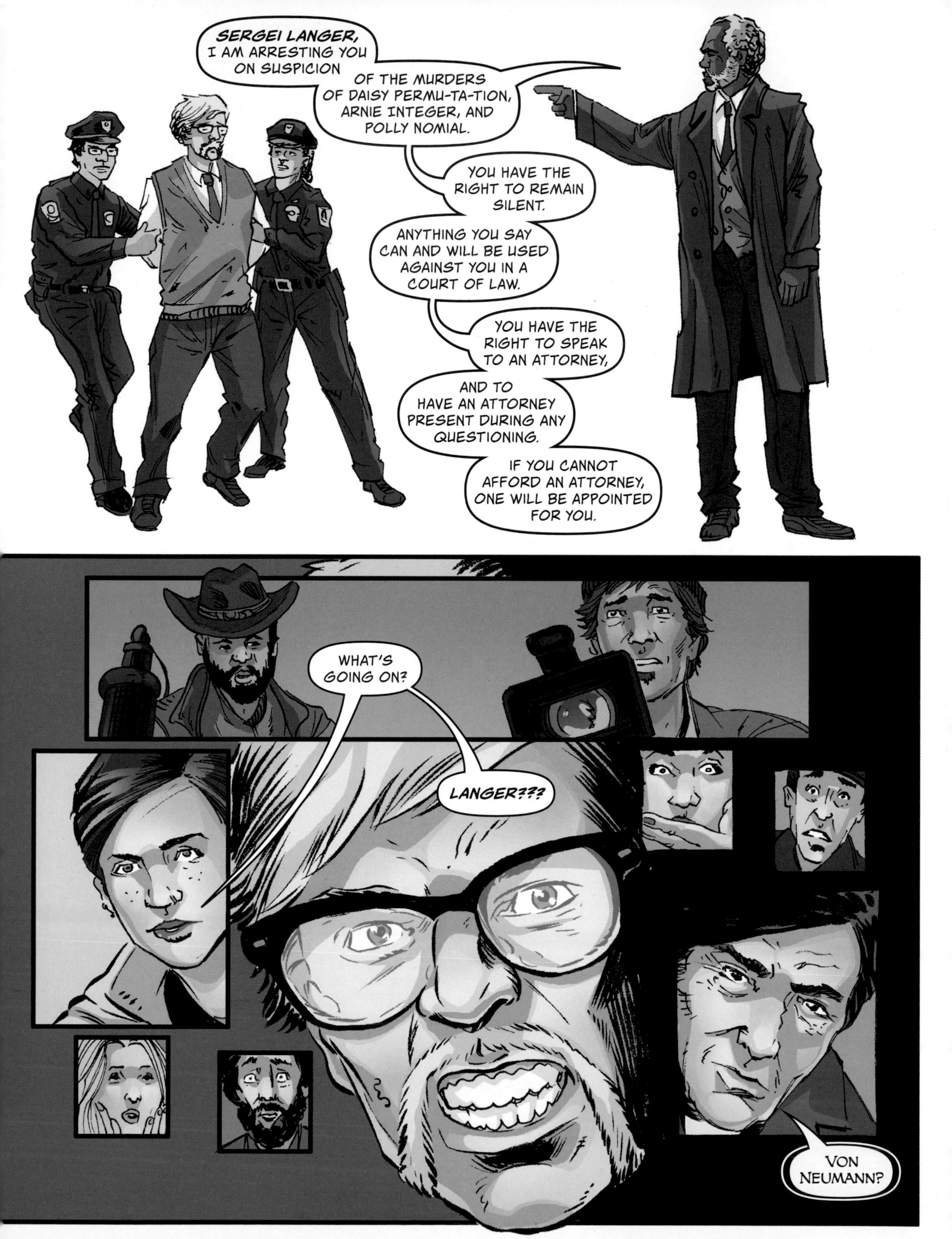
SERGEI LANGER, I AM ARRESTING YOU ON SUSPICION
OF THE MURDERS OF DAISY PERMU-TA-TION, ARNIE INTEGER, AND POLLY NOMIAL.
YOU HAVE THE RIGHT TO REMAIN SILENT.
ANYTHING YOU SAY CAN AND WILL BE USED AGAINST YOU IN A COURT OF LAW.
YOU HAVE THE RIGHT TO SPEAK TO AN ATTORNEY,
AND TO HAVE AN ATTORNEY PRESENT DURING ANY QUESTIONING.
IF YOU CANNOT AFFORD AN ATTORNEY, ONE WILL BE APPOINTED FOR YOU.
WHAT'S GOING ON?
LANGER???
VON NEUMANN?

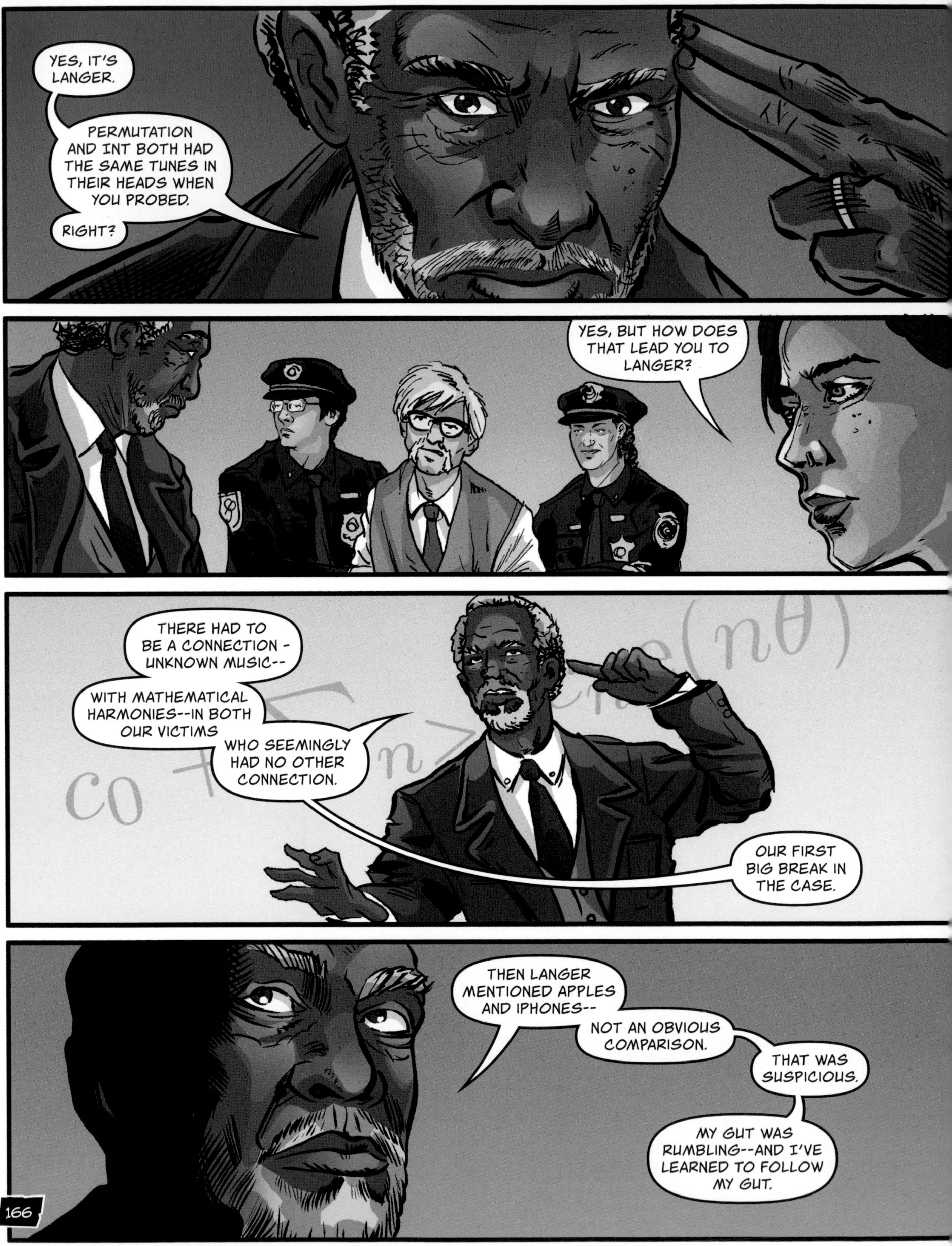
YES, IT'S LANGER.
PERMUTATION AND INT BOTH HAD THE SAME TUNES IN THEIR HEADS WHEN YOU PROBED.
RIGHT?
YES, BUT HOW DOES THAT LEAD YOU TO LANGER?
THERE HAD TO BE A CONNECTION - UNKNOWN MUSIC--
WITH MATHEMATICAL HARMONIES--IN BOTH OUR VICTIMS
WHO SEEMINGLY HAD NO OTHER CONNECTION.
OUR FIRST BIG BREAK IN THE CASE.
THEN LANGER MENTIONED APPLES AND IPHONES--
NOT AN OBVIOUS COMPARISON.
THAT WAS SUSPICIOUS.
MY GUT WAS RUMBLING--AND I'VE LEARNED TO FOLLOW MY GUT.

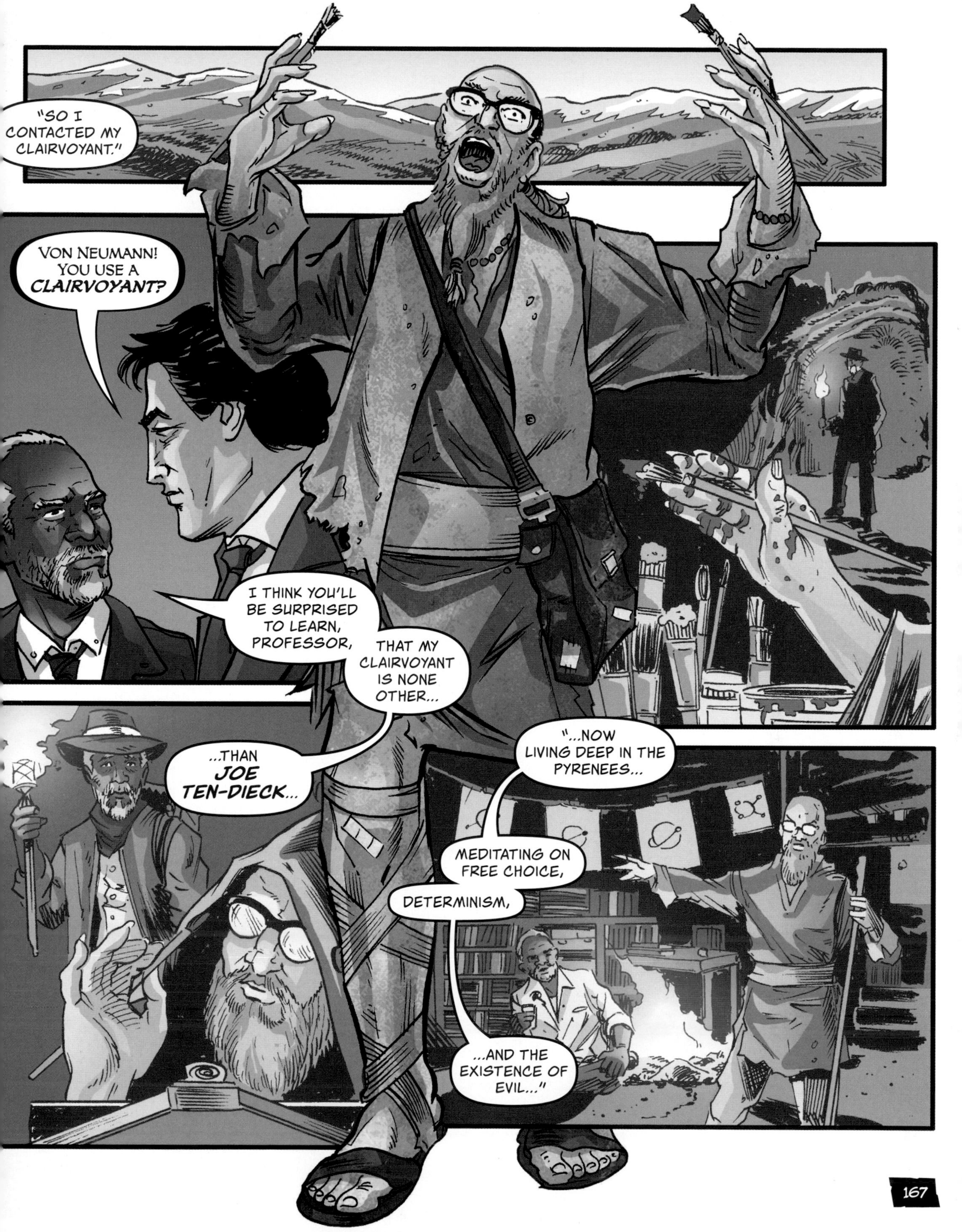
"SO I CONTACTED MY CLAIRVOYANT."
VON NEUMANN! YOU USE A CLAIRVOYANT?
I THINK YOU'LL BE SURPRISED TO LEARN, PROFESSOR,
THAT MY CLAIRVOYANT IS NONE OTHER...
...THAN JOE TEN-DIECK...
"...NOW LIVING DEEP IN THE PYRENEES...
MEDITATING ON FREE CHOICE,
DETERMINISM,
...AND THE EXISTENCE OF EVIL..."

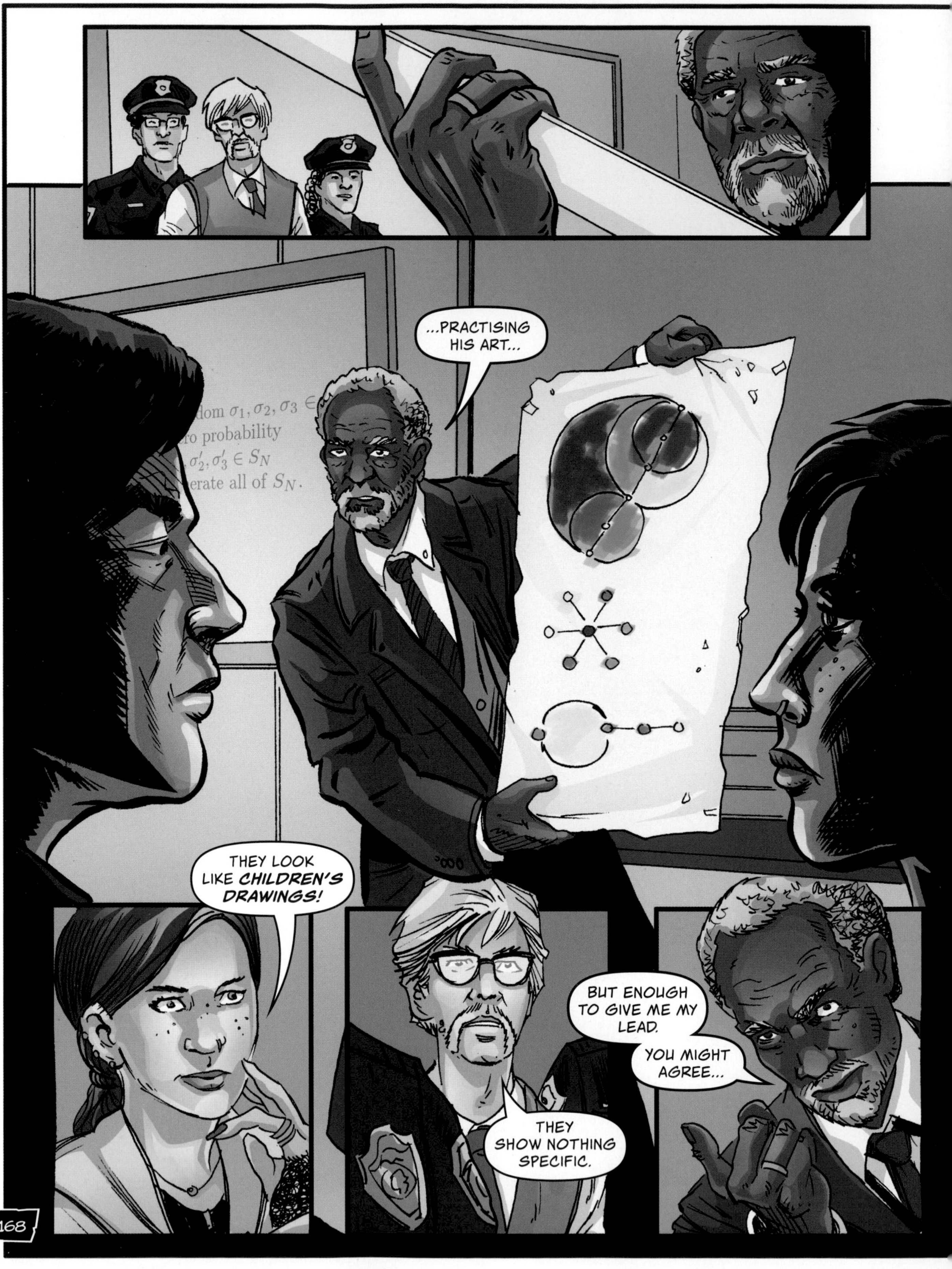

...PRACTISING HIS ART...
THEY LOOK LIKE CHILDREN'S DRAWINGS!
THEY SHOW NOTHING SPECIFIC.
BUT ENOUGH TO GIVE ME MY LEAD.
YOU MIGHT AGREE...

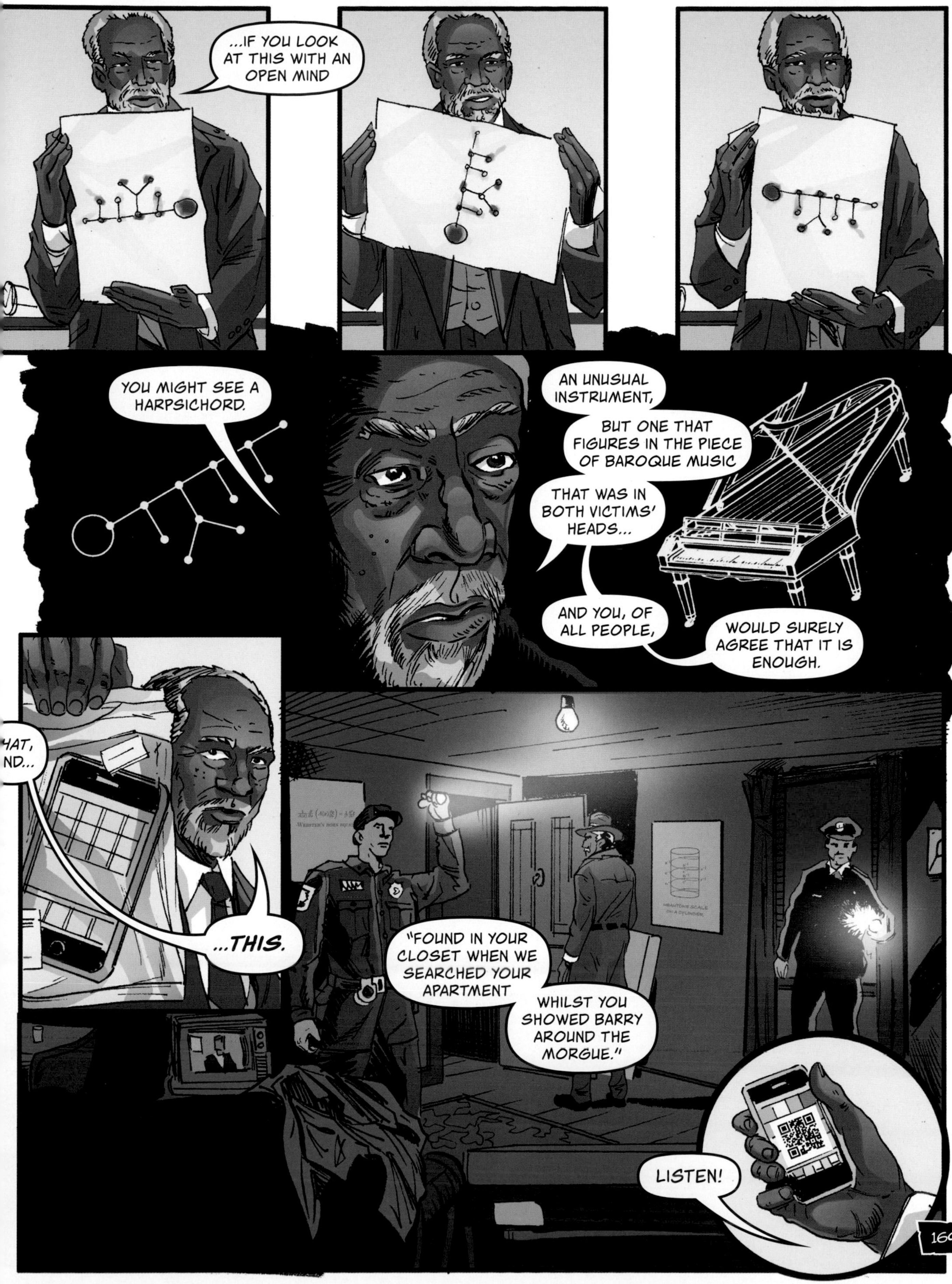
...IF YOU LOOK AT THIS WITH AN OPEN MIND
YOU MIGHT SEE A HARPSICHORD.
AN UNUSUAL INSTRUMENT,
BUT ONE THAT FIGURES IN THE PIECE OF BAROQUE MUSIC
THAT WAS IN BOTH VICTIMS' HEADS...
AND YOU, OF ALL PEOPLE,
WOULD SURELY AGREE THAT IT IS ENOUGH.
HAT, ND...
...THIS.
"FOUND IN YOUR CLOSET WHEN WE SEARCHED YOUR APARTMENT
WHILST YOU SHOWED BARRY AROUND THE MORGUE."
LISTEN!

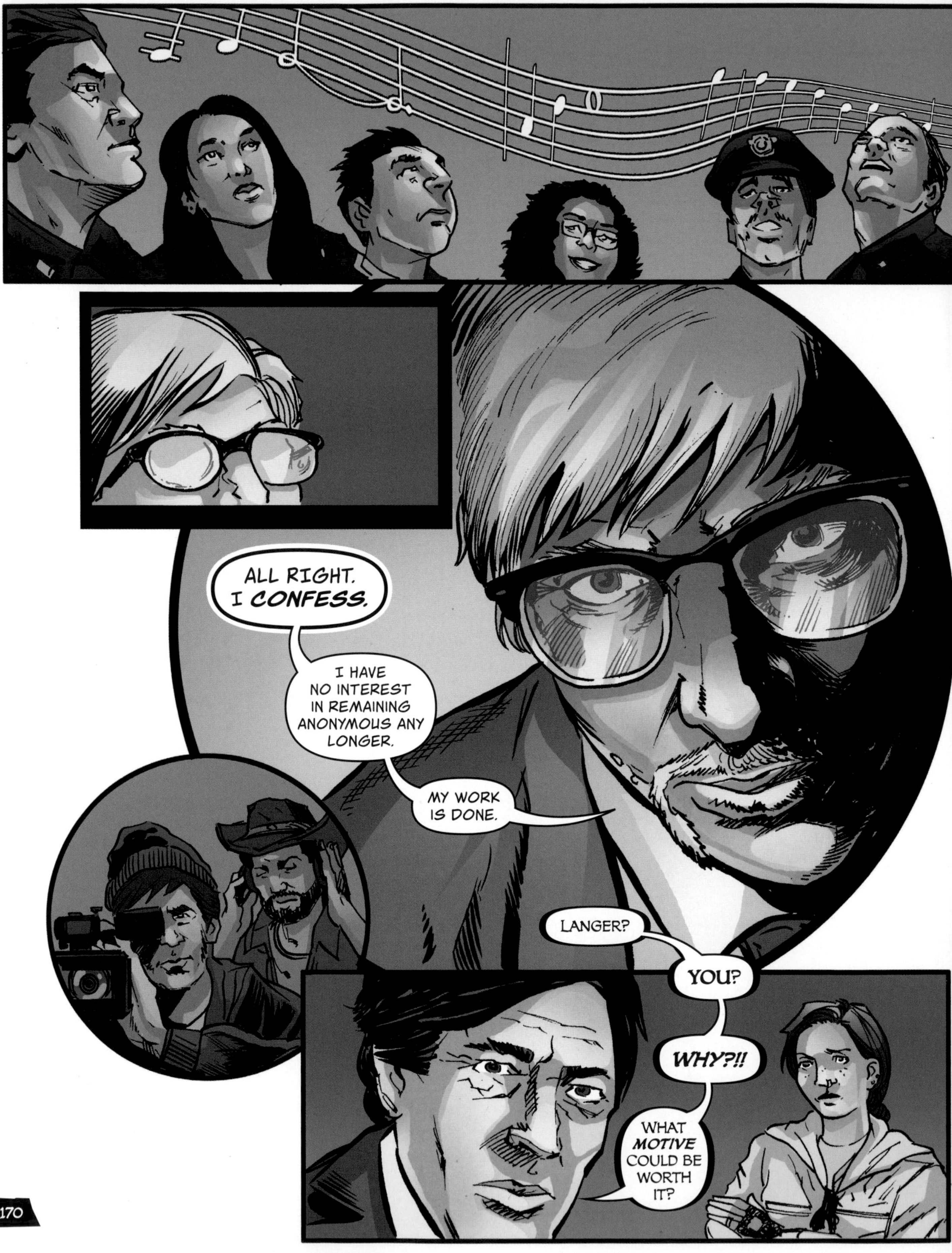
ALL RIGHT. I CONFESS.
I HAVE NO INTEREST IN REMAINING ANONYMOUS ANY LONGER.
MY WORK IS DONE.
LANGER?
YOU?
WHY?!!
WHAT MOTIVE COULD BE WORTH IT?

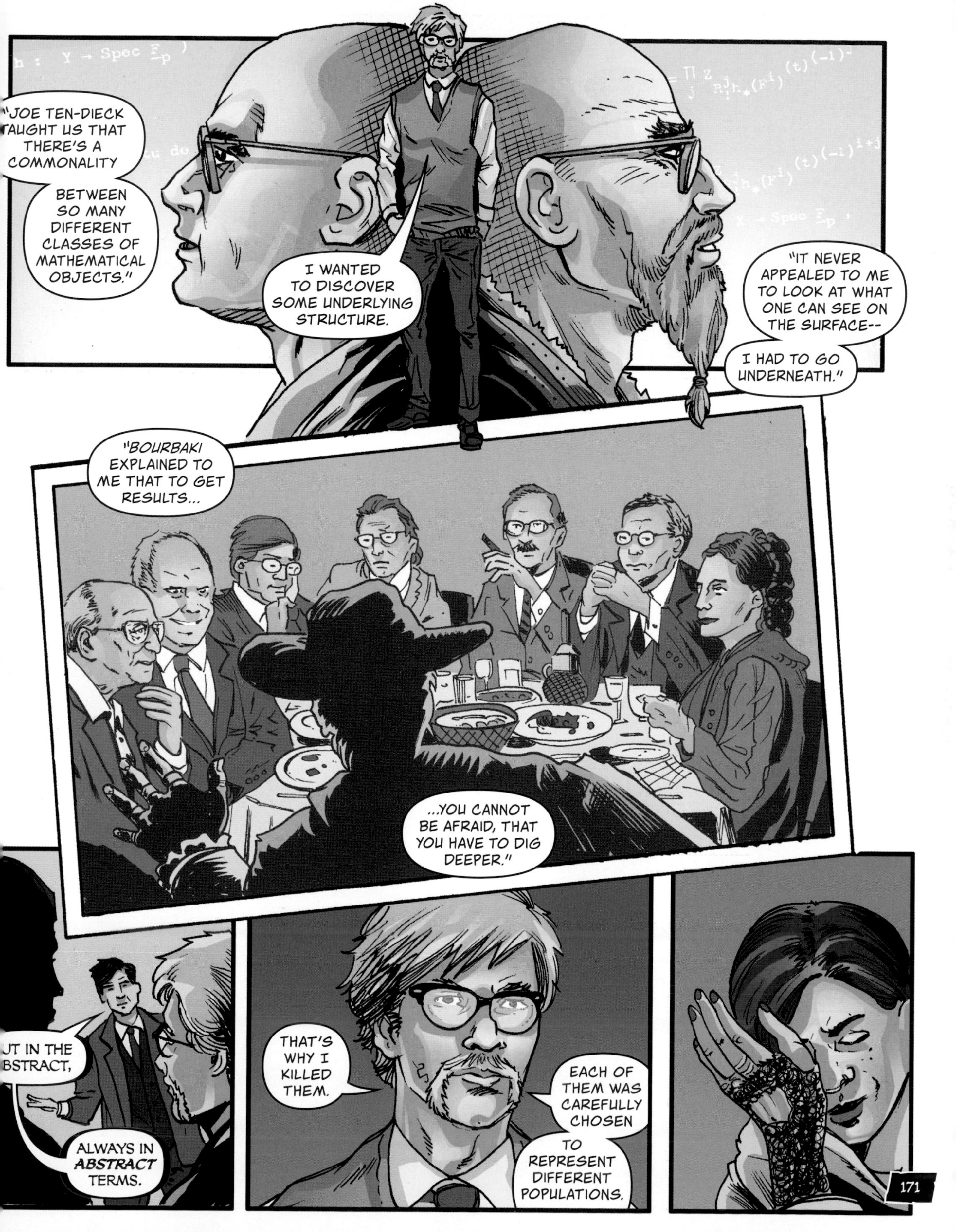
"JOE TEN-DIECK TAUGHT US THAT THERE'S A COMMONALITY
BETWEEN SO MANY DIFFERENT CLASSES OF MATHEMATICAL OBJECTS."
I WANTED TO DISCOVER SOME UNDERLYING STRUCTURE.
"IT NEVER APPEALED TO ME TO LOOK AT WHAT ONE CAN SEE ON THE SURFACE--
I HAD TO GO UNDERNEATH."
"BOURBAKI EXPLAINED TO ME THAT TO GET RESULTS...
...YOU CANNOT BE AFRAID, THAT YOU HAVE TO DIG DEEPER."
UT IN THE BSTRACT,
ALWAYS IN ABSTRACT TERMS.
THAT'S WHY I KILLED THEM.
EACH OF THEM WAS CAREFULLY CHOSEN
TO REPRESENT DIFFERENT POPULATIONS.

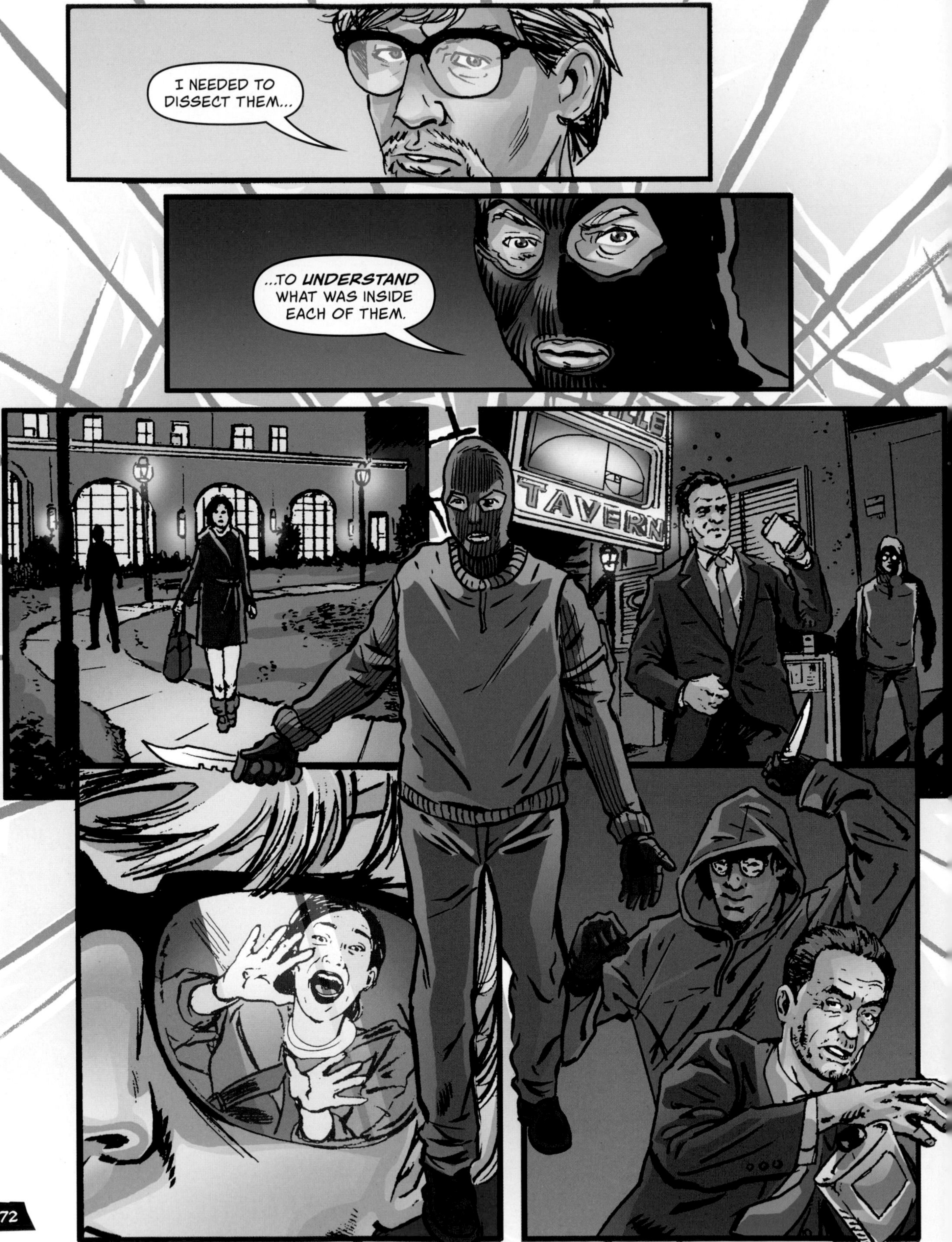
I NEEDED TO DISSECT THEM...
...TO *UNDERSTAND* WHAT WAS INSIDE EACH OF THEM.
TAVERN

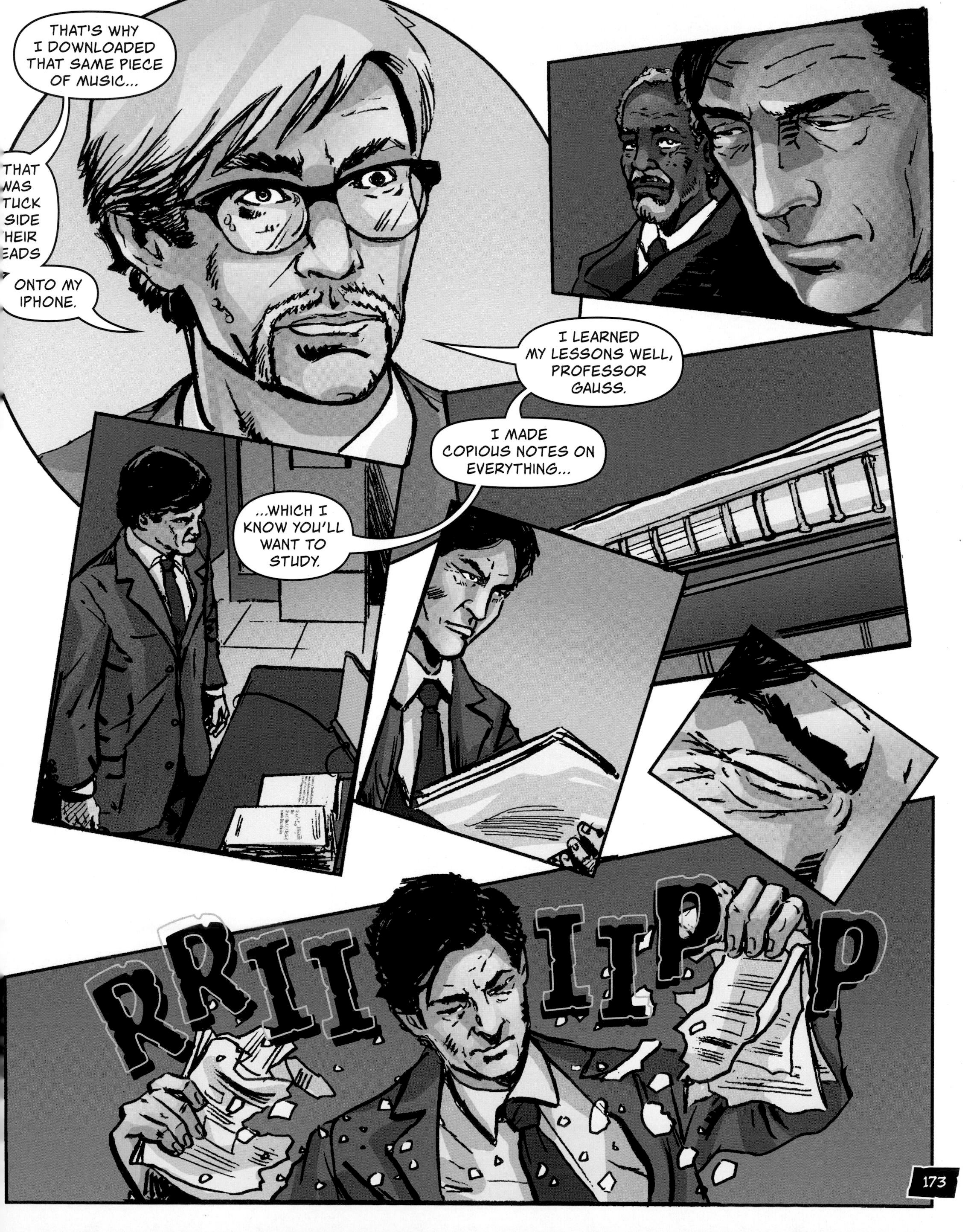
THAT'S WHY I DOWNLOADED THAT SAME PIECE OF MUSIC...
THAT WAS TUCK SIDE HEIR EADS
ONTO MY IPHONE.
I LEARNED MY LESSONS WELL, PROFESSOR GAUSS.
I MADE COPIOUS NOTES ON EVERYTHING...
...WHICH I KNOW YOU'LL WANT TO STUDY.
RRIIIIPP

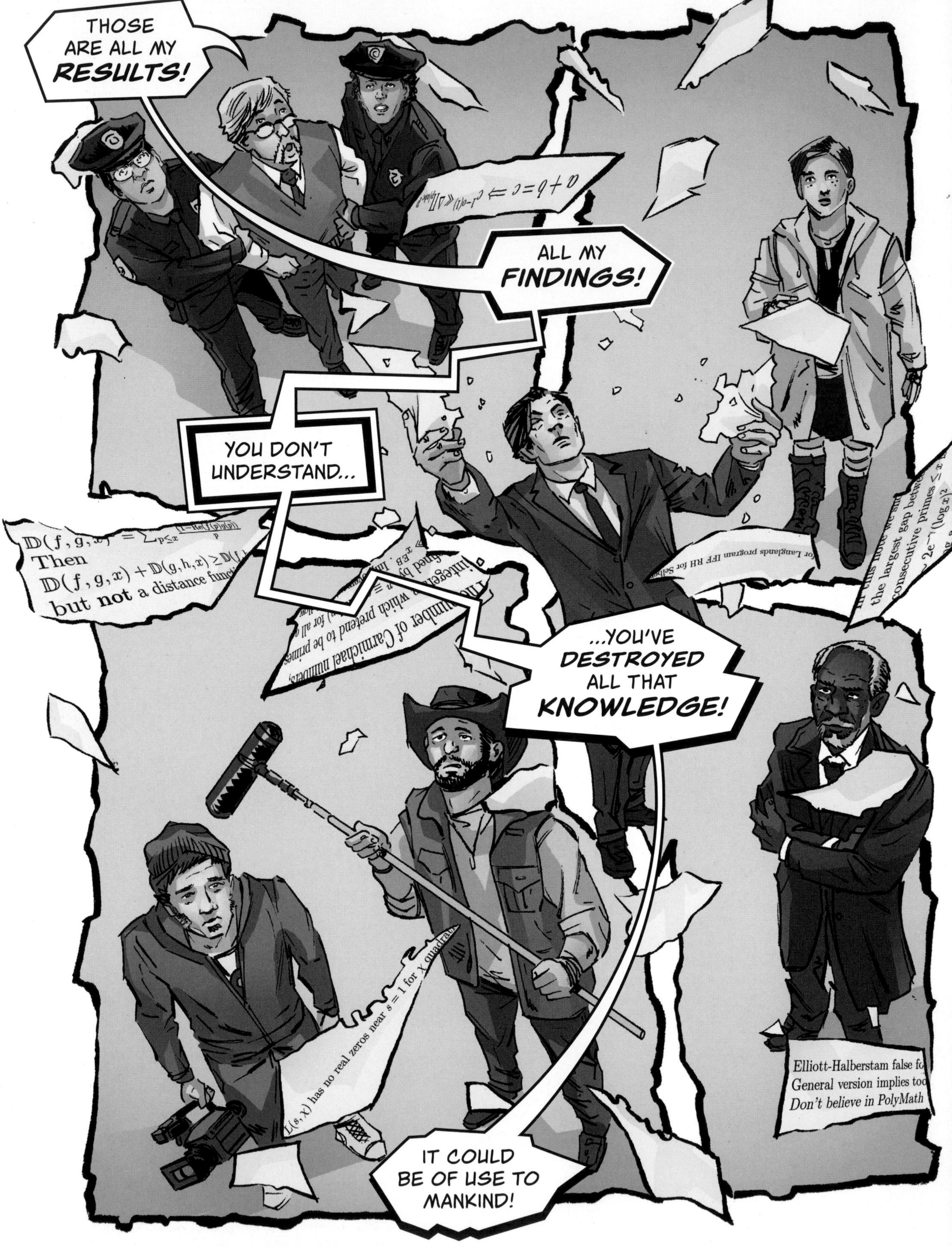
THOSE ARE ALL MY RESULTS!
ALL MY FINDINGS!
YOU DON'T UNDERSTAND...
...YOU'VE DESTROYED ALL THAT KNOWLEDGE!
IT COULD BE OF USE TO MANKIND!
Then
but not a distance
Elliott-Halberstam false fo
General version implies too
Don't believe in PolyMath

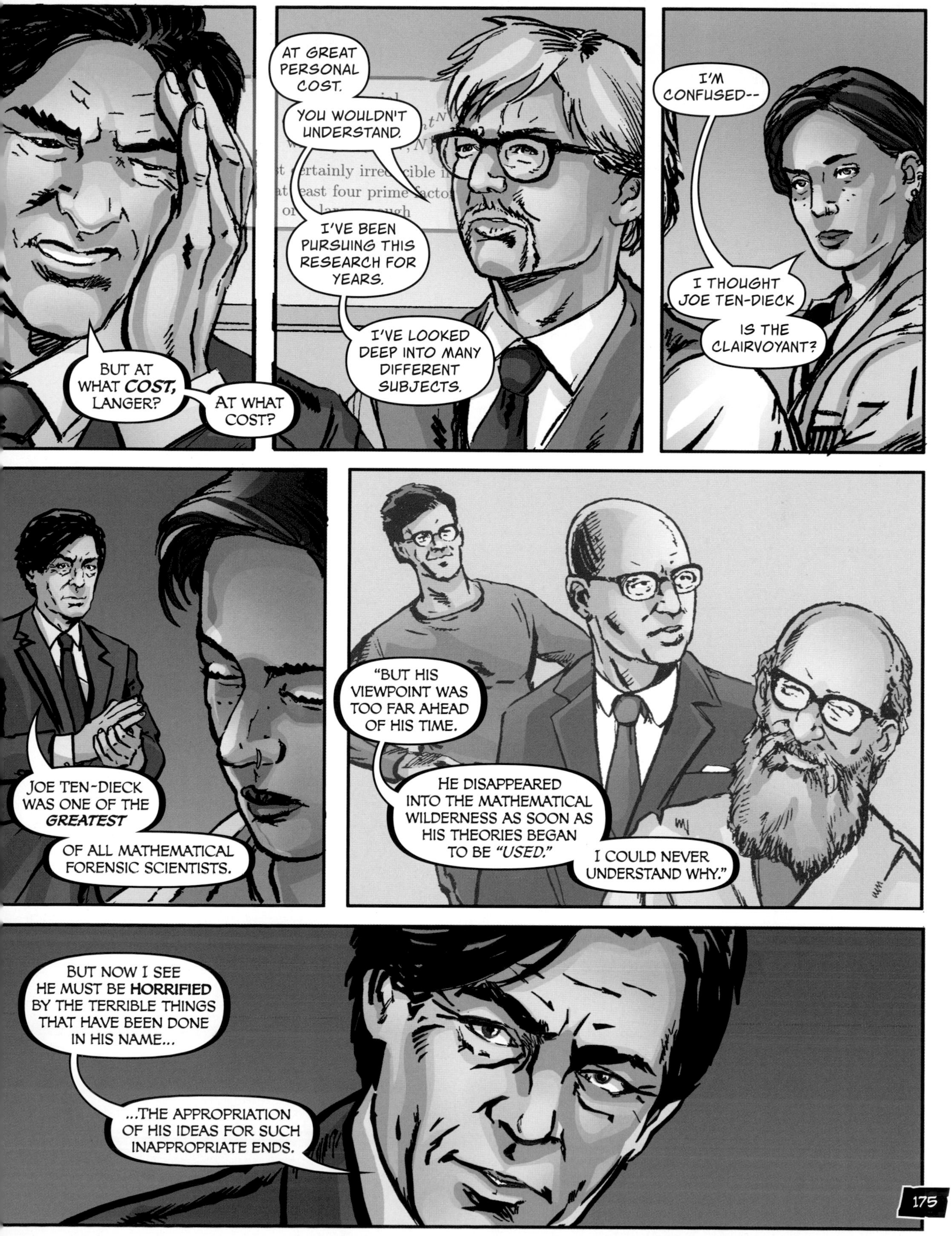
BUT AT WHAT COST, LANGER?
AT WHAT COST?
AT GREAT PERSONAL COST.
YOU WOULDN'T UNDERSTAND.
I'VE BEEN PURSUING THIS RESEARCH FOR YEARS.
I'VE LOOKED DEEP INTO MANY DIFFERENT SUBJECTS.
I'M CONFUSED--
I THOUGHT JOE TEN-DIECK
IS THE CLAIRVOYANT?
JOE TEN-DIECK WAS ONE OF THE GREATEST
OF ALL MATHEMATICAL FORENSIC SCIENTISTS.
"BUT HIS VIEWPOINT WAS TOO FAR AHEAD OF HIS TIME.
HE DISAPPEARED INTO THE MATHEMATICAL WILDERNESS AS SOON AS HIS THEORIES BEGAN TO BE "USED."
I COULD NEVER UNDERSTAND WHY."
BUT NOW I SEE HE MUST BE HORRIFIED BY THE TERRIBLE THINGS THAT HAVE BEEN DONE IN HIS NAME...
...THE APPROPRIATION OF HIS IDEAS FOR SUCH INAPPROPRIATE ENDS.

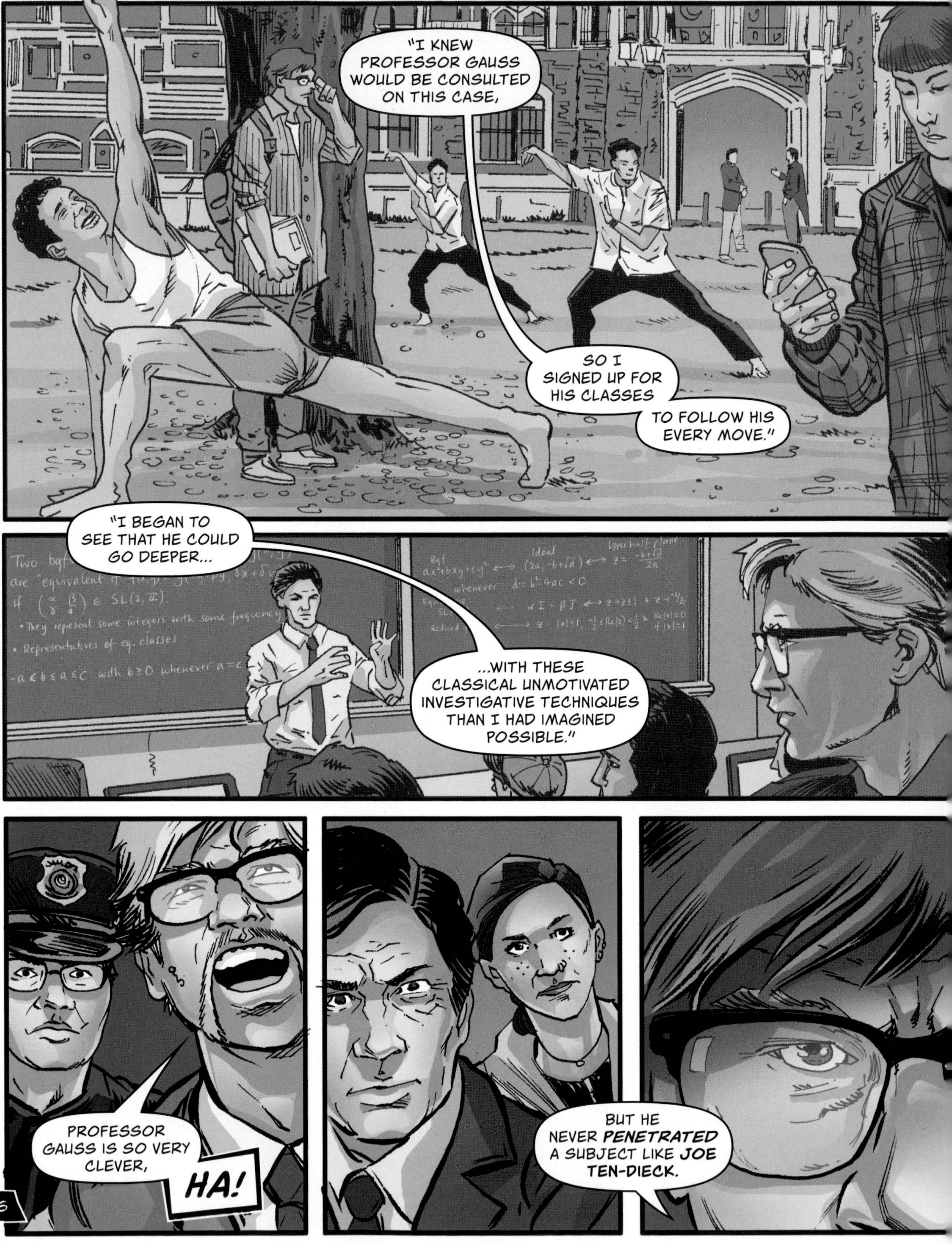
"I KNEW PROFESSOR GAUSS WOULD BE CONSULTED ON THIS CASE,
SO I SIGNED UP FOR HIS CLASSES
TO FOLLOW HIS EVERY MOVE."
"I BEGAN TO SEE THAT HE COULD GO DEEPER...
...WITH THESE CLASSICAL UNMOTIVATED INVESTIGATIVE TECHNIQUES THAN I HAD IMAGINED POSSIBLE."
PROFESSOR GAUSS IS SO VERY CLEVER,
HA!
BUT HE NEVER ***PENETRATED*** A SUBJECT LIKE **JOE TEN-DIECK.**

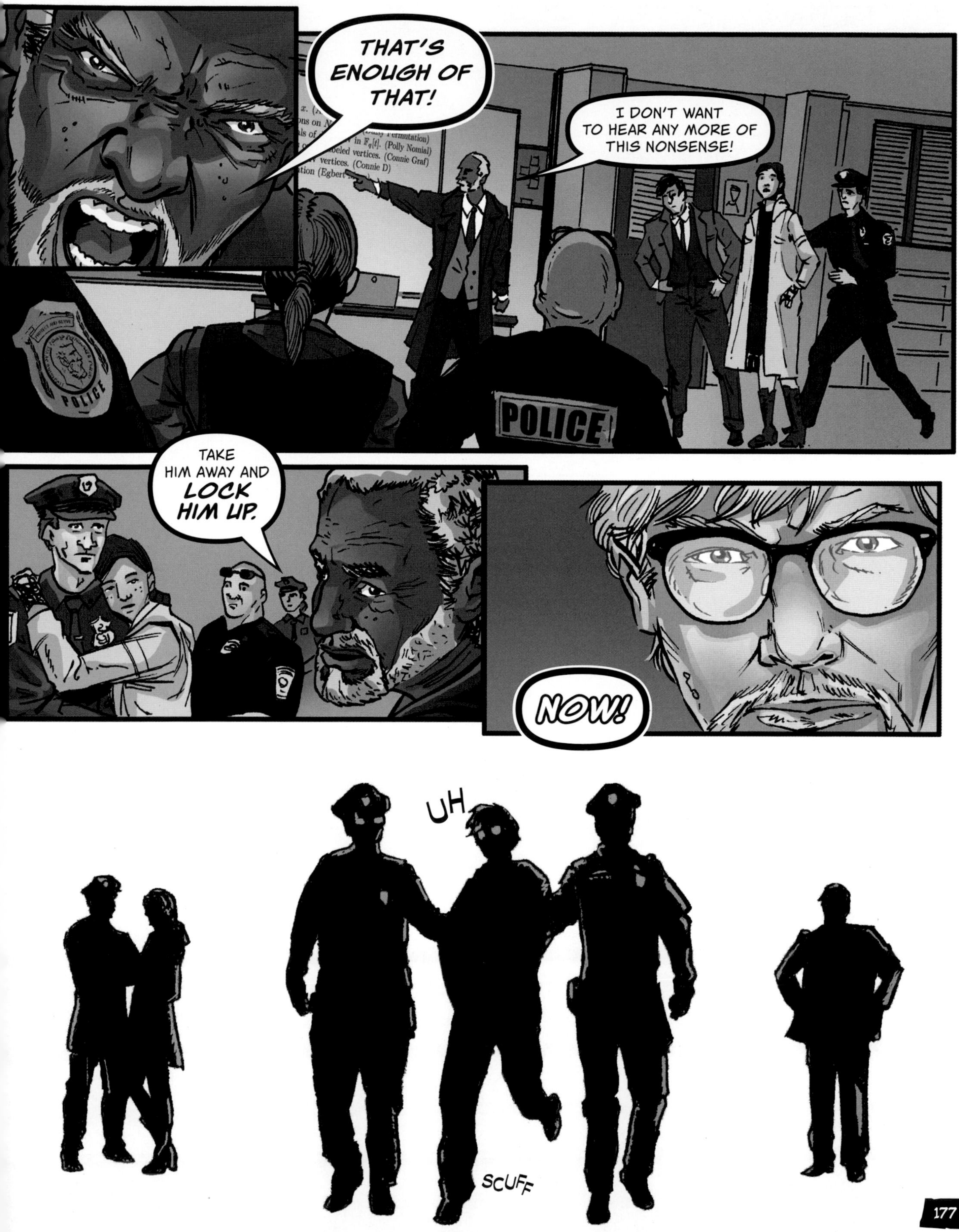
THAT'S ENOUGH OF THAT!
I DON'T WANT TO HEAR ANY MORE OF THIS NONSENSE!
POLICE
POLICE
TAKE HIM AWAY AND LOCK HIM UP.
NOW!
UH
SCUFF

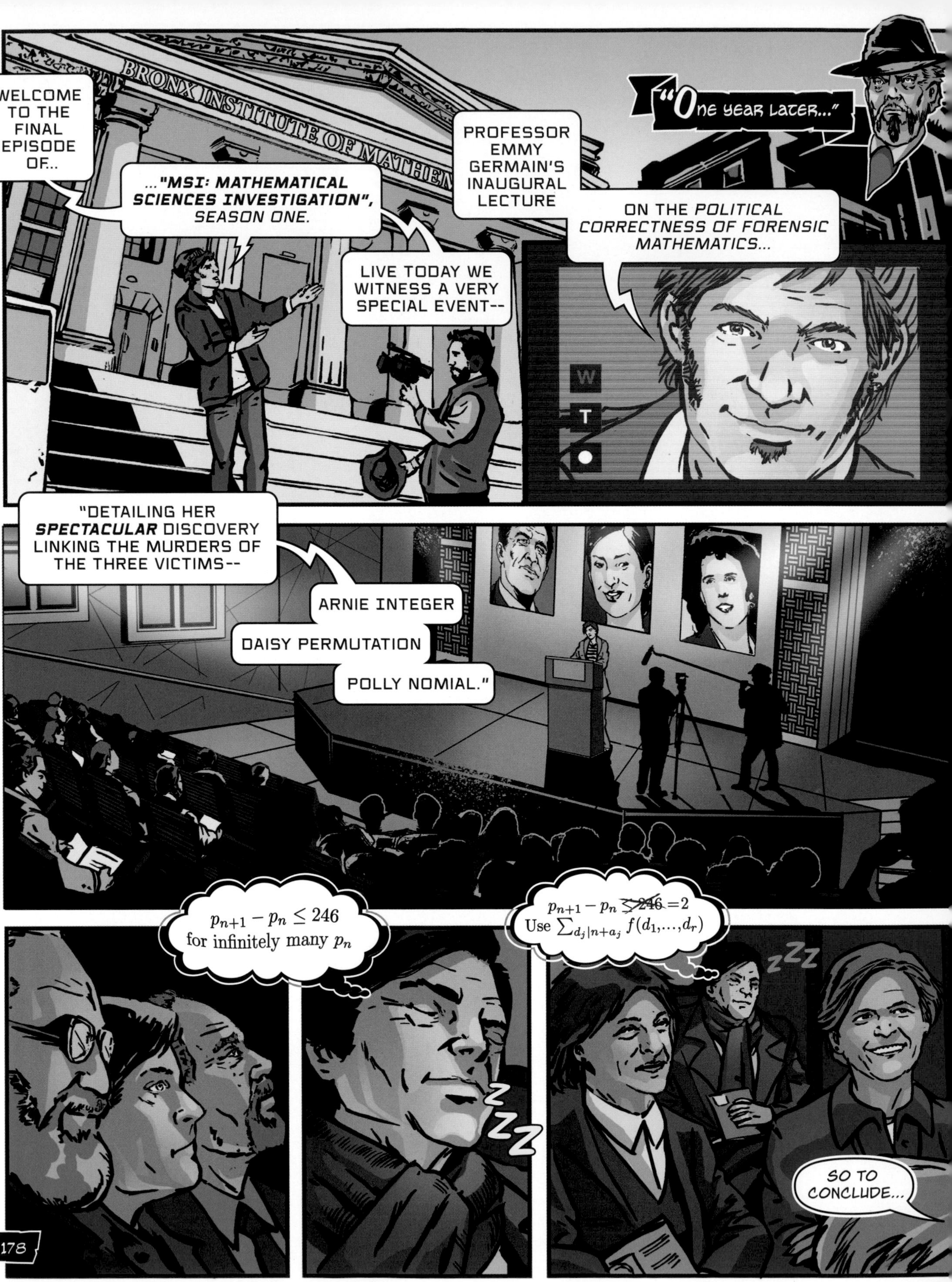

BRONX INSTITUTE OF MATHEM
WELCOME TO THE FINAL EPISODE OF...
..."MSI: MATHEMATICAL SCIENCES INVESTIGATION", SEASON ONE.
LIVE TODAY WE WITNESS A VERY SPECIAL EVENT--
PROFESSOR EMMY GERMAIN'S INAUGURAL LECTURE
"One year later..."
ON THE POLITICAL CORRECTNESS OF FORENSIC MATHEMATICS...
W
T
"DETAILING HER SPECTACULAR DISCOVERY LINKING THE MURDERS OF THE THREE VICTIMS--
ARNIE INTEGER
DAISY PERMUTATION
POLLY NOMIAL."
$p_{n+1} - p_n \leq 246$ for infinitely many $p_n$
Z Z Z
$p_{n+1} - p_n$ ~~$\leq 246$~~ $= 2$
Use $\sum_{d_j|n+a_j} f(d_1,\ldots,d_r)$
Z Z Z
SO TO CONCLUDE...

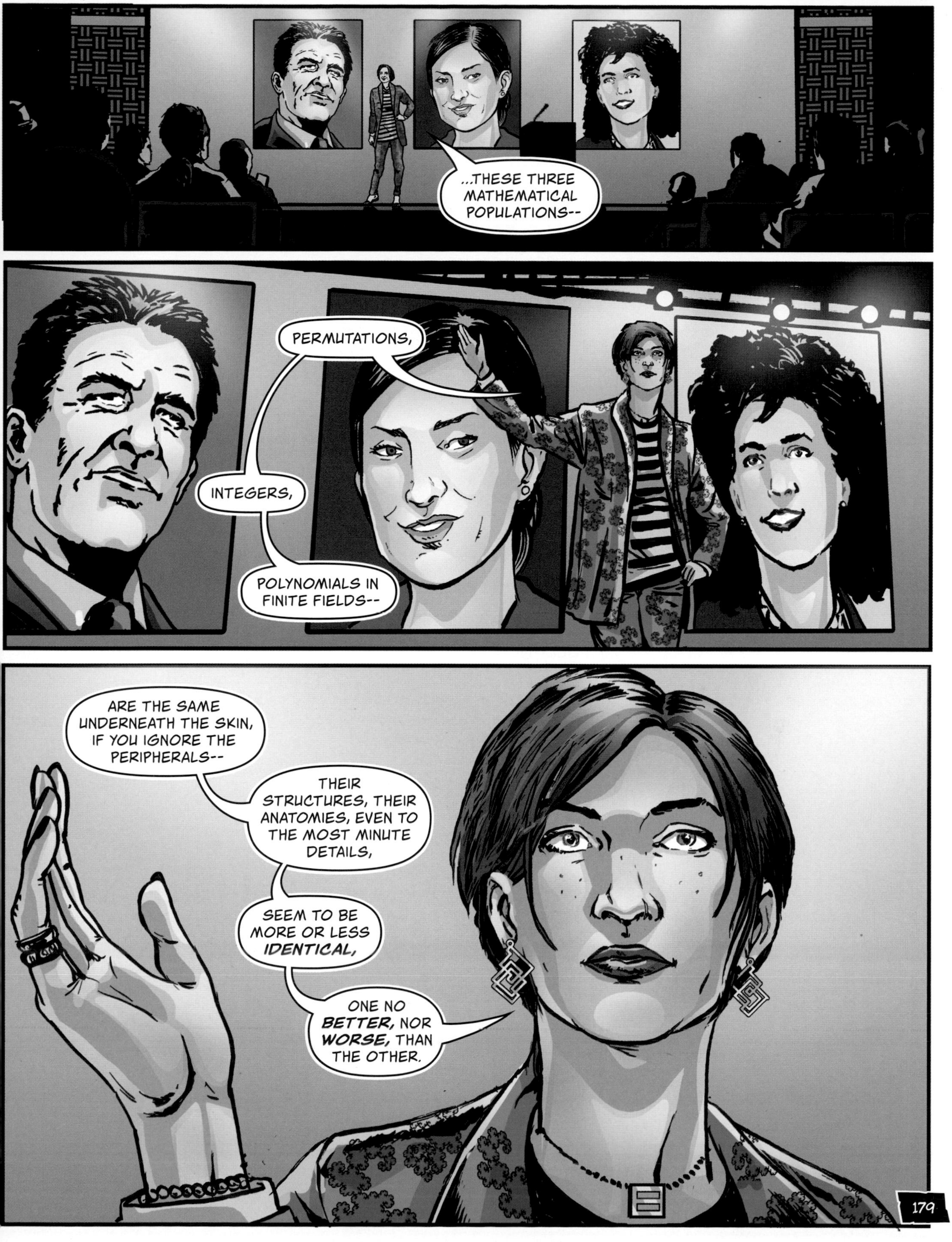
...THESE THREE MATHEMATICAL POPULATIONS--
PERMUTATIONS,
INTEGERS,
POLYNOMIALS IN FINITE FIELDS--
ARE THE SAME UNDERNEATH THE SKIN, IF YOU IGNORE THE PERIPHERALS--
THEIR STRUCTURES, THEIR ANATOMIES, EVEN TO THE MOST MINUTE DETAILS,
SEEM TO BE MORE OR LESS *IDENTICAL,*
ONE NO ***BETTER,*** NOR ***WORSE,*** THAN THE OTHER.

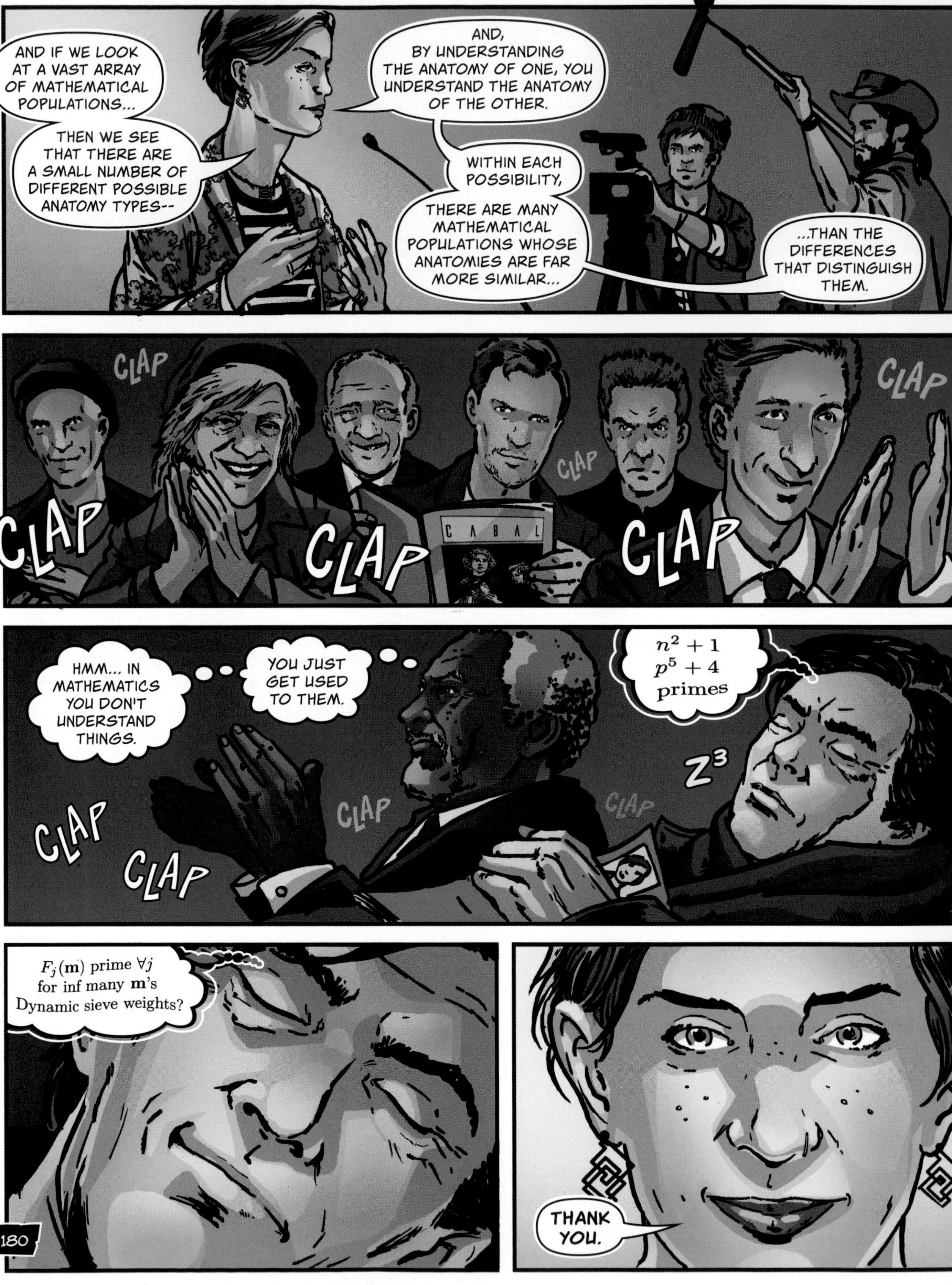
AND IF WE LOOK AT A VAST ARRAY OF MATHEMATICAL POPULATIONS...
THEN WE SEE THAT THERE ARE A SMALL NUMBER OF DIFFERENT POSSIBLE ANATOMY TYPES--
AND, BY UNDERSTANDING THE ANATOMY OF ONE, YOU UNDERSTAND THE ANATOMY OF THE OTHER.
WITHIN EACH POSSIBILITY,
THERE ARE MANY MATHEMATICAL POPULATIONS WHOSE ANATOMIES ARE FAR MORE SIMILAR...
...THAN THE DIFFERENCES THAT DISTINGUISH THEM.
CLAP
CLAP
CLAP
CLAP
CLAP
CLAP
CABAL
HMM... IN MATHEMATICS YOU DON'T UNDERSTAND THINGS.
YOU JUST GET USED TO THEM.
$n^2 + 1$
$p^5 + 4$
primes
$Z^3$
CLAP
CLAP
CLAP
CLAP
$F_j(\mathbf{m})$ prime $\forall j$
for inf many $\mathbf{m}$'s
Dynamic sieve weights?
THANK YOU.

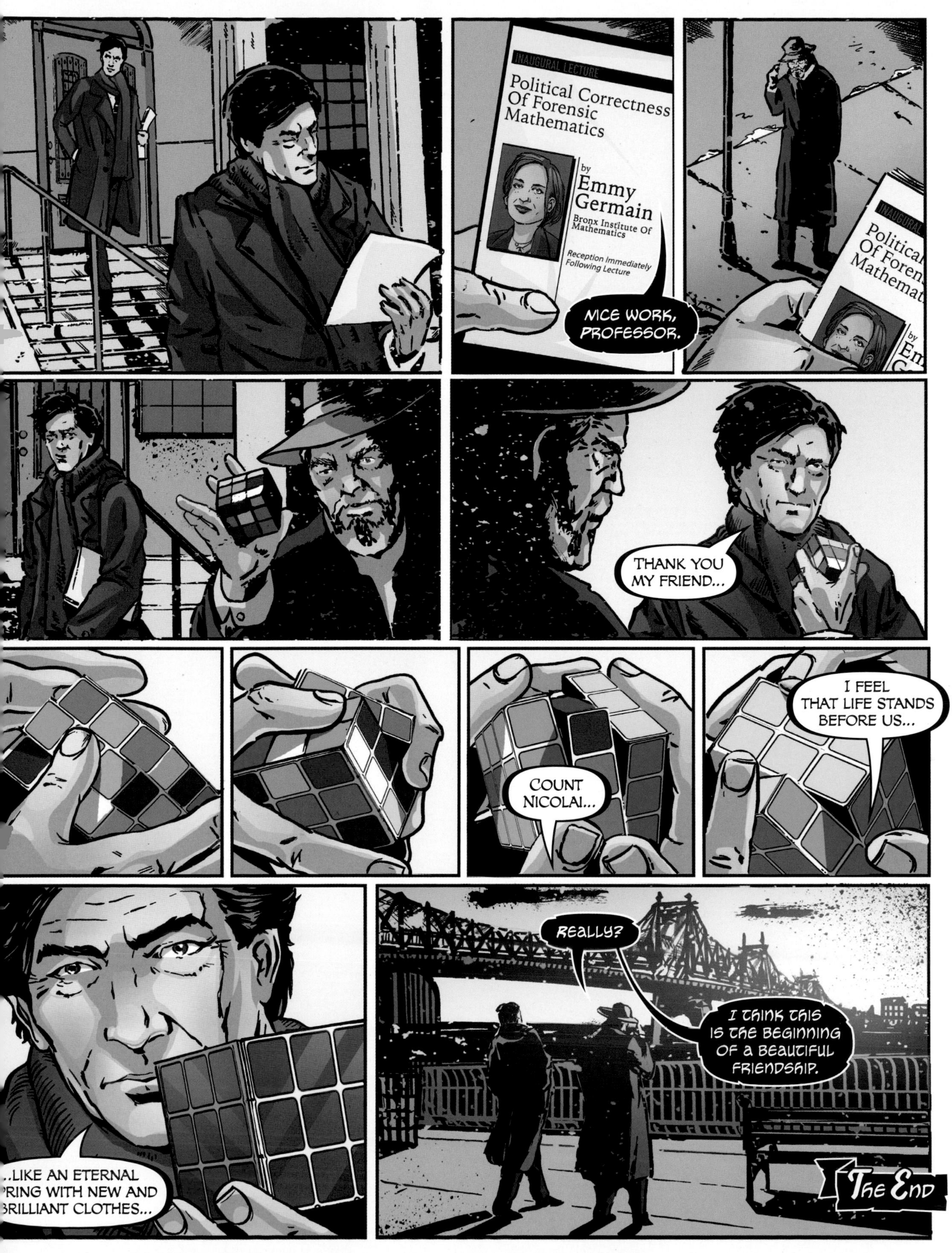
INAUGURAL LECTURE
Political Correctness Of Forensic Mathematics
by Emmy Germain
Bronx Institute Of Mathematics
Reception Immediately Following Lecture
NICE WORK, PROFESSOR.
THANK YOU MY FRIEND...
COUNT NICOLAI...
I FEEL THAT LIFE STANDS BEFORE US...
..LIKE AN ETERNAL
PRING WITH NEW AND
BRILLIANT CLOTHES...
REALLY?
I THINK THIS IS THE BEGINNING OF A BEAUTIFUL FRIENDSHIP.
THE END

# DRAMATIS PERSONAE

## Prime Suspects:
## The Anatomy of Integers and Permutations

### Jennifer and Andrew Granville

Integers and permutations are fundamental mathematical objects that inhabit quite distinct worlds though, under more sophisticated examination, one cannot help but be struck by the extraordinary similarities between their anatomies. This comic book stemmed from an experiment to present these similarities to a wider audience in the form of a dramatic narrative. In these after-pages, we will clarify some of the mathematical ideas alluded to in the comic book, giving the details of Gauss's lectures and Langer's presentation at the police precinct. We will also break down the content of some of the background artwork, explaining how some of it refers to breakthroughs in this area of mathematics, some of it to other vaguely relevant mathematics, while some content is simply our attempt at mathematical humor.

Our goal in *Prime Suspects* has been not only to popularize the fascinating and extraordinary similarities between the fine details of the structure of integers and of permutations, but also to draw attention to several key cultural issues in mathematics:

- How research is done, particularly the roles of student and adviser;
- The role of women in mathematics today; and
- The influence and conflict of deep and rigid abstraction.

Our original narrative was written in the form of a screenplay set in a metaphorical fantasy world, where detective mathematicians interpret several key notions, eventually discovering some of the more interesting theorems in this area. The screenplay has been performed as a live reading in Princeton, Berkeley, and Montreal. For each of these readings, the brilliant stage designer, Michael Spencer, created a site-specific production that offered intriguing and entertaining solutions for how to present the mathematics visually and viscerally.

The vision of the artist, Robert J. Lewis, has taken the screenplay and ideas from the site-specific productions and developed them in this graphic novel. The comic book format offers a world of different opportunities compared to those of a live performance. New characters were added, and new ways of presenting the mathematics emerged, while still incorporating key elements of the live performance, such as Robert Schneider's fascinating and beautiful score *Reverie in Prime Time Signatures*. The story of how each of these different creative processes intersect and collaborate—the writing of the original screenplay, staging the live performances, recording the score, and developing the graphic novel—have all become part of further art work: a documentary film by Tommy Britt.

The characters we created needed to be strong enough to drive the plot and bring color and life to the story—mathematical superheroes and villains! So we took the liberty of loosely basing our characters on the work, lives, and traits of famous mathematicians. To put the record straight, we now present brief biographies of our characters' namesakes.

**Dramatis Personae**

# Character References

## K. F. Gauss

> Archimedes, Newton and Gauss, these three, are in a class by themselves among the great mathematicians.
>
> E. T. BELL

Many central themes of modern mathematics, both pure and applied, came into their modern incarnation through the works of Karl Friedrich Gauss (1777–1855). Gauss's genius was spotted when he was a little boy. At age 10, his teacher told his class to add up all the whole numbers up to 100. Gauss ingeniously determined the correct answer, 5050, in a few moments. By the time he was 16, he had proved several deep mathematical theorems and had brilliantly interpreted data on prime numbers so as to make a guess as to their distribution, an insight that has motivated research ever since. At 23, he solved the ancient problem of what regular shapes could be constructed using only a ruler and compass, and by 24, he had written his first book, *Disquisitiones Arithmeticae*, which is still the model for all introductory books in number theory. Gauss went on to make discoveries in many different areas, notably astronomy and co-creating the telegraph.

Gauss did not publish much, preferring "few but ripe." Indeed, other mathematicians rediscovered his unpublished work on more than one occasion. He excelled at motivating students, such as Friedrich Bessel; Richard Dedekind; Sophie Germain; and Bernhard Riemann; and his junior colleague, Lejeune Dirichlet:

> Usually he sat in a comfortable attitude, looking down . . . with his hands folded above his lap. He spoke quite freely, very clearly, simply and plainly. . . . When he wanted to emphasize a new viewpoint . . . he lifted his head . . . and gazed . . . with his beautiful, penetrating blue eyes while speaking emphatically. . . . If he proceeded from an explanation of principles to the development of mathematical formulas, then he got up, and in a stately very upright posture he wrote on a blackboard . . . in his peculiarly beautiful handwriting: he always succeeded through economy and deliberate arrangement in making do with a rather small space. For numerical examples . . . he brought along the requisite data on little slips of paper.
>
> **DEDEKIND ON GAUSS**

**Emmy Germain.** There are still not enough women in research mathematics, although the situation has been slowly improving. Blatant sexism has largely been eliminated, but women still encounter obstacles to succeeding in this profession.

There have been many great female mathematicians in history, including Hypatia in ancient Alexandria; Sophie Germain in early nineteeth-century France; Emmy Noether in early twentieth-century Germany; and Julia Robinson, arguably the leading figure in solving Hilbert's tenth problem 40 years ago. There are more leading female mathematicians today than ever before, including Maryam Mirzakhani, the first woman to be awarded the Fields medal (which is the closest thing to a Nobel prize in mathematics).

Our heroine, Emmy Germain, is a young female mathematician, her name combining those of Emmy Noether (1882–1935) and Sophie Germain (1776–1831). Her background is based on that of Sophie Germain, who never formally attended any school or gained a degree. She taught herself geometry, algebra, and calculus, and then Latin and Greek to study Euler and Newton in the original. She was home-schooled! Germain began sending her ideas to Lagrange, and then to Gauss, using the pseudonym "Monsieur Le Blanc." Joseph-Louis Lagrange demanded to meet with this extraordinary correspondent and was shocked to find that she was not a "Monsieur." In 1806, when Napoleon's forces invaded Prussia, Germain asked a personal friend, General Pernety, to make sure of Gauss's safety. It was only then that Gauss discovered her secret, and wrote to her:

> But how to describe to you my admiration and astonishment at seeing my esteemed correspondent, Monsieur Le Blanc, metamorphose himself into this illustrious personage who gives such a brilliant example of what I would find it difficult to believe. A taste for the abstract sciences in general and above all the mysteries of numbers is excessively rare: one is not astonished at it: the enchanting charms of this sublime science reveal only to those who have the courage to go deeply into it. But when a person of the sex which, according to our customs and prejudices, must encounter infinitely more difficulties than men to familiarize herself with these thorny researches, succeeds nevertheless in surmounting these obstacles and penetrating the most obscure parts of them, then without doubt she must have the noblest courage, quite extraordinary talents and superior genius. Indeed nothing could prove to me in so flattering and less equivocal manner that the attractions of this science, which has enriched my life with so many joys, are not chimerical, [than] the predilection with which you have honored it.
>
> GAUSS'S LETTER TO SOPHIE GERMAIN

Emmy Noether was strongly supported from an early age by leading male mathematicians, in her case David Hilbert and Felix Klein. Unfortunately, her appointment to the faculty at the University of Göttingen was initially blocked by the philologists and historians, so when she taught, the classes had to be advertised under Hilbert's name. After the First World War, she obtained a full faculty position, only relinquishing that when the Nazis banned Jewish faculty in the early 1930s. In 1934, Noether visited the Institute for Advanced Study in Princeton, although she was not welcome at Princeton University, calling it "the men's university, where nothing female is admitted." She finished her career on the faculty at Bryn Mawr.

**DETECTIVE JACK (J. J.) VON NEUMANN.** John von Neumann (1903–1957) was a major figure in twentieth-century mathematics and in such applications as the development of the H-bomb and of the modern computer. He had an enormous impact on many fields of pure mathematics as well as developing the ideas behind important applications of mathematics. He was an inspiration to many mathematicians of his time with his ability to penetrate different fields with new perspectives.

von Neumann was another child prodigy: At the age of six, he was able to exchange jokes with his father in classical Greek, and would entertain family by quickly memorizing random pages of the Budapest phone directory. Despite von Neumann's brilliance at school in mathematics, his dad did not want him studying a subject that would not bring him wealth, so they compromised on chemistry. Von Neumann nonetheless attended math classes:

> Johnny was the only student I was ever afraid of. If in the course of a lecture I stated an unsolved problem, the chances were he'd come to me as soon as the lecture was over, with the complete solution in a few scribbles on a slip of paper.
>
> GEORGE POLYA ON VON NEUMANN

In the same year that he got his diploma in chemical engineering, he also obtained his doctorate in mathematics, developing the theory of "ordinals," an understanding of the different possible types of infinities. For the next three years, von Neumann taught and researched in Berlin. Besides his prodigious mathematical output, he was a denizen of the Cabaret-era Berlin nightlife circuit, enjoying the clubs and the parties.

In 1930, von Neumann came to Princeton, and in 1933, he was one of the original six professors at the Institute for Advanced Study.[1] He worked well with others—indeed, he could present deep mathematics with amazing lucidity—but was not so good in the classroom. He was notorious for writing complicated equations on the blackboard and erasing them before students had a chance to copy them down!

After marrying in 1938, Von Neumann continued to enjoy good company and parties, but now the parties centered around their house—frequent, famous, and long! Von Neumann was always a snappy dresser and well groomed, the antithesis of the stereotypical scruffy mathematician. He had as lively views on international politics and practical affairs as on mathematical problems.

JOE TEN DIECK AND COUNT NICHOLAS BOURBAKI.   Alexander Grothendieck (1928–2014) was the most controversial mathematician of the second half of the twentieth century. Grothendieck took an abstract, broad perspective on mathematics in a way that paid extraordinary dividends. He took the view that one should only work with very general methods and only derive concrete results that follow obviously, never allowing oneself to stretch far from the abstract to deduce the tangible. Most mathematicians find it difficult to think this way. For example, Gauss, by contrast, used poignant examples to motivate most of his theoretical discoveries.

Groethendieck has many ardent followers, some of whom have an almost religious adherence to this way of looking at things, refusing to acknowledge that other techniques could yield anything interesting.[2] Groethendieck's point of view fits well with the French Bourbaki movement, whose participants sought to rewrite everything in mathematics from an abstract perspective, all "in the correct order." They would write their books collaboratively: One person would write a chapter and then give it up to the next person to edit, removing any hint of joy or personality

---

1.   Along with Einstein, Hermann Weyl, Oswald Veblen, Marston Morse, and James Alexander.
2.   Indeed if, say, combinatorics resolves what had been considered an interesting question, then resolute adherents would infer that the original interest must have been a misperception.

from the writing. Given that no one person was supposed to dominate the writing of any one book, they decided to publish all of their output under the (mythical) collective name, Count Nicolas Bourbaki.

The malign influence of Bourbaki has caused rifts in the research and teaching of mathematics. Too often, leaders of the Bourbaki school have refused to acknowledge the top-quality work of non-Bourbakiists. Moreover, the idea that one must learn mathematics in such a rigid way has diminished the appeal of our subject to many fine young minds. It is only recently that the Bourbaki influence is diminishing—a sure sign is the welcome rise of probabilty theory and combinatorics, popular subjects worldwide.

Groethendieck was not well equipped to minister to his following, dropping out of full-time mathematics in the late 1980s. In 1984, he had proposed a scientific program that included a very simple type of graph to study Riemann surfaces and the absolute Galois group, which he called a "Child's Drawing." In 1991, he went into hiding, completely breaking ties with other mathematicians. He lived in a remote farming village in the Pyrenees, writing long political and personal tracts.

We wanted a flamboyant, larger-than-life, slightly mysterious persona for our Narrator, who helps us navigate the complicated time-twists in our story, so we decided to bring Count Nicholas Bourbaki to life (courtesy of Orson Welles). This allowed us to include our own versions of famous photos of Bourbaki group meetings—including some of our own characters.

**SERGEI LANGER.** Serge Lang (1927–2005) was the most prolific author of research-level textbooks in mathematics of the second half of the twentieth century (writing more than 60 books!). Even though they could be carelessly written and somewhat unmotivated, the sheer volume, and the fact that he often wrote the first textbook in a new and exciting area, meant that he had an enormous influence on the development of mathematics. Lang said that the best way to learn a topic was to write a book on it, and he did so quickly; someone once called the Yale math department to speak to Lang and was told by the receptionist: "He can't talk to you right now, he is busy writing a book. Can I put you on hold?"

In 1939, Lang's family fled France for Los Angeles, where he graduated from Caltech in physics at age 19. Moving to Princeton to do a doctorate in philosophy, he fell under the influence of Emil Artin (and the new developments in algebra), gaining a PhD in mathematics in 1951. By 1955, he was on the faculty at Columbia, resigning in 1971 to protest their treatment of anti-war protestors. In 1972, he found a position at Yale, where he settled down and became a prolific author and tireless campaigner on many issues.

He was a great believer in putting mathematics in a more abstract context, in the Bourbakiist mold of trying to give the big picture, regardless of whether the material was suited to this approach:

> When I first saw Lang's Diophantine geometry I was disgusted with the way in which my own contributions to the subject had been disfigured and made unintelligible. . . . The whole style of the author contradicts the sense for simplicity and honesty which we admire in the works

**Dramatis Personae**

of the masters in number theory—Lagrange, Gauss,. . . . Now Lang has published another book on algebraic numbers which, in my opinion, is still worse than the former one. I see a pig broken into a beautiful garden and rooting up all flowers and trees. . . . I am afraid that mathematics will perish . . . if the present trend for senseless abstraction . . . cannot be blocked up.

C. L. SIEGEL ON LANG'S WRITING (1962)

Lang fought many vigorous campaigns against perceived injustices. He kept copious files on everything that bothered him, in mathematics, in other sciences, and in politics, which he would happily share! He had a particular disgust of the shoddy use of mathematics in social science research:

I don't like the nonsense that passes for rational discourse so often in our society. I am very much bothered by the inaccuracies, ambiguities, code words, slogans, catch phrases, public relation devices, sweeping generalizations, and stereotypes, which are used (consciously or otherwise) to influence people. I am bothered by the inability of many to recognize these for what they are.

I am bothered by the misinformation which can get disseminated uncritically through the media and by the obstructions which prevent correct information from being disseminated. These obstructions come about in many ways—personal, institutional, through self-imposed inhibitions, through external inhibitions, through outright dishonesty, through incompetence. . . . I am bothered by the way misinformation, disguised as scholarship, is used in social, political, and educational contexts to affect policy decisions. I am bothered by the way misinformation is accepted uncritically, and by the way people are unable to recognize it or reject it.

SERGE LANG

**BARRY BELL AND SILENT BOB.** E. T. Bell (1883–1960) was the greatest of all mathematical biographers. In *Men of Mathematics*, Bell crafted romantic histories of the great mathematicians of the seventeenth, eighteenth, and nineteenth centuries, inspiring many generations of prodigies to see a bright future in the subject. Mathematically, the book is fairly accurate, though some of the life stories are ambitious reconstructions based on little documentary evidence.

There have been a few mathematical biographers in recent times, most notably Constance Reid and recently the extraordinary Siobhan Roberts. Current mainstream reporters include Barry Cipra, Dana Mackenzie, Keith Devlin, and the wonderful Erica Klarreich, among others. Our reporter's name is an all-American amalgam of Barry Cipra and E. T. Bell.

Barry's quiet boom operator and sound recordist plays a pivotal role in our story, but it was not until Robert J. Lewis drew him that he looked a lot like the popular character "Silent Bob," who is film director Kevin Smith's cameo role in many of his own movies. At this co-incidence, we approached Kevin, who kindly accommodated our request to use his beloved character in our comic book.

Ben Green, Terry Tao, and Tamar Ziegler. As we developed and adapted the original screenplay into a graphic novel, we realized that we needed extra characters. Two beat cops seemed like a good idea, but who? In 2004, Green and Tao had made the most extraordinary breakthrough on prime patterns, introducing revolutionary techniques. They would do. They formulated how to get the ultimate consequence of their ideas but, to complete the proof (in 2012), they brought in Tamar Ziegler, a top ergodic theorist. Thus, Tamar comes in (on p. 164) toward the end of our story, to join the team just as the going gets tough!

Minor characters. Page 60 is an homage to one of the most interesting mathematicians on the planet, Persi Diaconis, who combines wonderful mathematical input to probability theory, both pure and applied, with mathematical magic tricks! Read about him on Wikipedia and obtain *Magical Mathematics: The Mathematical Ideas That Animate Great Magic Tricks* by Diaconis and Ron Graham.

Enrico Bombieri (p. 23) is a leading analytic number theorist, Fields medalist, and very generous mentor, who freely shares ideas with the many young people around him, including co-author Andrew in the early 1990s. Gauss's mentoring style is based, to some extent, on Enrico's. However, Enrico is Italian (though less gregarious than our Gauss), so when we needed a Godfather, the natural choice was Enrico, with Andrew there to kiss his ring as a sign of respect!

Page 133 is based on the front gate of Fort George G. Meade, home of the National Security Agency NSA, with a cameo of Gauss being checked out by Alan Turing (a historical mish-mash).

The Bourbaki group meets on pp. 115, 116, and 171. The participants include likenesses of Alexandroff, Bourbaki (of course), Henri Cartan, Deuring, Jean Dieudonne, Grothendieck, Herbrand, Hermann, Hopf, Jacobson, Krull, our Langer, Saunders Maclane, Pierre Samuel, F. K. Schmidt, Laurent Schwarz, Shoda, Olga Taussky, Tsen, van der Waerden, Andre Weil, and Witt. Jean-Pierre Serre appears both as his young self and how he is now—a trim 92-year-old. Andrew's PhD advisor, Paulo Ribenboim, is standing (he is a Bourbaki fan), and Robin Hartshorne sits at the next table.

Serre (another Fields medallist) is one of the most creative and brilliant mathematicians of the twentieth century, a charming, urbane Parisian. Paris streets have been named after the great French intellectuals since the time of Napoleon. More than 100 streets are named after great mathematicians, and Jean-Pierre Serre Boulevard (p. 85) will surely be a wide, tree-lined boulevard, where great ideas breeze in and refresh the passers-by.

On p. 123, the two concert musicians are Bryna Kra on violin and Robert Schneider (the composer of the music in the comic book) on piano.

Supporting characters. When it came to populating the many scenes in the book that require 'extras,' we decided to give Robert a list of colleagues and friends from the world of mathematics

David Eisenbud is the crime scene photographer on p. 6, as he had taken photos of our performance at Berkeley. Hugh Williams is visiting precinct 1093 (on p. 8), which is why he is wearing his Canadian Mountie uniform. Dan Everett refused to use a car, even in rural Georgia, which

**Dramatis Personae**

led to him becoming Gauss's chauffeur (p. 18). Spain Rodriguez and his wife, Susan Stern, who had some part in the development of this comic book, get a cameo on p. 79. The other bikers are Bombieri and his close collaborators Friedlander and Iwaniec, their jackets emblazoned with the slogan "Truth and Beauty" of the Institute for Advanced Study (and a great paraphrasing of what we strive for in mathematical research).

Andrew's main research collaborator is K. Soundararajan, winner of the Simons Foundation "genius award." Fed up with his name being intimidating and mispronounced by his western colleagues, he decided to be known as "Sound" (one name worked for Socrates and Pele, so why not Sound?). Given that sound plays a major role in our story, Sound's cameos come in terms of what is said, not what is seen (p. 39).

Various police officers are cameos for mathematician colleagues (just the page number of the first appearance is given here): Andras Biro (p. 6), Tim Gowers (p. 58), Roger Heath-Brown (p. 9), Jacques Hurtubise (p. 160), Veronique Hussin (p. 9), Kevin James (p. 9), Kaisa Matomaki (p. 56), Barry Mazur (p. 56), Carl Pomerance (p. 8), Yvan St Aubin (p. 8), Anitha Srinivasan (p. 9), and Endre Szemeredi (p. 93).

Other cameos include: Javier Cilleruelo (p. 41), Octav Cornea (p. 74), HSM Coxeter (p. 73), Ernie Croot (p. 68), Chantal David (p. 41), Krzysztof Dzieciolowski (p. 57), Yasha Eliashberg (p. 74), Mo Hendon (p. 121; and his juggling pattern is an art exhibit on the wall of the concert hall), Jorge Jimenez-Urroz (p. 87), Dimitris Koukoulopoulos (p. 41), Matilde Lalin (p. 121, doing Rubik's cube behind her back), François Lalonde (p. 74), Benoit Larose (p. 41), Dusa McDuff (p. 74), Alekos Moisiadis (p. 71), the real John von Neumann (p. 57), Mel Nathanson (p. 119, checking his watch at the airport—where he often collected co-author Andrew in 2010—Newton and his apprentice Einstein (p. 72), Ken Ono (p. 65), and Iosif Polterovich (p. 68)

The book is an homage not only to film noir and TV police dramas but also to individual favorite movies. For example, the images and dialogue of the opening sequence reference the opening of *Pulp Fiction*, and the final scene nods to *Casablanca*. There are many more in-between as well as a multitude of characters from films and TV shows, writers, people, and ideas that we want to celebrate.

# THE MATHEMATICS OF *PRIME SUSPECTS*

Andrew Granville

## THE SUBJECTS TO BE STUDIED

### Integers and their prime factors

We all know the *integers*, the whole numbers

$$\ldots, -5, -4, -3, -2, -1, 0, 1, 2, 3, 4, 5, \ldots,$$

going off to infinity in either direction, positive and negative. They come up throughout mathematics in all sorts of different guises.

*Prime numbers* are those integers greater than 1 that cannot be broken down, multiplicatively, into smaller parts. That is, they are *not* the product of two smaller positive integers. Describing primes in terms of what they are not is a rather negative definition, but it is much harder to describe them as what they *are* (which is one reason they turn out to be difficult to count and even, at times, to identify).

Key to our understanding of positive integers is that they factor into primes, for example,

$$50 = 2 \times 5 \times 5, \text{ and } 1001 = 7 \times 11 \times 13.$$

The Fundamental Theorem of Arithmetic tells us that every positive integer can be written as a product of primes in a unique way (unique except for the order in which one writes them down). For example, $60 = 2 \times 2 \times 3 \times 5$, and and each of 2, 3, and 5 is prime. One can write the primes in a different order like $60 = 2 \times 5 \times 3 \times 2$; the same primes, the same number of each, but in a different order. Put another way, if we multiply together some prime numbers (say, 7, 11, and 13), then we get a positive integer (in this case, $7 \times 11 \times 13 = 1001$), and this is the only integer that is the product of these primes. So they identify each other, the primes and the quantity of each prime, giving you a specific integer, and that integer can only be broken down into primes in one way. The "primes are the fundamental constituent parts of integers; their genetic code, if you like. Any integer can be identified by the primes it contains, which ones and how many of each type." [Emmy, p. 29]

### Permutations and their cycles

When we reorganize some objects, we *permute them*. For example, if you shuffle a deck of cards, then they mix and end up in a different order. That reordering is a *permutation*. It is most convenient if we record where each object goes. To simplify things, suppose that the cards, from a deck of seven, are marked

> $a, b, c, d, e, f, g$

in order, and after your shuffle they are ordered

> $c, f, b, d, g, a, e.$

Then the card marked "*a*" that started in position 1 (the top of the deck) has moved to position 6 (the sixth card down), the card marked "*b*" that started in position 2 has moved to position 3, and so forth. We might therefore write the shuffle as

> $1 \rightarrow 6, \ 2 \rightarrow 3, \ 3 \rightarrow 1, \ 4 \rightarrow 4, \ 5 \rightarrow 7, \ 6 \rightarrow 2, \ 7 \rightarrow 5,$

or we could write this in a more structured way as

> $1 \rightarrow 6 \rightarrow 2 \rightarrow 3 \rightarrow 1, 4 \rightarrow 4,$ as well as $5 \rightarrow 7 \rightarrow 5.$

These are three *cycles*, the positioning of the cards cycling between them. For example, the card that was in first place moved to sixth place, the card that was in sixth place moved to second place, and so forth, until the card that was in third place moved to the vacant first place. We have a simple notation to describe these cycles:

> $(1, 6, 2, 3)(4)(5, 7).$

That is, the shuffle that permuted the order of our cards can be described as the *product of cycles* $(1, 6, 2, 3)(4)(5, 7)$. This is the only way that the permutation can be described as a product of cycles, and any given product of cycles gives a unique permutation. "Cycles are the fundamental constituent parts of a permutation, just like primes are the fundamental constituent parts of an integer. You can always break a permutation up into cycles." [Langer, p. 38]

On p. 42, Gauss asks how many permutations can be formed out of 12 balls. In our example here, but now playing with a full deck, we might ask: How many different orders are there of a deck of 52 playing cards? Emmy's explanation on p. 43 for the number of orderings of 12 balls can be modified to determine the number of orderings of 52 playing cards. There are 52 possibilities for the top card. Once that has been chosen, there are only 51 possibilities for the second card. Once the top two cards have been selected, only 50 possibilities remain for the next card, and so forth. So the total number of possible orderings of 52 cards (which is the same as the number of permutations on a set of 52 objects) is

> $52 \times 51 \times 50 \times \ldots \times 3 \times 2 \times 1.$

This product is denoted "52!" by mathematicians and is an enormous number, something like 8 followed by 67 zeros. If you used all the computers in the world from now until the end of time,

**The Mathematics of *Prime Suspects***

calculating one permutation each nanosecond, you would not come close to running through each of these orderings. So a good shuffle can really mix up the cards.[1]

People sometimes ask: What is the reason for being interested in pure mathematics? Why is it useful? We have just seen that "When we try to understand the number of possible choices in a given situation, even taking into account certain restrictions, then the theory of permutations comes to the fore." [Langer, p. 39] It does not matter what we are choosing and organizing—billiard balls, playing cards, hats at a hat check, a team sports schedule—the same mathematical ideas describe what we need to understand. Over many years, people have realized that the same basic ideas, the same patterns of organization, run through many different endeavours, and when they abstract the essence of those ideas, they find themselves involved in *pure mathematics*.

## Forensics

In *Prime Suspects*, the "forensic team" does an in-depth analysis of the basic constituent parts of integers and of permutations, comparing their *anatomies* and determining how those constituent parts—the prime factors and the cycles—are laid out. But how can we compare such different-looking objects? "We're agreed that it makes sense to calibrate the primes with the cycles. But how?" [Emmy, p. 48] We need to find some feature or measurement of the two that we can sensibly compare. It is Gauss who makes that leap of logic on p. 50, asking himself: What proportion of each has just one fundamental component?

### Proportion of permutations that are a cycle

We denote the set of permutations on $N$ letters by $S_N$. We have seen that the number of permutations on $N$ letters is $|S_N| = N!$. What about the number of cycles on $N$ letters, using all $N$ letters? To begin with, the first letter, which I will call $a$, can be sent to any of the other $N - 1$ letters; say the choice is the letter $q$. Then $q$ can be sent to any of the other $N - 2$ letters (that is, other than $a$ and $q$), say the letter $x$. We continue on like this; the last letter must be sent back to $a$. So the total number of cycles on $N$ letters is

$$(N - 1) \times (N - 2) \times \cdots \times 2 \cdot 1 \cdot 1 = (N - 1)!.$$

As a proportion of the elements of $S_N$ this gives $(N - 1)!/|S_N| = (N - 1)!/N! = 1/N$. In other words, "exactly one in every $N$ permutations on $N$ letters is a cycle." [Gauss, p. 50]

### Proportion of integers that are a prime

For prime numbers, this is a *much* more difficult question. Indeed, resolving this question is considered perhaps the greatest achievement of nineteenth-century mathematical research. The story begins at the end of the eighteenth century, when the child prodigy, Gauss, studied tables

---

1.   A perfect riffle shuffle is when one cuts the deck into two exact halves and then perfectly interlaces the cards from the two halves. After eight perfect riffle shuffles, the deck returns to its original order. This is why you want to be wary of someone who shuffles too well!

of primes to find patterns (see the top right panel of p. 54). As a 15-year-old in 1792, he guessed that

> *At around x, roughly 1 in every $\log x$ integers is prime,*

a prediction that took more than one hundred years to fully justify. Gauss did not propose how to prove this claim, though he set his best student, Lejeune Dirichlet, on this problem, who went some way but did not resolve it. The ten-page masterpiece, written by Bernhard Riemann in 1859, proposed a most extraordinary plan to prove Gauss's guess, bringing in ideas from a much wider circle of mathematical concepts. Riemann's plan was not completed until 1896, and his paper still inspires much of the work done in pure mathematical research today (and certainly my own).

Gauss's logarithm, "$\log x$," is the logarithm of $x$ in base $e$, the so-called *natural logarithm*, often written "$\ln x$" when first encountered. The reader might not have encountered the concept of logarithms before, but it is a key tool in mathematics. We begin with the example $2^3 = 8$, so that $\log_2 8 = 3$, that is, 3 is the logarithm of 8 in base 2. Similarly, the logarithm of 100 in base 10 is 2, because $100 = 10^2$. The great thing about logarithms is that they magically change multiplication into addition, so if, for example, we want to multiply 8 and 16, which equal $2^3$ and $2^4$, respectively, then their product is $8 \times 16 = 2^3 \times 2^4 = 2^{3+4} = 2^7$. So to do the multiplication, we need only to *add* 3 and 4. There is also a connection between the logarithm and the number of digits you need to write down a given number. For example, 78356 is a 5-digit number, and $\log_{10} 78536 = 4.894\ldots$, which rounds up to 5, so the logarithm gives an idea of the size of the number, measured in terms of its length.

What is the number $e$, the base of the natural logarithm? This is most easily given by an infinite sum

$$e = 1 + \frac{1}{1!} + \frac{1}{2!} + \frac{1}{3!} + \frac{1}{4!} + \frac{1}{5!} + \cdots = 2.718281828459045235\ldots.$$

It is not obvious why this number is "natural" (indeed, the digits of the numbers seem to have no pattern) but there are all sorts of sophisticated reasons. It is often presented as the basis on which one computes compound interest and understands exponentiating (because $\lim_{n \to \infty} (1 + \frac{1}{n})^n = e$), it has a very surprising property in calculus (that $\frac{d}{dx} e^x = e^x$), but most important of all, it yields the logarithm that provides the right way to describe the density of primes around $x$!

## CALIBRATION

1 in $N$ permutations (of $N$ letters) are cycles, and 1 in $\log x$ of the integers around $x$ are prime, so if we are to compare permutations and integers, perhaps the natural way to do so is to equate this scarcity. So in a formula for permutations that is given in terms of the number $N$, perhaps we can replace $N$ by $\log x$ to obtain the analogous formula for integers? Does this make any sense? A priori, "Integers . . . and permutations . . . are about as similar as apples and iPhones" [Langer, p. 34], but how do we know that there are no similarities in their anatomies until we try to find

some? There is not necessarily an analogy of each question about permutations in the theory of integers, but if we find such analogies, we can test how deeply they lie. Arguably the most obvious statistic to compare is the number of indecomposable components in a typical member of each population.

## The number of indecomposable parts that a "typical" permutation/integer has

A typical permutation has about $\log N$ cycles. By the word "typical," I mean that "almost all" permutations have $\log N$ cycles plus or minus something small, much smaller than $\log N$, at least when $N$ is large enough. These are usual notions in *arithmetic statistics*, what happens "almost all the time," and how well things are approximated. One can be more precise about quantifying these things, but we will not do so here.

"If our hypothesized 'calibration' is to make sense, then we should be able to simply swap out the '$N$' for a '$\log x$,' and guess that a typical integer has about $\log\log x$ distinct prime factors." [Gauss, p. 57] This is the 1919 theorem of HARDY and RAMANUJAN, which inspired the subject of *additive number theory*.[2] Fantastic! The calibration seems to make sense! Let's test it further.

## The number of indecomposable parts, more precisely

Typically about $\log N$ cycles occur in a permutation on $N$ letters, but how often are there somewhat more, or somewhat fewer? Statisticians use the *Normal distribution* to model many things. "Data that seems chaotic often organizes itself into certain recognizable patterns. Any one roll of the dice is random, but a casino banks on patterns emerging over many rolls. The most common pattern can be seen when you graph the data—the data plot is shaped like a bell around the average. All the bells have the same basic shape, though the center may appear in different places, and some may be fatter than others. We measure the width of the bell by its variance." [Langer, pp. 60–61] So there are two measures: the average and the variance.

For permutations, GONCHAROV showed in 1942 that as we vary over the permutations on $N$ letters, the number of cycles is given by a Normal distribution with mean and variance about $\log N$. This means that the number of cycles is $\log N$ plus or minus a not-too-large multiple of $\sqrt{\log N}$ (this is the *standard deviation*, which is the square root of the variance). The exact formula is a little complicated: The proportion of permutations in $S_N$ for which the number of cycles lies between

$$\log N + a\sqrt{\log N} \text{ and } \log N + b\sqrt{\log N},$$

where $a < b$, is

$$\frac{1}{\sqrt{2\pi}} \int_a^b e^{-t^2/2} dt.$$

---

2. Ramanujan's inspiring yet sad life story is told in the recent movie *The Man Who Knew Infinity*.

In 1940, the great Paul ERDŐS and Marc KAC showed that, as we vary over the integers $\leqslant x$, the number of prime factors of an integer is given by a Normal distribution with mean and variance around $\log \log x$. Again we see the calibration works, now on a more difficult statistic. Let $\omega(n)$ denote the number of distinct prime factors of a given integer $n$. Then the proportion of integers $n$ up to $x$ for which

$$\log \log x + a\sqrt{\log \log x} < \omega(n) \leqslant \log \log x + b\sqrt{\log \log x},$$

where $a < b$, is also

$$\frac{1}{\sqrt{2\pi}} \int_a^b e^{-t^2/2} dt.$$

### The size of the different parts

We know that a typical permutation on $N$ letters has about $\log N$ cycles, but not how long those cycles are. Are they mostly short? Or mostly long? Or is there a more involved way to lay them out? Let's suppose that we have a permutation on $N$ letters with $\ell$ cycles, and that these cycles have lengths $d_1 \leqslant d_2 \leqslant \ldots \leqslant d_\ell$. We first note that $d_1 + d_2 + \cdots + d_\ell = N$, and that they are each between 1 and $N$.

Our best approach is to take the logarithm of each of these lengths (which is far from obvious). Then we have a set of $\ell$ real numbers, each between 0 and $\log N$, namely,

$$0 \leq \log d_1 \leq \log d_2 \leq \cdots \leq \log d_\ell \leq \log N.$$

For a typical permutation, $\ell$ is about $\log N$, so we have roughly $\log N$ points lying on interval $[0, \log N]$, of length $\log N$. Therefore, we might naively guess that the points, the $\log d_j$, are evenly spaced out, with average spacing 1. There are many possibilities for how they might be distributed with average spacing 1, so what is typical? Statisticians have noted a certain pattern that often describes very well how events spread out over time; this occurs in many situations, for example the frequency of traffic on a freeway or the timing of clicks on a webpage. These patterns are called *Poisson point processes* and deal with the appearance of random variables over time. The key is to note that the average spacing between our points, $\log d_{j+1} - \log d_j$ is 1, and so for a typical permutation, we expect that the proportion of intervals of length $\lambda$ (inside the interval $[0, \log N]$) that contain exactly $m$ of the $\log d_j$, is about

$$\frac{e^{-\lambda}\lambda^m}{m!}$$

(and the proportion gets closer to this value as $N$ gets larger). For such an insightful statistic, we should not be surprised that this formula is more complicated to describe than much of what has appeared before.

How about the size of the prime factors of the integers? Our calibration replaces $N$ by $\log x$, and so $\log N$ by $\log \log x$ in all of our formulas. As a best guess, we should perhaps replace the $j$th smallest cycle length $d_j$ in our permutation by $\log p_j$, the logarithm of the $j$th smallest prime factor in our integer. By this measure, if

$$n = p_1 p_2 \cdots p_k, \text{ where } p_1 \leqslant p_2 \leqslant \cdots \leqslant p_k$$

are all prime, then the $\log \log p_j$ usually give about $\log \log x$ values in the interval $[\log \log 2, \log \log x]$, which has length about $\log \log x$. Again, one can prove that these values are distributed, for typical integers $n$, like a Poisson point process.

### Convincing similarities?

Our two organisms, permutations and integers, seem to satisfy the same rules for the size and lay out of their indecomposable components. Is this surprising? The Poisson and Normal distributions appear in many situations in mathematics, so perhaps these successful comparisons are not too surprising. After all, the cycle lengths and prime factor sizes have to be laid out somehow, according to some rules, so one's first guess would probably be something random; hence the Poisson and Normal distributions. "Pretty well ubiquitous in such situations," [Langer, p. 90] which leads one to ask whether there are measures of permutations or integers that involve rather unusual functions, so that it would be more surprising if our two organisms calibrate so well. "To get something interesting, we need to look at unusual aspects of the anatomies of permutations and integers that are much less likely to be identical." [Emmy, p. 92]

## UNUSUAL ASPECTS OF THEIR ANATOMIES

### The smallest part; finding the right question

Every integer has a unique smallest prime factor. 2 is the smallest prime factor of half of the integers, 3 of one-sixth of the integers, 5 of one-fifteenth of the integers, and so forth.

However, permutations do not necessarily have a unique shortest cycle. For example, if the positions of two different letters are fixed by some given permutation, then that permutation has two cycles of length 1. These cycles are different, even though they have the same length, so this situation is not directly comparable with what can possibly happen with the integers. In this case, if 4 divides an integer (that is, 2 divides the integer twice), then it is the same prime that divides twice. The frequency with which the smallest cycle length is 1 is rather more complicated:

$$\frac{1}{1!} - \frac{1}{2!} + \frac{1}{3!} - \frac{1}{4!} + \cdots \pm \frac{1}{N!}.$$

(As $N$ gets larger, this quantity gets ever closer to $1 - 1/e$, where $e$ is again the base of the natural logarithm.) Therefore, we see that no obvious analogy exists between these results for integers and for permutations, and the reason seems to be that we do not quite know how to compare the very smallest components. "How can I compare the smallest primes to the smallest cycles?" [Emmy, p. 106] They're just such different objects.

It seems we should simply ignore these smallest parts, as they seem hard to compare. Selection of what to compare, and how to measure the comparisons is as much an art as a science. "Part of the art of . . . mathematics is to decide what evidence to ignore, if any. . . . We found that the smallest components . . . have only a tiny influence on the key comparisons, so we will ignore these peripheral artifacts." [Langer, p. 93]

### How can we compare the smallest parts?

We can get around such objections, by taking a broader perspective. "Instead of specifying the smallest component, we could look at all of the integers, and permutations, that have no components smaller than a given size." [Gauss, p. 117] More precisely, instead of asking how many integers have a smallest prime factor $p$, we can ask how many integers have a prime factor $\leqslant y$ (for some well chosen $y$) and pose an analogous reformulation for the smallest components of a permutation. This works, and in an interesting and surprising way, as I will explain. "At a stroke, we've not only gotten rid of the issue that the smallest component happens rarely, but we can also work in the range where we know our organisms are comparable." [Emmy, p. 117]

### The size of the smallest part

A permutation on $N$ letters cannot have two cycles of length $> N/2$, and so if its smallest cycle has length $> N/2$, then it must be a cycle of length $N$. If the smallest cycle of a permutation on $N$ letters has length $> N/3$, then the permutation breaks down into one or two cycles and no more. The correct question is therefore: For each fixed $u > 1$, how many permutations on $N$ letters have all cycles of length $> N/u$? We have seen that for $u \in (1, 2]$, the answer is the number of cycles on $N$ letters, namely, $(N - 1)!$, which is the proportion $1/N$ of all permutations. In fact, for any fixed $u > 0$, the proportion is $c(u)/N$ for some constant $c(u)$ that depends on $u$ but not $N$, though we typically write $\omega(u) = c(u)/u$, so that the proportion of permutations whose cycles all have length $> N/u$ is $\omega(u)/(N/u)$. What are the values of these constants $\omega(u)$? It starts off easily enough: $\omega(u) = 0$ for $0 < u < 1$, then $\omega(u) = 1/u$ for $1 \leqslant u < 2$; but after that, $\omega(u)$ becomes a mess to describe. The simplest way to calculate it for $u \geqslant 2$ is via the formula

$$\omega(u) = \frac{1}{u}\left(1 + \int_0^{u-1} \omega(t)\, dt\right).$$

This is *Buchstab's function*; it "is something that can only really be usefully described as a certain average of its history. It's self-referential, and thus hard to compute." [Gauss, p. 129] This is not the sort of function one meets at school and rarely at university. It has a *transcendental definition*, and there is no easy way to describe it.

Remarkably, the analogous result for integers was proved by Buchstab in 1949: By our calibration, we replace the $d_j$ by $\log p_j$ and the $N$ by $\log x$, and so having all $d_j > N/u$ is analogous to each $\log p_j > (\log x)/u$, which is better written as each $p_j > x^{1/u}$. Buchstab showed that the proportion of integers up to $x$ that have all their prime factors $> x^{1/u}$ is also about $u\,\omega(u)/\log x$ for large $x$.

One can graph the function $\omega(u)$ (as we see at the top of p. 128), and, as $u$ gets larger, it oscillates on either side of the value $e^{-\gamma}$ (a constant that appears in many places but is difficult to describe). These oscillations diminish in size very rapidly.

One can develop this further and show that the proportion of permutations that have their $k$ smallest cycles of lengths at least $N/u_1, N/u_2, \ldots, N/u_k$, respectively, is the same (after calibrating) as the proportion of integers that have their $k$ smallest prime factors at least $x^{1/u_1}, x^{1/u_2}, \ldots, x^{1/u_k}$, respectively.

No one (in their right mind) would expect to see as peculiar a function as Buchstab's function appearing twice in analogous questions without there being a fundamental underlying reason. Moreover, if strange things happen when comparing the smallest indecomposable components, how about when we compare the largest? "Why not compare the organisms from each population with no components larger than a given size?" [Emmy, p. 132]

## The size of the largest part

Again fix $u \geqslant 1$, and ask what proportion of permutations have all their cycle lengths $< N/u$. In 1944, Goncharov showed that this proportion is roughly a constant, $\rho(u)$, when $N$ is large, a constant that depends only on $u$ and not on $N$. For $0 < u \leqslant 1$, this includes all permutations, so $\rho(u) = 1$. For larger $u$, this function $\rho(u)$ does not appear to have any simple definition (like a closed formula). The most palatable way to describe it is through the integral delay equation:

$$\rho(u) = \frac{1}{u} \int_{u-1}^{u} \rho(t) \, \mathrm{d}t \text{ for } u > 1.$$

This is another function that is a certain average of its history, though in this case, it is rapidly diminishing (as in the graph on p. 132).

The integer version asks for the proportion of integers up to $x$ all of whose prime factors are $\leqslant y$. Such integers are called the *y-smooth integers*, and there are roughly $\rho(u)x$ of them when $y = x^u$. For example, the number of integers up to $x$ whose prime factors are all $\leqslant \sqrt{x}$ is roughly $\rho(2)x$ (and $\rho(2) = 1 - \log 2$). This result has proved essential to our understanding of the running times of many computer algorithms (particularly those used in cryptography). Moreover, recent advances in our understanding of the properties of integers come from appreciating what happens to their *y*-smooth parts. In 1930, Dickman observed that such a result should be true, and his remarks were made rigorous and expanded on by de Bruijn in the 1950s. The function $\rho(u)$ is called the *Dickman–de Bruijn function*.

In 1976, the great computer scientist Donald Knuth and his collaborator Luis Trabb Prado observed that one also gets the same proportions for the largest $k$ cycles of a permutation and the largest $k$ prime factors of an integer. That is, for any given $1 \leqslant u_1 \leqslant u_2 \leqslant \cdots \leqslant u_k$, the proportion of permutations on $N$ letters whose $j$th largest cycle has length $\leqslant N/u_j$ for $j = 1, 2, \ldots, k$ is more or less the same as the proportion of integers $\leqslant x$ that have their $j$th largest prime factor $\leqslant x^{1/u_j}$ for $j = 1, 2, \ldots, k$.

Probability theorists have found that these proportions come up in many situations. They can be described by what is called a *Poisson-Dirichlet* process and, among other things, this process is used to model industrial rock-crushing procedures!

# AND WHEN THERE ARE EXACTLY $\ell$ PARTS

"We created our calibration by comparing the proportion of integers that are primes with the proportion of permutations that are cycles. But what if we now compare the proportion of integers that have exactly two prime factors with the proportion of permutations that have exactly two cycles—will they be the same? Or three? Or more?" [Langer, p. 110]

In 1947, Jordan showed that, for any fixed integer $\ell \geqslant 1$, the proportion of permutations on $N$ letters with exactly $\ell$ cycles is

$$\sim \frac{1}{N} \frac{(\log N)^{\ell-1}}{(\ell-1)!}.$$

We already know this for $\ell = 1$, but it is new information when $\ell$ is larger. In fact, Jordan's formula works as long as $\ell$ is significantly smaller than $\log N$. It is worth noting that one gets this quantity if one goes back to the Poisson formula, $\frac{e^{-\lambda}\lambda^m}{m!}$ (which we saw earlier in a different context) and replaces $m$ by $\ell - 1$ and $\lambda$ by $\log N$. The Poisson formula truly does arise in many different contexts, and we might expect that this formula holds for all values of $\ell$, even large values (technically, because we might guess that the number of parts of a permutation satisfies a *Poisson distribution* with mean $\log N$).

However, if $\ell$ is comparable in size to $\log N$, then there is an unexpected extra factor: The proportion of permutations with exactly $\ell$ cycles is

$$\sim \frac{1}{\Gamma(r+1)} \cdot \frac{1}{N} \frac{(\log N)^{\ell-1}}{(\ell-1)!},$$

where $r := (\ell-1)/\log N$. The "Gamma function," $\Gamma(\cdot)$, interpolates the factorial function, meaning that $\Gamma(r+1) = r!$ when $r$ is an integer $\geqslant 0$, and it provides the optimal way to go in between integer values. Our first estimate above follows from this as $\Gamma(1) = 1$, and $\Gamma$ is continuous. This new estimate holds as long as $\ell$ is significantly smaller than $(\log N)^2$.

Because $\Gamma(2) = 1$ also, we find that the Poisson formula is also valid if $\ell$, the number of indecomposable parts, is close to $\log N$. Interestingly, these are the only two points at which $\Gamma(x) = 1$. Otherwise, $\Gamma(x)$ is not 1, and we do not get the expected Poisson distribution. For example, $\Gamma(3/2) = \sqrt{\pi/4} = .886\ldots$, so the number of permutations with about $\frac{1}{2}\log N$ cycles is about $1/.886 \approx 1.13$ times what we would guess from the Poisson distribution.

What about *integers* with exactly two prime factors? Or three or more? Can we simply replace $N$ by $\log x$ in the above results? In 1909, Landau showed that, for any fixed integer $\ell \geqslant 1$, the proportion of integers up to $x$ with exactly $\ell$ prime factors is

$$\sim \frac{1}{\log x} \frac{(\log \log x)^{\ell-1}}{(\ell-1)!},$$

as we, by now, expect. This formula again works if the ratio $r := (\ell-1)/\log \log x$ is very small, and it is the Poisson formula with mean $\lambda = \log \log x$. When $r$ (and so $\ell$) is larger, we need to again introduce the factor $1/\Gamma(r+1)$, but there is also another factor:

$$\prod_{p \text{ prime}} \left(1 + \frac{r}{p-1}\right)\left(1 - \frac{1}{p}\right)^r,$$

which has no analogy when looking at permutations. Even this extra factor equals 1 at the two most interesting values, $r = 0$ and $r = 1$ (which is worth verifying for yourself), that is, where $\ell$ is small and where $\ell$ is close to what is typical. So the extra factors do not contribute in the most interesting ranges for the variables.

Nonetheless, the second factor provides a worrying difference between the two anatomies. It is typically close to 1, and its value is only really affected by the primes $p \leqslant r^2$; that is, the small primes. But we agreed earlier that we should consider the small primes to be a peripheral artifact that can be ignored.

This formula for the number of integers with exactly $\ell$ prime factors was first established by L. G. Sathé in 1954 by a long, complicated argument. The then-young Norwegian genius, Atle Selberg, was the first person to review Sathé's breakthrough and realized that by developing Riemann's old ideas in a new direction, he could re-prove this formula in just a few pages. Anyone looking to research into this area of number theory could not do better than begin by reading and fully appreciating Selberg's stunningly clever argument.

### When $\ell$ is large

There is no precise, yet simple, formula for the number of permutations with more than $(\log N)^2$ cycles. However, we can get accurate formulas if we compare the answers to two questions of this type. For example, let $\ell$ be an integer that is neither too small nor too large: Specifically, suppose that $\ell$ and $N/\ell$ are large, and let $v = \frac{N}{\ell}\log(\frac{N}{\ell})$. If $m$ is a positive integer that is significantly smaller than both $N$ and $v$ then

$$\frac{\text{Proportion of permutations on } N - m \text{ letters with } \ell - 1 \text{ cycles}}{\text{Proportion of permutations on } N \text{ letters with } \ell \text{ cycles}} \approx \frac{\ell}{\log v}.$$

This result follows from deep estimates of Moser and Wyman (1958).[3]

Analogously, for integers up to $x$, if $\ell$ and $(\log x)/\ell$ are large enough, we let $v = \frac{\log x}{\ell}\log(\frac{\log x}{\ell})$. If $d$ is a positive integer for which $\log d$ is significantly smaller than $\log x$ and $v$, then

$$\frac{\text{Proportion of integers up to } x/d \text{ with exactly } \ell - 1 \text{ prime factors}}{\text{Proportion of integers up to } x \text{ with exactly } \ell \text{ prime factors}} \approx \frac{\ell}{\log v},$$

which follows from the 1988 results of Hildebrand and Tenenbaum.

---

3. In fact, they estimated the *Stirling numbers of the first kind*, which give precisely the number of permutations on $N$ letters with exactly $\ell$ cycles.

### *The size of the $\ell$ different parts*

We wish to understand how the indecomposable parts are laid out when there are $\ell$ of them, and $\ell$ is not too close to the average. We begin by taking the same approach as before. For permutations, there are $\ell$ values of the form $\log d_j$ in the interval $[0, \log N]$, so they have average spacing $\frac{1}{\ell}\log N$. The largest part can be given in terms of all the smaller ones (because $\sum_j d_j = N$), and so it makes sense to restrict our attention to all but the largest part (more on the largest part in the next section). One can show that the points

$$\frac{\log d_j}{\frac{1}{\ell}\log N}, \quad 1 \leqslant j \leqslant \ell - 1,$$

are distributed on $[0, \ell]$ like $\ell - 1$ independent randomly chosen numbers, provided $\ell$ is small (that is, $\ell \leqslant \frac{1}{2}\log\log N$). Similarly, if $\ell \leqslant \frac{1}{2}\log\log\log x$, then the points

$$\frac{\log\log p_j}{\frac{1}{\ell}\log\log x}, \quad 1 \leqslant j \leqslant \ell - 1,$$

are distributed on $[0, \ell]$ like $\ell - 1$ independent randomly chosen numbers, for integers $n \leqslant x$ with exactly $\ell$ prime factors.

For larger $\ell$, we need to restrict the range in which we consider cycle lengths for two reasons: A vanishing proportion of the cycles are much larger than $N/\ell$, and too many cycle lengths much smaller than $\ell$ are repeated.[4] So then it is true that the values

$$\frac{\log d_j}{\frac{1}{\ell}\log(N/\ell^2)}, \quad \ell \leqslant d_j \leqslant N/\ell,$$

satisfy a Poisson point process in $[0, \ell]$ with average gap size 1 for permutations on $N$ letters with $\ell$ cycles.

The extra logarithm, when working with the prime factors of integers, means that the restrictions on small indecomposable parts are not applicable. Therefore, we find that for integers up to $x$ with exactly $\ell$ prime factors, the values

$$\frac{\log\log p_j}{\frac{1}{\ell}\log\log(x^{1/\ell})}, \quad p_j \leqslant x^{1/\ell},$$

satisfy a Poisson point process in $[0, \ell]$ with average gap size 1.

### *The largest of the $\ell$ irreducible components*

When there are significantly fewer parts than usual, then the largest component is usually more or less as large as it can be. For example, if $\ell$ is significantly smaller than $\log N$ (when the parameter $r = \frac{\ell-1}{\log N}$ is close to 0), then the longest cycle in a permutation with $\ell$ cycles typically has length close to $N$. And if $\ell$ is significantly smaller than $\log\log x$ (when the parameter $r$ is

4. The "cutoffs" are more precisely at $(N/\ell)\log(N/\ell)$ and $\ell/\log \nu$, respectively.

close to 0), then for the largest prime factor $p$ of an integer $\leqslant x$ with exactly $\ell$ prime factors, we typically have $\log p$ close to $\log x$.

Things are more difficult when $\ell$ is significantly larger than is typical: "How large is the largest component of an integer or a permutation if it has more components than is usual? Perhaps if there are more components than usual then there is more chance to have a particularly large one? Or perhaps, since their average size is smaller, the largest component is smaller than is typical?" [Langer, p. 146]

To describe the results on this question, we need to bring in a new parameter. So far, for permutations, we have used $r = \frac{\ell-1}{\log N}$ and $\nu = \frac{N}{\ell} \log(\frac{N}{\ell})$. Now we replace $r$ by the more precise $\xi := \frac{\ell-1}{\log \nu}$. For small $\ell$, there is no significant difference between the values of $r$ and $\xi$, but there is for large $\ell$. One makes the analogous definition for integers with $N$ replaced by $\log x$. In what follows, we assume that $\ell$ is large, but $\ell \leqslant \sqrt{N}$.

We find that for typical permutations on $N$ letters with exactly $\ell$ cycles, the longest cycle has length about $\frac{\log \xi}{\xi} N$. Thus the longest cycle of a permutation typically gets smaller the more cycles the permutation has. The analogous result is true for integers.

If we fix $\xi$, then we find a similar theory to that of the Dickman–de Bruijn function. The proportion of permutations on $N$ letters with exactly $\ell$ cycles that all have length $\leqslant N/u$ is about $\rho_\xi(u)$ (this is also the proportion of integers $n \leqslant x$ with exactly $\ell$ distinct prime factors that are all $\leqslant x^{1/u}$). Here, $\rho_\xi(u) = 1$ for $0 \leqslant u \leqslant 1$, and $\rho_\xi(u) = 1 - \xi \int_1^u (u-t)^{\xi-1} \frac{dt}{t}$ for $1 \leqslant u \leqslant 2$, with

$$\rho_\xi(u) = \frac{\xi}{u} \int_{u-1}^u \rho_\xi(t) \left(\frac{t}{u}\right)^{\xi-1} dt \ \text{ for all } u \geqslant 2.$$

(One can easily verify that $\rho_1(u) = \rho(u)$.)

It seems that there are "exactly the same proportions of any specific anatomy type in each of these two populations, no matter how peculiar these precise anatomical characteristics." [von Neumann, p. 154]

## AN UNDERLYING CAUSE?

"It seems that no matter what statistic we look at, the two yield the same proportions. Could this be a coincidence? With so much evidence, it seems increasingly unlikely that it could be, and we surely must suspect that there is an underlying cause." [Langer, p. 147]

Why are the anatomies of integers and permutations so similar? There are two proposed explanations for why their anatomies are so similar, one from probability theory, the other from analytic combinatorics (as handed out on p. 162), though I find neither explain "why," but rather "how." Moreover, it could be that these two viewpoints are really the same but in different guises. It is often difficult to penetrate the different languages of mathematics and, in this case, one feels that many elements are in common without it being clear whether the differences are fundamental.

These frameworks suggest that one should find similar anatomies in other classes of mathematical objects. We will discuss a few of these in a subsequent subsection.

## The probabilistic model

Arratia, Barbour, and Tavaré (1997) developed a probabilistic model that yields a good approximation to the structure of randomly chosen permutations and randomly chosen integers, so that the properties of the model give accurate predictions for their anatomies. This model considers the joint distribution of $(z_1, z_2, \dots)$, where each $z_i$ is an independent random variable having a Poisson distribution with parameter $1/i$. This distribution (subject to the side condition $\sum_i i z_i = N$) is very close to the joint distribution of $(c_1(\sigma), c_2(\sigma), \dots)$, where we run through the permutations $\sigma$ on $N$ letters, and $c_i(\sigma)$ denotes the number of cycles of length $i$ in $\sigma$. This builds on the "Poisson-Dirichlet process" as developed by Sir John Kingman; it is known to be very applicable, for example, to the Ewens sampling formula in population genetics.

When we look at the whole population of permutations, this model mostly predicts things very well. However, it is not clear to me whether it works so well when we restrict to subpopulations with $\ell$ parts, because the probability questions that arise when we add in the condition $\sum_i z_i = \ell$ seem to be somewhat more delicate.

Several authors have proved some remarkable results using the probabilistic model. My two favorite are due to Eric Bach, who designed a computer algorithm that rapidly outputs a random integer, fully factored (more on this later), and to Alexander Smith, who estimated the largest gap, $\log\log p_{j+1} - \log\log p_j$, between consecutive prime factors of a typical integer in his groundbreaking work [20] on ranks of elliptic curves.

## Analytic combinatorics

In 2001, Panario and Richmond noted that many of these statistics are true for a fairly general class of combinatorial objects for which the generating function takes the following form: The number of objects of size $m$ is given by the coefficient of $z^m$ in a generating function of the form $a(1 - z/\rho)^{-b} \exp(E(z))$, where $|E(z)| \leqslant |z - \rho|^\epsilon$, if $z$ is sufficiently close to $\rho$. Their work appears to me to be more the development of an efficient calculating tool to prove that certain qualified organisms have similar anatomies rather than an explanation of why they are so similar.

## Fundamental structures

In his invited lecture at the prestigious 1994 International Congress of Mathematicians in Zurich,[5] Anatoly Vershik provided perhaps the best explanation for the similarity of so many phenomena for different objects: Fundamental mathematical structures should be organized in a natural way. There are a few outstanding possibilities for this "natural anatomy" (seven are listed in Vershik's paper [22]), including the structure we see here. What is perhaps new in this article is the amazing amount of detail that these different anatomies share.

The idea that wildly different objects should be organized along very similar lines has emerged recently in an area on the boundary between quantum chaos in mathematical physics and the

---

5.  This lecture kindled my interest in these similarities. The first page of Vershik's handwritten slides from that talk can be seen on p. 163.

**The Mathematics of *Prime Suspects***

theory of zeta functions in analytic number theory: Sets of eigenvalues of various naturally arising operators (for example, in quantum chaos) and zeros of $L$-functions, also seem to always be organized in very similar ways, according to the distribution of the eigenvalues of matrices randomly selected from certain groups. In 1999, Katz and Sarnak showed that only a small set of possible groups seem to arise (see the bulletin board notice on p. 125). I don't think anyone can say why. Indeed, it all seems unreasonably convenient, begging for a unifying explanation.

### Polynomials in finite fields

Toward the end of the graphic novel, we met a third victim, Polly Nomial, with a similar anatomy to integers and permutations. More precisely, polynomials mod $p$ (or in $\mathbb{F}_p[t]$) factor into irreducible polynomials mod $p$, the indecomposable components. There are $p^N$ monic polynomials of degree $N$ in $\mathbb{F}_p[t]$. Emmy finds a clue on p. 148, the first line of which reads that the number of monic irreducible polynomials in $\mathbb{F}_q[t]$ of degree $N$ is

$$\pi_q(N) = \frac{1}{N} \sum_{d|N} \mu(d) q^{N/d}.$$

That is, roughly 1 in $N$ of the monic polynomials of degree $N$ are irreducible. This is the calibration (see the whiteboard behind Gauss on p. 163), and from it, one can write a host of analogous properties for these polynomial factorizations from what we noted above; most have by now been proved by my students.

### Other organisms with similar anatomies

Vershik identified many other classes of objects with similar anatomies:

- The connected components of the 2-regular graphs on $N$ labeled vertices (that is, the vertices should all appear in a set of nontrivial disjoint cycles);
- The connected components of the directed graphs given by the edges $(i, f(i))$ of any map $f : \{1, 2, \ldots, N\} \to \{1, 2, \ldots, N\}$ (which typically have about $\frac{1}{2} \log N$ components);
- The equivalence classes of mappings $\{1, 2, \ldots, N\} \to \{1, 2, \ldots, N\}$, where $\pi_1, \pi_2$ are equivalent if there exist permutations $\sigma, \tau$ such that $\pi_2 = \sigma \pi_1 \tau$; and
- Additive arithmetic semigroups and other algebraic objects generalizing the rational integers.

A version of this list can be found behind von Neumann on p. 177.

### Constructing integers with the probabilistic model

Arratia, Barbour, and Tavaré [2] gave a probabilistic model to represent the integers, which can be described as a process to randomly select an integer $n \leqslant x$, so that it will be given fully factored: For each prime $p \leqslant x$, let $e_p$ be a random variable for which

$$\operatorname{Prob}(e_p = k) = \frac{1}{p^k} \left(1 - \frac{1}{p}\right) \quad \text{for each integer } k \geqslant 0.$$

Let $Y := \prod_{p \leqslant x} p^{e_p}$. For each integer $n \leqslant x$, define independent random variables $u_n$ so that $\text{Prob}(u_n = 1) = n/x$, and $\text{Prob}(u_n = 0) = 1 - n/x$; for $n > x$, let $u_n = 0$. Then

$$\text{Prob}(Y = n \text{ and } u_n = 1 \mid Y \leqslant x \text{ and } u_Y = 1) = \frac{1}{x}.$$

So the process proceeds by selecting the random variables as described. The algorithm fails if $Y > x$ or if $u_n = 0$, which occur with probability $1 - c_x$, where $c_x \approx 1/(e^{\gamma} \log x)$. Otherwise, the algorithm succeeds, and we obtain a fully factored, randomly chosen integer $\leqslant x$. We expect this algorithm to succeed 1 in every $1/c_x \approx e^{\gamma} \log x$ times it is run.

This algorithm would run slowly on a computer, because it requires the calculation of $e_p$ for each prime $p \leqslant x$: Determining each $e_p$ may be quick, but because there are so many of them, this part of the algorithm is very slow. One can simplify by using Vershik's 1995 observation that a random integer up to (integer) $x$ can be constructed by what is, in essence, a Markov chain, picking at each step a random prime factor of our integer. Hence, our algorithm to select a random factored integer $n \leqslant x$ runs as follows. The probability that we select $n = 1$ is $1/x$; if we do so, then the algorithm terminates. Otherwise, the probability that prime $p$ divides $n$ is $\frac{1}{x-1} \left[ \frac{x}{p} \right]$. So we select prime $p$ with probability

$$\left( 1 - \frac{1}{x} \right) \frac{1}{L} \left[ \frac{x}{p} \right], \text{ where } L := \sum_{\substack{p \text{ prime} \\ p \leqslant x}} \left[ \frac{x}{p} \right].$$

If we have selected prime $p$, then we obtain $n = mp$, where $m$ is a randomly selected integer $\leqslant [x/p]$. Now we repeat the process for $m$. Because the range in which we search gets at least halved each time we run this process, we will not have to run it more than $\frac{\log x}{\log 2}$ times. Thus, the algorithm is fast, provided we can select $p$ rapidly, and this was achieved by Bach in his 1998 PhD thesis [4], proving that random, fully factored integers up to $x$ can be found extremely quickly. Bach's algorithm for selecting primes $p$ quickly is clever but complicated.

We take another approach to producing a random factored integer $\leqslant x$, based on the fact that permutations and integers have such similar anatomies. Let $N = \log x$, and select a random permutation $\sigma$ on $N$ letters. Writing $\sigma$ as a product of cycles of lengths $d_1 \leqslant \cdots \leqslant d_k$, we then select random prime numbers $p_i \in \left( e^{d_i(\sigma)}, e^{d_i(\sigma)+1} \right)$ and consider the product $p_1 p_2 \cdots p_k$. With this algorithm, the probability that integer $n \leqslant x$ is produced is close to $1/x$ (up to a constant multiple), which is not quite what was required. What we want is that the probability that integer $n$ is produced is exactly $1/x$. To fix this, one can import Bach's ideas [4] to "doctor the odds" and make our algorithm work as claimed.[6]

---

6. 1 should clarify that selecting a random permutation and a random prime, as described, can be done easily, quickly, and correctly.

**The Mathematics of *Prime Suspects***

# DIVISORS

A positive integer $m$ is a *divisor* of $n$ if $m$ divides $n$. Determining the number and distribution of divisors of $n$ is a broad subject involving much research, past and present. At first sight, there is no obvious analogy of a divisor when talking about permutations, but one can find a useful analogy by working from a certain perspective: If $n = p_1 \cdots p_k$, the factorization of $n$ into primes, then each divisor $m$ can be written as $p_{j(1)} \cdots p_{j(\ell)}$ for some subset $\{p_{j(1)}, \ldots, p_{j(\ell)}\}$ of $p_1, \ldots, p_k$ (and each such product gives a divisor of $n$). In this language, the analogy for permutations would therefore be: If $\sigma = C_1 \cdots C_k$, the factorization of $\sigma$ into cycles, then each divisor $\mu$ can be written as $C_{j(1)} \cdots C_{j(\ell)}$ for some subset $\{C_{j(1)}, \ldots, C_{j(\ell)}\}$ of $C_1, \ldots, C_k$. But now we have to ask ourselves: What does $C_{j(1)} \cdots C_{j(\ell)}$ mean? For example, if $\sigma = (a)(bh)(cdg)(ef)$ is a permutation on 8 letters, then what is the subproduct of cycles $(bh)(cdg)$? This evidently describes a permutation on the 5 letters $b, c, d, g, h$ and has no effect on the 3 letters $a, e, f$. Its partner $(a)(ef)$ permutes the 3 letters $a, e, f$ and has no effect on the 5 letters $b, c, d, g, h$. So one way to think of this "divisor" is that $\sigma$ fixes the subsets $\{b, c, d, g, h\}$ and $\{a, e, f\}$ in that each gets mapped by $\sigma$ to itself (though the elements of each subset might well be permuted). Therefore, if $\sigma$ permutes $N$ letters and fixes a subset $D$ of those $N$ letters, then we say that $\sigma_D$ ("$\sigma$ restricted to $D$") is a *divisor* of $\sigma$. If the $N$ letters partition into the two sets $D$ and $M$, then $\sigma = \sigma_D \sigma_M$. If $\sigma$ is a cycle, then its only divisors are $\sigma_\emptyset$ and itself, very much in analogy with how we define primes.

The average number of divisors of an integer $\leqslant x$ is

$$\frac{1}{x} \sum_{n \leqslant x} \sum_{d \mid n} 1 = \frac{1}{x} \sum_{d \leqslant x} \sum_{\substack{n \leqslant x \\ d \mid n}} 1 \approx \frac{1}{x} \sum_{d \leqslant x} \frac{x}{d} = \sum_{d \leqslant x} \frac{1}{d} \approx \log x.$$

Both steps with an "$\approx$" need some justification, but that is not too difficult.

The average number of divisors of permutations of a set $\Lambda$ of $N$ letters is

$$\frac{1}{N!} \sum_{\sigma \in S_N} \sum_{\substack{D \subset \Lambda \\ \sigma \text{ fixes } D}} 1.$$

Here $\sigma$ fixes $D$ precisely for those $\sigma$ that are a permutation on both $D$ and $\Lambda \setminus D$. If $|D| = k$, then there are $k! \cdot (N - k)!$ such permutations. Therefore the above equals

$$\frac{1}{N!} \sum_{k=0}^{N} \sum_{\substack{D \subset \Lambda \\ |D| = k}} k! \cdot (N - k)! = \frac{1}{N!} \sum_{k=0}^{N} \binom{N}{k} \cdot k! \cdot (N - k)! = \sum_{k=0}^{N} 1 = N + 1.$$

Another good analogy!

These results are a little surprising, because we know that almost all integers $n \leqslant x$ have about $\log \log x$ prime factors, few of them repeated, and so $2^{\{1 + o(1)\} \log \log x} = (\log x)^{\log 2 + o(1)}$ divisors.

It is one of those strange situations in which the average differs markedly from what happens typically. Similar remarks apply to divisors of permutations.

### Erdős's multiplication table

When you were very young, you might have had to learn the multiplication table by heart. Perhaps you memorized all the values of $a$ times $b$ for $1 \leqslant a, b \leqslant 12$, written in a grid like this:

| × | 1 | 2 | 3 | 4 | 5 | 6 | 7 | 8 | 9 | 10 | 11 | 12 |
|----|----|----|----|----|----|----|----|----|-----|-----|-----|-----|
| 1 | 1 | 2 | 3 | 4 | 5 | 6 | 7 | 8 | 9 | 10 | 11 | 12 |
| 2 | 2 | 4 | 6 | 8 | 10 | 12 | 14 | 16 | 18 | 20 | 22 | 24 |
| 3 | 3 | 6 | 9 | 12 | 15 | 18 | 21 | 24 | 27 | 30 | 33 | 36 |
| 4 | 4 | 8 | 12 | 16 | 20 | 24 | 28 | 32 | 36 | 40 | 44 | 48 |
| 5 | 5 | 10 | 15 | 20 | 25 | 30 | 35 | 40 | 45 | 50 | 55 | 60 |
| 6 | 6 | 12 | 18 | 24 | 30 | 36 | 42 | 48 | 54 | 60 | 66 | 72 |
| 7 | 7 | 14 | 21 | 28 | 35 | 42 | 49 | 56 | 63 | 70 | 77 | 84 |
| 8 | 8 | 16 | 24 | 32 | 40 | 48 | 56 | 64 | 72 | 80 | 88 | 96 |
| 9 | 9 | 18 | 27 | 36 | 45 | 54 | 63 | 72 | 81 | 90 | 99 | 108 |
| 10 | 10 | 20 | 30 | 40 | 50 | 60 | 70 | 80 | 90 | 100 | 110 | 120 |
| 11 | 11 | 22 | 33 | 44 | 55 | 66 | 77 | 88 | 99 | 110 | 121 | 132 |
| 12 | 12 | 24 | 36 | 48 | 60 | 72 | 84 | 96 | 108 | 120 | 132 | 144 |

(as seen on p. 153). The easily bored young Paul Erdős asked himself: How many different integers are there in the table? There is the obvious symmetry down the diagonal, meaning that one only need look at the upper triangle for distinct entries. One spots other coincidences, like $3 \times 4 = 2 \times 6$ and $4 \times 5 = 2 \times 10$, and wonders how many there are. We ask the following precise question: What proportion of the integers up to $N^2$ appear in the $N$-by-$N$ multiplication table, that is, how many equal $ab$, where $1 \leqslant a, b \leqslant N$? For $N = 6$, we have 18 distinct entries, that is, $1/2$ of $N^2$; for $N = 10$, we have 42 distinct entries, that is $.42$ of $N^2$; for $N = 12$ we have 59, just below 41%. If $p(N) = \#\{\text{Distinct entries in } N\text{-by-}N \text{ table}\}/N^2$ is the proportion of integers with such a representation, then $p(25) = .36$, $p(50) = .32$, $p(75) \approx .306$, $p(100) \approx .291$, $p(250) \approx .270$, $p(500) \approx .259$, $p(1000) \approx .248$. Can one guess what the limit is, as the table gets bigger? Erdős proved that the limit is 0, and his proof is from The Book:[7]

A typical integer up to $N$ has about $\log \log N$ prime factors, and so the product $ab$ of two such integers $a$ and $b$ has about $2 \log \log N$ prime factors. However, almost all integers up to $N^2$ have about $\log \log N^2 = \log \log N + \log 2$ prime factors, very different from $2 \log \log N$, the number

---

7. True mathematicians are motivated by elegant proofs, none more so than the great Paul Erdős. Erdős claimed that "the supreme being" keeps a book that contains all of the most beautiful proofs of each theorem, and just occasionally are we mortals allowed to glimpse this book. Erdős liked to say "You don't have to believe in God, but you should believe in The Book." See Aigner and Ziegler's Book collection [1].

**The Mathematics of *Prime Suspects***

of prime factors of $ab$ for typical $a, b \leqslant N$. Therefore, a typical integer $\leqslant N^2$ cannot equal the product of two typical integers $\leqslant N$. And that's the proof!

This proof is alluded to on the corkboard on p. 9, which advertises the "Distinguishing features seminar," with the text "How to tell the difference between one, and two that are half the size and joined together." We next see the same corkboard on p. 155, on which the Erdős proof is given:

DISTINGUISHING FEATURES

(from *The Book*)

Typical integers of size $x$ have

about $\log \log x$ components

So the product of two of size $\sqrt{x}$

have about $2 \log \log x$ components.

The question is raised on p. 154, as well as the analogous question for permutations. Erdős's proof can be easily adapted to show that the number of integers $\leqslant x$ that are the product of two integers $\leqslant \sqrt{x}$ is more like

$$\frac{x}{(\log x)^\delta}, \text{ where } \delta = 1 - \frac{(1 + \log \log 2)}{\log 2}.$$

In 2008, Kevin Ford gave the much more accurate answer

$$\asymp \frac{1}{(\log x)^\delta (\log \log x)^{3/2}},$$

as seen on p. 157, where the notation "$\asymp$" means "up to some bounded multiplicative constant."

The analogous question for permutations asks for the number of permutations on $N$ letters that fix some subset of $[N/2]$ letters. The argument in the previous section shows that on average, a permutation fixes one such subset, but this does not imply that they all do. Indeed, some permutations may fix far more than one such subset. On p. 156 we see the answer: Just $\asymp N!/N^\delta (\log N)^{3/2}$ have a fixed subset of size $[N/2]$, in analogy to the integer case. The analogous result also holds for polynomials (the proof of these last two cases appears in Meisner [16]).

## Generating the full permutation group

In some perspectives on group theory, permutations $\sigma$ are hard to distinguish from their *conjugates*, the permutations given by the formula $\tau^{-1} \sigma \tau$, as one runs through the elements $\tau$ of $S_N$. A set $A$ of permutations *generates* $S_N$ if every element of $S_N$ can be written as a product of elements of $A$. For example, if $A = \{a, b\}$ then products include "words" like $aabbbbabbaab$ as well as all three of $aba, aab, baa$, as a different order may yield a different permutation. The set $A$ *transitively generates* $S_N$ if every set of conjugates of the elements of $A$ *generate* $S_N$. Our story starts with a recent paper by Pemantle, Peres, and Rivin [19], who showed that four randomly chosen permutations transitively generate $S_n$ with probability bounded away from 0 (see the panel behind Emmy on p. 164). Inspired by their understanding of divisors of integers, and the analogy between the anatomies of integers and permutations, Eberhard, Ford, and the real Ben Green

[8] showed by contrast that three randomly chosen permutations of $S_N$ almost always **do not** transitively generate $S_n$ (see the whiteboard, partially obscured by Gauss's head, on p. 168). They show this by establishing that three permutations will almost always each have a fixed set of size $k$.

# MORE MATHEMATICS OF *PRIME SUSPECTS*

## Rings and groups

Integers can be added, subtracted, and multiplied in a straightforward, consistent way, and when we do so, the result is an integer. Such a structure is called a *ring*, which includes an *additive group* and is found throughout mathematics.

In the sleuthing world, when a male body is first found, it is traditionally called "John Doe" until it is identified. In the math world, "Doe" becomes "$\rho$," which is pronounced "roe." In their lab tests to identify the corpse, our lab technicians have entered (on a monitor on p. 20) "Not part of a multiplicative group. In an additive group? An integer?"

Each shuffle of a deck of cards permutes the order of the cards, and so when we shuffle twice, we get a further permutation. Any permutation can be reversed, and so the permutations form a *group*. One note of caution: When we do two shuffles in a row, the resulting permutation usually depends what order we do them in. For example, on 4 letters, let's take first the permutation $(1, 2, 3)(4)$ and then $(12)(34)$; this gives the permutation $(1)(2, 4, 3)$, whereas these permutations in the reverse order give $(1, 3, 4)(2)$. A little unsettling at first, which is why the *center* is useful—those permutations that when combined with others, always give the same result, no matter what order we take them in.

The even integers form a *subgroup* of the additive group of integers (which form a group when adding them)—this is a subgroup, because when you add and subtract even integers, you get back an even integer. The *alternating group* is an important subgroup of the permutations, and also consists, like the even integers, of half of the permutation group. A given permutation on $N$ letters is in the *alternating group* if the number of cycles is even or odd, as $N$ is even or odd. In our story, Daisy is an odd permutation, so she lies outside the alternating group.

## The mathematics of music

Music plays a big part in this murder mystery, specifically, Robert Schneider's beautiful *Reverie in Prime Time Signatures* discussed by him in a later chapter in this book. The mathematics of harmony was explored back in ancient Greek times; see David Benson's terrific book [6], which provided us with the equations we needed. On pp. 97 and 166, we see general Fourier series, which can be used to break down functions into a sum of waves, and in particular, music into its sounds with different wavelengths. On p. 98, we see, half-obscured, FM (frequency modulation) synthesis using Bessel functions:

$$\sin(\phi + z \sin \theta) = \sum_{n=-\infty}^{\infty} J_n(z) \sin(\phi + n\theta).$$

On p. 103, we encounter Sophie Germain's equation for a gong, and on p. 104, the wave equation for a string. Finally, on p. 169, we see, on the whiteboards in Langer's apartment, both Webster's horn equation and the meantone scale on a cylinder (sure signs of villainy!).

In 1966, Marc Kac published the wonderful "Can One Hear the Shape of a Drum?" (see pp. 63, 64), in which one tries to determine the shape of a drum from the frequencies of the sound it makes. In 1992, Carolyn Gordon, David Webb, and Scott Wolpert constructed two regions with different shapes (as seen on Gauss's monitor on p. 63) that create the same frequencies. Later, in the mall, we find that these mathematicians are now drum makers.

### The ancient Greeks

Modern mathematics owes a great deal to the ancient Greeks, in terms of philosophy (what do concepts, like a whole number, really mean?), rigor (by providing incontrovertible proofs of basic facts), and even depth (building up from basic ideas to formulate and prove deep facts). The sieve of Eratosthenes plays a prominent role in the story (p. 104), and we also honor several other immortal Greek mathematicians in a gallery in our mall on p. 67.[8] Here we find portraits (from left to right) of Euclid, Ptolemy, Pythagoras, Diophantus, and Hypatia, as well as a statue of Archimedes, widely considered, alongside Gauss and Newton, as one of the three greatest mathematicians of all time. There is also a portrait of Eratosthenes on p. 104.

### The Hardy-Ramanujan upper bound

On p. 58, we see a homework problem from Professor Gauss's class, written on the blackboard, peeking out from under the projector screen with the photos of Hardy and Ramanujan. They showed that there exist constants $c_0, c_1 > 0$ such that the number of integers $\leqslant x$ with exactly $\ell$ prime factors is

$$\leqslant \frac{c_0 x}{\log x} \frac{(\log \log x + c_1)^{\ell-1}}{(\ell-1)!}.$$

The homework problem is to use this bound to show that almost all integers have about $\log \log x$ prime factors. One can do this by bounding, using the Hardy-Ramanujan inequality, the number of integers with $< .99 \log \log x$ or $> 1.01 \log \log x$ prime factors (and then replacing .99 and 1.01 by any constants that are on either side of 1).

### Obscure references

Police lab techs would surely determine a cadaver's height and weight. The only object in number theory with an associated height and weight is a *modular form*, so we gave some particulars of a modular form on a monitor on p. 21.

---

8. The spiral staircase and piano are based on the Fields Institute in Toronto, where your artist and authors had a creative meeting. That same day, Frank Morgan, a leading expert on the mathematics of the shapes of soap bubbles, was there, blowing and discussing soap bubbles. He was watched by the Institute's then director, Ed Bierstone.

Emmanuel Kowalski, Guillaume Ricotta, and Emmanuel Royer recently graphed the evolution of *Kloosterman sums*, and in each case, they obtained a bat-like drawing (no one understands why!). Because these sums have been recently discovered, we find them in the lab on p. 25. And because Kloosterman sums are the next step forward after *Gauss sums*, we also find an Andy Warhol-esque version of these diagrams on the mezzanine wall of Gauss's apartment on p. 37.

Klaus Roth showed how to find three-term arithmetic progressions in any reasonably dense set in the 1950s. One of the greatest breakthroughs of the past 20 years (and most inspiring to me personally) was Tim Gowers's adaptation of Roth's arguments to find arbitrarily long arithmetic progresssions. On p. 30, we find a poster purportedly about pensions (a Roth IRA being a tax- and penalty-free pension fund in the US) but with lots of puns alluding to the mathematical breakthrough.

In about 1637, Pierre de Fermat wrote a note to himself in the margin of his copy of Diophantus's *Arithmetic* that he had "a truly marvelous proof" that $x^n + y^n = z^n$ has no solutions in positive integers $x, y, z, n$ with $n \geqslant 3$, but that "this margin is too small to contain it." Fermat's son published Fermat's private notes after his death, and there were several such enigmatic pronouncements. All but this one had been resolved within about a hundred years of publication, so that a proof of **Fermat's Last Theorem** became the greatest of all mathematical mysteries. Attempts to prove Fermat's Last Theorem led to some of the most extraordinary progress in mathematics. For example, E. E. Kummer's attempts in the mid-nineteenth century provided enormous impetus to the early development of algebra. In the twentieth century, a more concerted attempt was made to understand integer and rational solutions to equations with few variables. Pierre Deligne used Alexander Groethendieck's ideas to do so in finite fields ("misappropriated" according to Groethendieck, who "disappeared into the mathematical wilderness as soon as his theories began to be 'used.'" [Gauss, p. 175]) In 1983, Gerd Faltings showed that any such equation of high enough degree can only have finitely many solutions. Then came some ultimately futile attempts to use "geometric inequalities" to bound the size of solutions to equations and so prove Fermat's Last Theorem. But it was only in the mid-1990s that Andrew Wiles finally proved Fermat's Last Theorem, in what can only be described as a tour de force of modern concepts in mathematics. This explains the "cold case" bulletin on p. 117.

After Wiles's work, the next great quest has been to prove the *abc*-conjecture, which literally gives bounds on the prime powers that can divide solutions to $a + b = c$ in positive, coprime integers. In 2012, the respected Shinichi Mochizuki announced that he had proved the *abc*-conjecture using his own "inter-universal Teichmüller theory," which few others have been able to appreciate. Usually such obscure mathematics has blatant mistakes, but not in this case, which has led to the uncomfortable position that the mathematical community could neither believe nor dismiss this work. The warning notice on p. 31 refers to this situation. At the time of publication, top mathematicians Peter Scholze and Jakob Stix have announced that they have found a serious flaw in Mochizuki's argument.

Perhaps the oldest unresolved great problem in mathematics is the *Riemann Hypothesis*, enunciated by Riemann back in 1859, that the interesting roots of the Riemann zeta-function all

**The Mathematics of *Prime Suspects***

lie on the line $x = \frac{1}{2}$ on the complex plane. A lot of evidence suggests that this is true: We know that well over 40% of the (infinitely many) zeros do lie on this line, and thus the "missing" poster on p. 137. The \$1 million reward is real, offered by the Clay Foundation for the resolution of this or any one of six other great mathematical questions.

Actually, one of those six great questions, the Poincaré conjecture, *was resolved* by Grigori Perelman in 2002. Perelman turned down mathematics most prestigious prize, the Fields medal, as well as the \$1 million prize, preferring to live on his mother's meager pension. The reports of his reasons vary: Like many mathematicians, he felt like his achievements rested on the foundation of others' work (Newton said, "If I have seen further it is by standing on the shoulders of Giants"), and they deserved the prizes as much as he did. He also bemoaned the ethics of a few mathematicians taking too much credit, and the community for tolerating that behavior, and so insisted that he stay out of the spotlight. The rumored reason that I liked best was that he did not feel that all members of the prize committees fully appreciated his accomplishments and so would not accept their prizes until they had.[9] This explains the slogan in the ad for "Perelman & Mum" on p. 72. The "one size fits all" paraphrases the actual Poincaré conjecture, which can be roughly stated as: if each part of a 3-d shape looks like a 3-d sphere, then it is a 3-d sphere.

The shop names in the mall include references to all of the Clay Millennium prize problems, as well as to Hilbert's 23 problems announced in 1900 for that new century.

The concept that mathematicians stand on each other's shoulders is explored on p. 163 in a series of nested Russian dolls, who are supposed to be the great Russian mathematicians Vershik, Perelman, Matijasevic, Gelfand, and Gromov (following the dolls in descending order of size). This nesting is not a true reflection of mathematical precedents, but rather artistic license!

We decorated Gauss's luxury penthouse with symbols of the real Gauss's life and career; for example, the massive Warhol-esque portrait of Euler, and smaller ones of Fermat, Weber (with whom he developed the telegraph), and Dirichlet. I also tried to imagine whom he would admire from subsequent works. Riemann obviously, and Manjul Bhargava (see below). He would like Klein bottles, and he has a plaque celebrating Weyl's equidistribution law over his billiard table: Supposing a rectangular billiard table has no pockets, is frictionless, and the bounce on each side is perfect, one can wonder whether a moving ball would eventually visit every part of the table. Weyl's law tells us that this depends only on the angle at which the ball was struck. Finally, Gauss could not help but be intrigued by tiling circles with circles, and he would be amused to accept a Fields' medal. In the bar scene (p. 41), we see portraits of the Bolyais, father and son, Gauss's close friends.

I have long been inspired by Gauss's work on binary quadratic forms, which appeared in his book *Disquisitiones Arithmeticae* written in Latin in 1798 when he was just 21. I am writing a "modern translation," which Gauss has in his apartment on p. 38, while we see Langer flicking through the original on pp. 37–38. In fact, the modern Gauss is introduced on p. 3, lecturing

---

9. All of Perelman's rumored complaints have some truth to them. Nowadays, pure mathematics is so broad and so deep that even the world's best mathematicians can only fully appreciate a fraction of what is out there. Prize committees tend to be selected to cover all areas, so only a few people on such a committee can truly understand each particular work under consideration.

from the original page in *Disquisitiones* on binary quadratic forms. He then gives a more modern lecture on the same subject in the flashback on p. 176. Manjul Bhargava's new approach to binary quadratic forms earns his portrait on Gauss's wall. Euler's polynomial $n^2 + n + 41$ takes prime values when you substitute in $n = 0$ or $n = 1$ or any integer up to $n = 39$. This is perhaps the weirdest of all such polynomials, and Gauss was one of the first to understand why it takes on so many prime values. Euler's polynomial has *discriminant* given by the familiar formula $b^2 - 4ac$, which is this case equals $1^2 - 4 \cdot 1 \cdot 41 = -163$, Gauss's vanity license plate that we see on p. 18.

*Elliptic curves* are the quintessential equations of degree 3 in two variables and are probably the most researched topic in number theory today. Their only mention comes in Emmy's thoughts on p. 100 as she envisions the beauty of their group law.

On p. 72, Einstein makes a cameo appearance working as an apprentice at Newton's fruit stand. Throughout Langer's presentation, Gauss is reading *Who Got Einstein's Office?*, a slightly controversial account of the history of the Institute for Advanced Study (this is the institute created in the 1930s in Princeton to house the great scientific minds who needed to escape Nazi persecution in Europe, like Einstein and von Neumann). On p. 160, he tosses the book aside while protesting, and some of Einstein's most famous equations come bouncing out.

What is the work of Langer that gets ripped up toward the end? On p. 173, we see reference to my own research work on Carmichael numbers, and then on pretentious distance, and these are ripped into shreds on p. 174 along with papers on various hot topics: the *abc*-conjecture, the nonexistence of Siegel zeros, the identification of Langlands and Selberg $L$-functions, the largest gaps between primes, and the general Elliott-Haberstam conjecture.

Biomathematicians like Mark Lewis in Alberta model the process of how coyotes and wolves create and maintain their territories, for instance, by scent marking. This helps conservationists protect species, combat bio-invasions, and control habitat for species survival. Andrea Bertozzi and her research group at UCLA have used this research, and mathematical models of cooperative behavior, to predict urban crime in Los Angeles and even gang territory via graffiti. Good predictions mean that police departments like the LAPD can have officers on the ground at the right locations, to fight and prevent crime. There is even a commercial product, *Predpol*, and our precinct has just been briefed (as we see on the flip chart on p. 9) that they are going to start testing it the following week.

The book also highlights a couple of beautiful recent, vaguely relevant, results:

- Behind Langer on p. 175 one sees a result of Bary-Soroker, Koukoulopoulos, and Kozma [5] that a random polynomial with positive integer coefficients from $1, \ldots, N$ is almost certainly irreducible if $N \geqslant 84$ (ideally, we would like to be able to take $N = 2$). Remember Polly Nomial's coefficients lie in a finite field, so this is a little off topic, but this is a terrific result with an answer that is not analogous to the finite-field case.
- On the back wall of the squad room (p. 164), we see reference to a new situation, in which a probabilistic model tells us how often the number of points on an elliptic curve take a certain value (see the paper [7] by *seven* authors).

**The Mathematics of *Prime Suspects***

During Emmy's final presentation, Gauss dreams about twin primes and other patterns of primes. On p. 178, he thinks of the latest, the Maynard-Tao result, showing that there are infinitely many pairs of primes that differ by no more than 246; he then dreams about how to improve that to the conjectured gap 2, the twin prime conjecture, by using sieve weights in a similar way. On p. 180, he dreams of polynomials that represent primes and then imagines proving this using "dynamic sieve weights." Isn't this how we all dream?

### *Rending a recursive infinity*

The artwork provided several opportunities to discuss infinity, and the intriguing dynamics of recursion and self-similarity. The paint job on the side of Barry's van is a developing psychedelic fractal as it goes through its scenes in the book. It encounters a "Dynamic Movers" van outside Emmy's place (p. 86), which has the slogan "working the same at any scale." Of course, we had to include our own version of Escher's ever-uphill staircase (p. 75), as well as some attempts by the cheese shop attendants to carve Escher-type shapes (p. 70). We were exhilarated by Robert J. Lewis's brilliant rendition of the Droste effect on the I-Pad on p. 93.

# BIBLIOGRAPHY

[1] MARTIN AIGNER and GÜNTHER M. ZIEGLER (1998) *Proofs from The Book*, Springer, Berlin.

[2] RICHARD ARRATIA, A.D. BARBOUR, and SIMON TAVARÉ (1997) "Random combinatorial structures and prime factorizations," *Notices Amer. Math. Soc.* **44**, 903–910.

[3] RICHARD ARRATIA, A.D. BARBOUR, and SIMON TAVARÉ (2003) *Logarithmic Combinatorial Structures: a Probabilistic Approach*, EMS Monographs in Mathematics **1**, European Mathematical Society, Zuri.

[4] ERIC BACH (1988) "How to generate factored random numbers," *SIAM J. Comput.* **17**, 179–193.

[5] LIOR BARY-SOROKER, DIMITRIS KOUKOULOPOULOS, and GADY KOZMA (2018) "Irreducible random polynomials of bounded height" (preprint).

[6] DAVID J. BENSON (2006) *Music: A Mathematical Offering*, Cambridge University Press, Cambridge.

[7] ALINA BUCUR, CHANTAL DAVID, BROOKE FEIGON, NATHAN KAPLAN, MATILDE LALIN, EKIN OZMAN, and MELANIE MATCHETT WOOD (2016) "The distribution of $\mathbb{F}_q$-points on cyclic $\ell$-covers of genus $g$," *International Mathematics Research Notices* **14**, 4297–4340.

[8] SEAN EBERHARD, KEVIN FORD, and BEN GREEN (2017) "Invariable generation of the symmetric group," *Duke Math. J.* **166**, 1573–1590.

[9] WILLIAM FELLER (1968) *An Introduction to Probability Theory and Its Application* (third ed.), Wiley, New York.

[10] ANDREW GRANVILLE (2006) "Cycle lengths in a permutation are typically Poisson distributed," *Electr. J. Combinatorics* **13**, 23.

[11] ANDREW GRANVILLE (2007) "Prime divisors are Poisson distributed," *Int. J. Number Theory* **3**, 1–18.

[12] G. H. HARDY and E. M. WRIGHT (1932) *Introduction to the Theory of Numbers*, Oxford University Press, Oxford.

[13] ADOLF HILDEBRAND and GÉRALD TENENBAUM (1988) "On the number of prime factors of an integer," *Duke Math. J.* **56**, 471–501.

[14]   NICK M. KATZ and PETER SARNAK (1999) "Zeroes of zeta functions and symmetry," *Bull. Amer. Math. Soc* **36**, 1–26.

[15]   DONALD E. KNUTH and LUIS TRABB PRADO (1976) "Analysis of a simple factorization algorithm," *Theoret. Comput. Sci.* **3**, 321–348.

[16]   PATRICK MEISNER (2018) *Erdős' Multiplication Table Problem for Function Fields and Symmetric Groups* (to appear). Available at https://arxiv.org/abs/1804.08483.

[17]   LEO MOSER and M. WYMAN (1958) "Asymptotic development of the Stirling numbers of the first kind," *J. London Math. Soc.* **33**, 133–146.

[18]   DANIEL PANARIO and BRUCE RICHMOND (2001) "Smallest components in decomposable structures: Exp-log class. Average-case analysis of algorithms," *Algorthmica* **29**, 205–226.

[19]   ROBIN PEMANTLE, YUVAL PERES, and IGOR RIVIN (2016) "Four random permutations conjugated by an adversary generate $S_n$ with high probability," *Random Structures & Algorithms* **49**, 409–428.

[20]   ALEXANDER SMITH (2017) "$2^\infty$-Selmer groups, $2^\infty$-class groups, and Goldfeld's conjecture" (to appear). Available at https://arxiv.org/abs/1702.02325.

[21]   ANATOLY M. VERSHIK (1987) "The asymptotic distribution of factorizations of natural numbers into prime divisors," *Soviet Math. Dokl.* **34**, 57–61.

[22]   ANATOLY M. VERSHIK (1995) "Asymptotic combinatorics and algebraic analysis," *Proc. ICM Zurich (1994)*, 1384–1394.

# THE MUSIC OF
# *PRIME SUSPECTS*

Robert Schneider

## REVERIE IN PRIME TIME SIGNATURES

My score, *Reverie in Prime Time Signatures*, is used as a clue in the graphic novel *Prime Suspects*. This music was written for, and first performed at, an experimental reading of the original script at the Institute for Advanced Study in Princeton, New Jersey, on December 12, 2009. I had the privilege of performing alongside cellist Heather McIntosh and clarinetist Alex Kontorovich, with me on an analog synthesizer. My program notes were largely incorporated into the graphic novel's script:

"A high note pulses on every beat—rising in pitch at the perfect squares" (Emmy, p. 102) "And the cello plays a note on every other beat, the clarinet every third beat, the keyboard every fifth beat, and every fifth beat is marked by a chord instead of a single note" (Silent Bob, p. 102). "It's a nod to the golden ratio?" (Emmy, p. 102).

"The golden ratio is related to the square root of 5. It's historically considered a model of aesthetic perfection. . . . The piece only incorporates prime-numbered time signatures: 2/4, 3/4, 5/4, and most of it's in 7/4!" (Emmy, p. 103). That is to say, there is a prime number of beats in each measure, so time signature 7/4 indicates 7 beats per measure. From the constraints imposed by these rhythmic patterns, melodies emerged naturally as I composed, special to each prime.

"The second interlude happens in 29/4 time at the 29th measure of the composition . . . it's a musical rendition of the sieve of Eratosthenes!" (Emmy, p. 103).

The cello marks beats that are multiples of 2, the clarinet marks multiples of 3, and the chords mark multiples of 5. Clearly, the beats on which none of these instruments play must not be multiples of 2, 3, or 5, which is enough to identify them as primes among the integers relevant to the composition, each accompanied by only the high pulse until the cello, clarinet, and keyboard chords sound together on the thirtieth beat (as 30 is a multiple of all three primes 2, 3, and 5), resolving before returning to the main theme.

In this tangled interlude, not quite random, our ears experience the formation of the sequence of the primes.

I have read that Leonardo da Vinci may have hidden a musical composition in his painting *The Last Supper*, and that Roslyn Chapel in Scotland has musical notation encoded in the masonry. As a variation on this theme, I sought to encode a hint of real mathematics in the musical composition: Eratosthenes' first step toward understanding the primes. These musings evidently inspired Silent Bob on pp. 99 and 100.

# Reverie in Prime Time Signatures

Robert Peter Schneider
Copyright © 2009 The Elephant 6 Publishing Co.

The Music of *Prime Suspects*

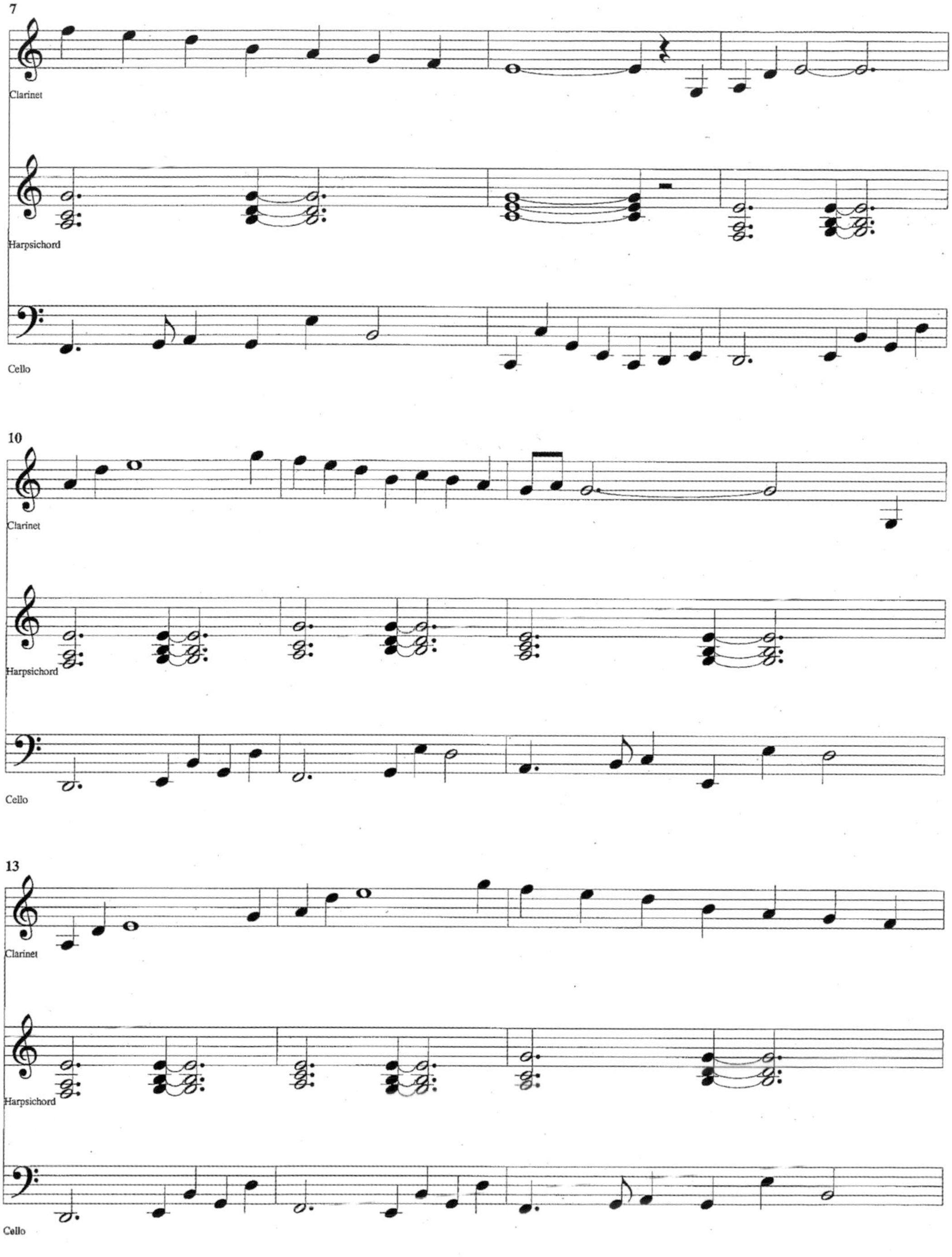
7
Clarinet
Harpsichord
Cello
10
Clarinet
Harpsichord
Cello
13
Clarinet
Harpsichord
Cello

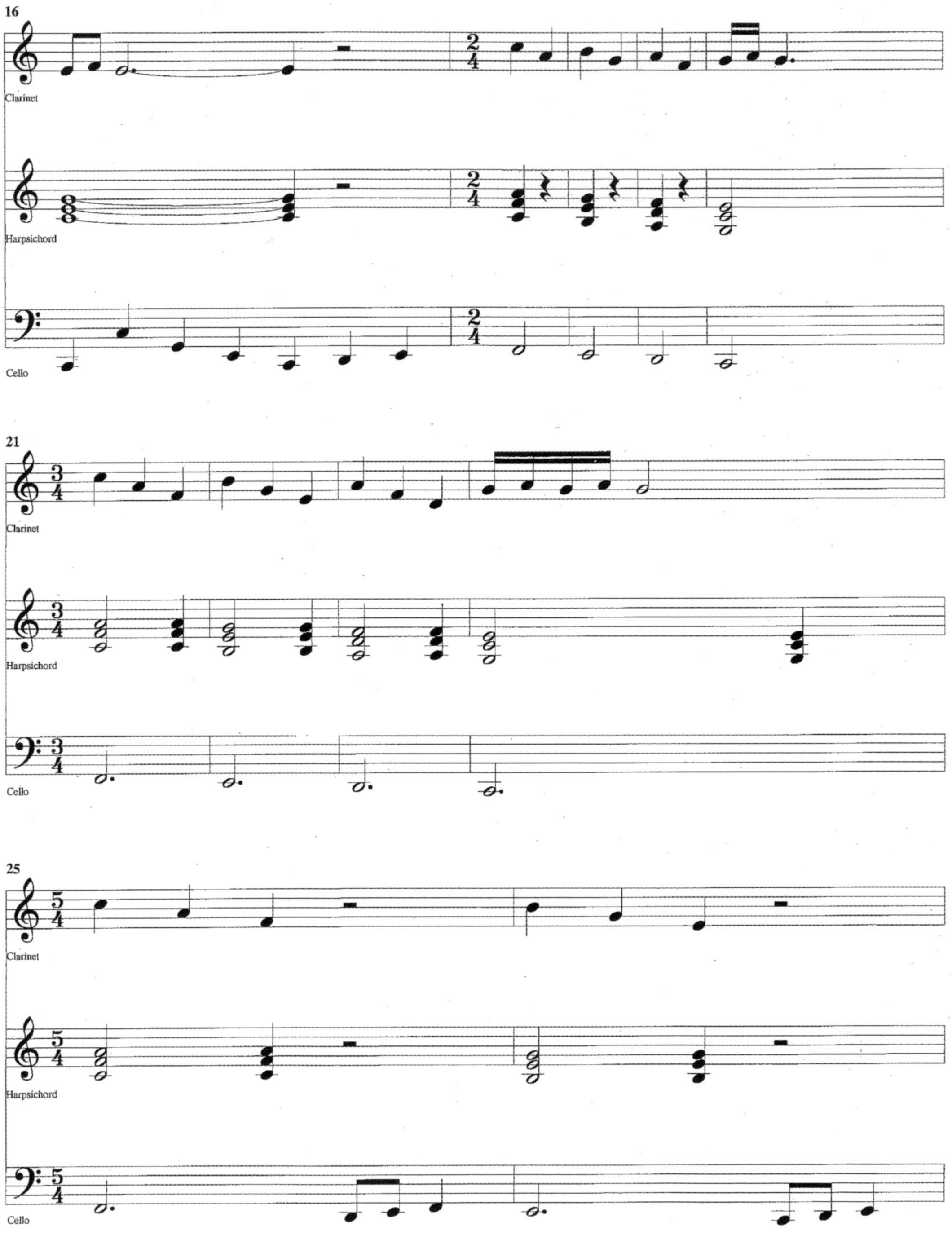
16
Clarinet
Harpsichord
Cello
21
Clarinet
Harpsichord
Cello
25
Clarinet
Harpsichord
Cello

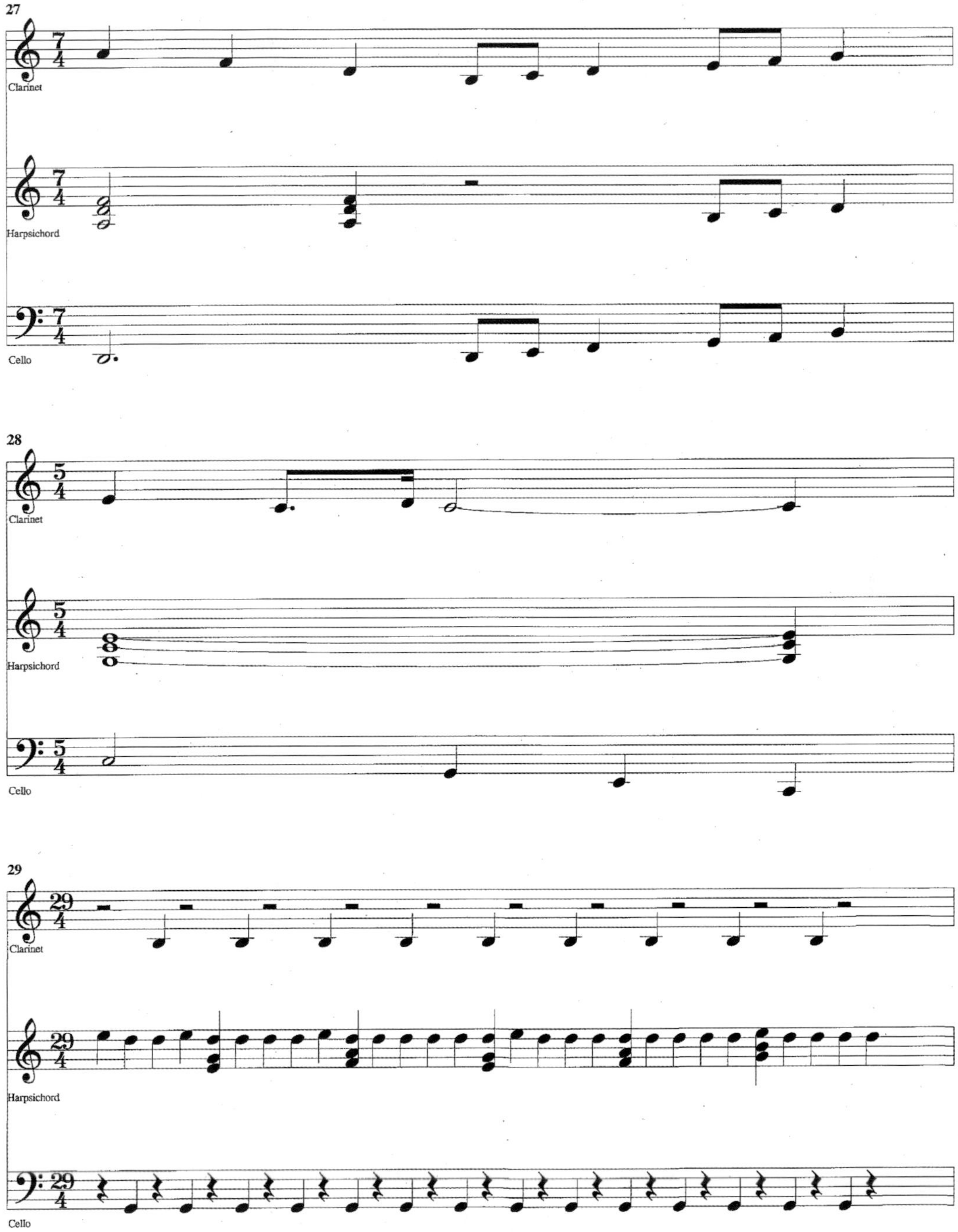
27
Clarinet
Harpsichord
Cello
28
Clarinet
Harpsichord
Cello
29
Clarinet
Harpsichord
Cello

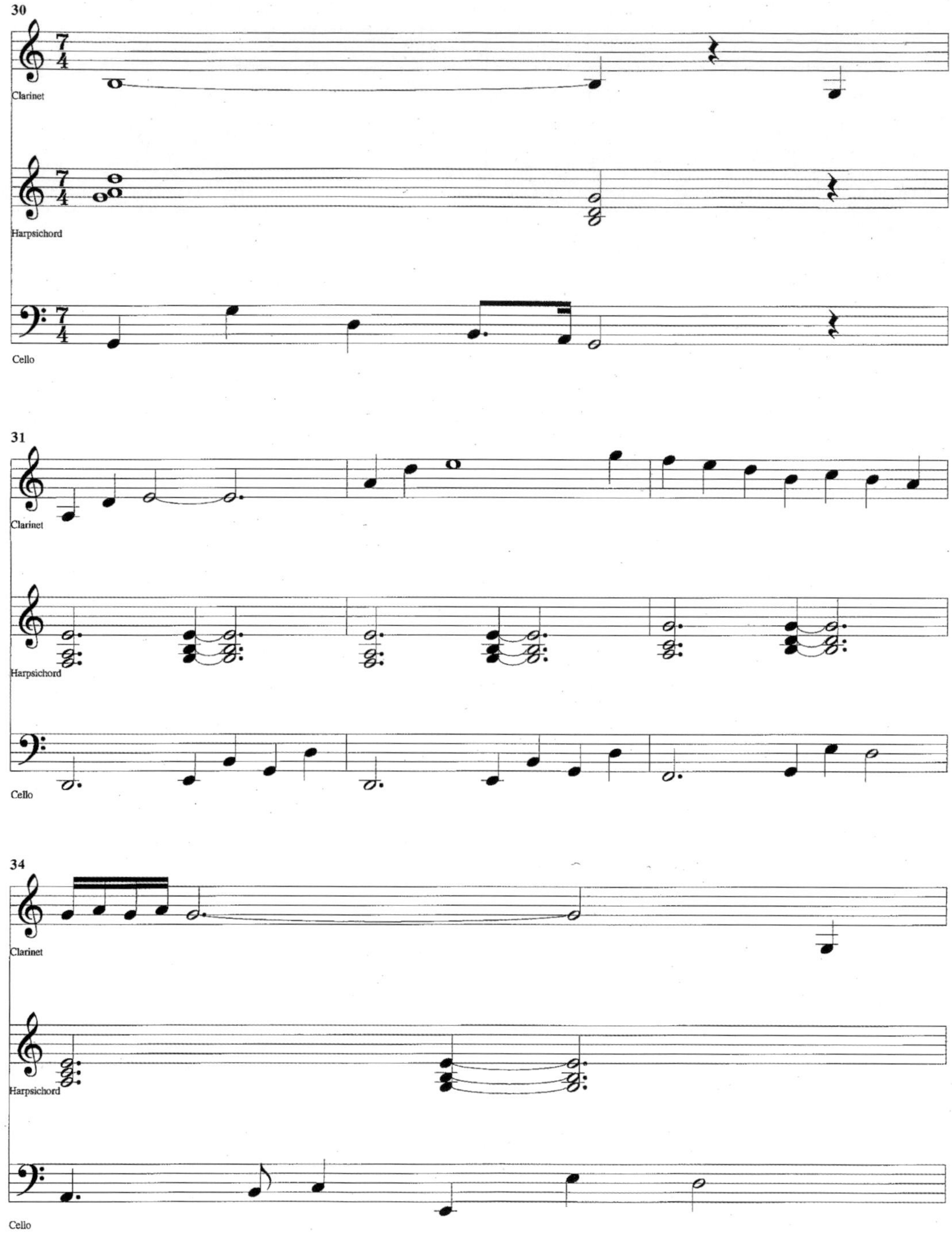
30
Clarinet
Harpsichord
Cello
31
Clarinet
Harpsichord
Cello
34
Clarinet
Harpsichord
Cello

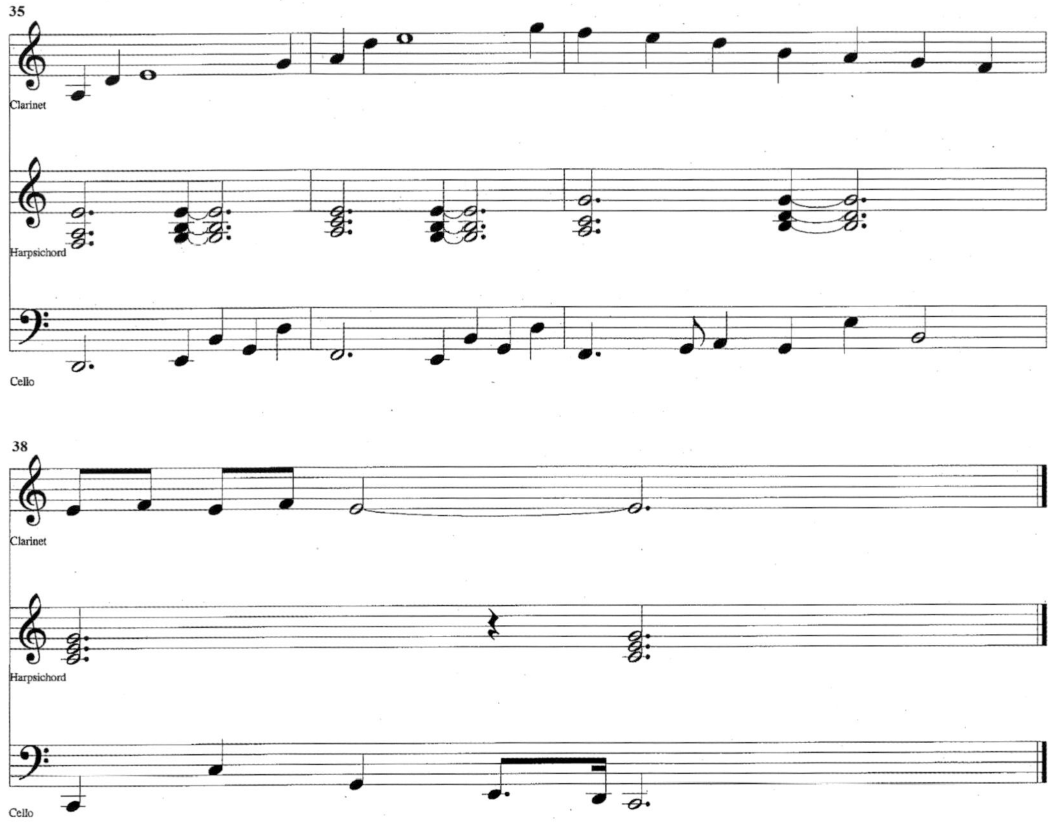
35
Clarinet
Harpsichord
Cello
38
Clarinet
Harpsichord
Cello

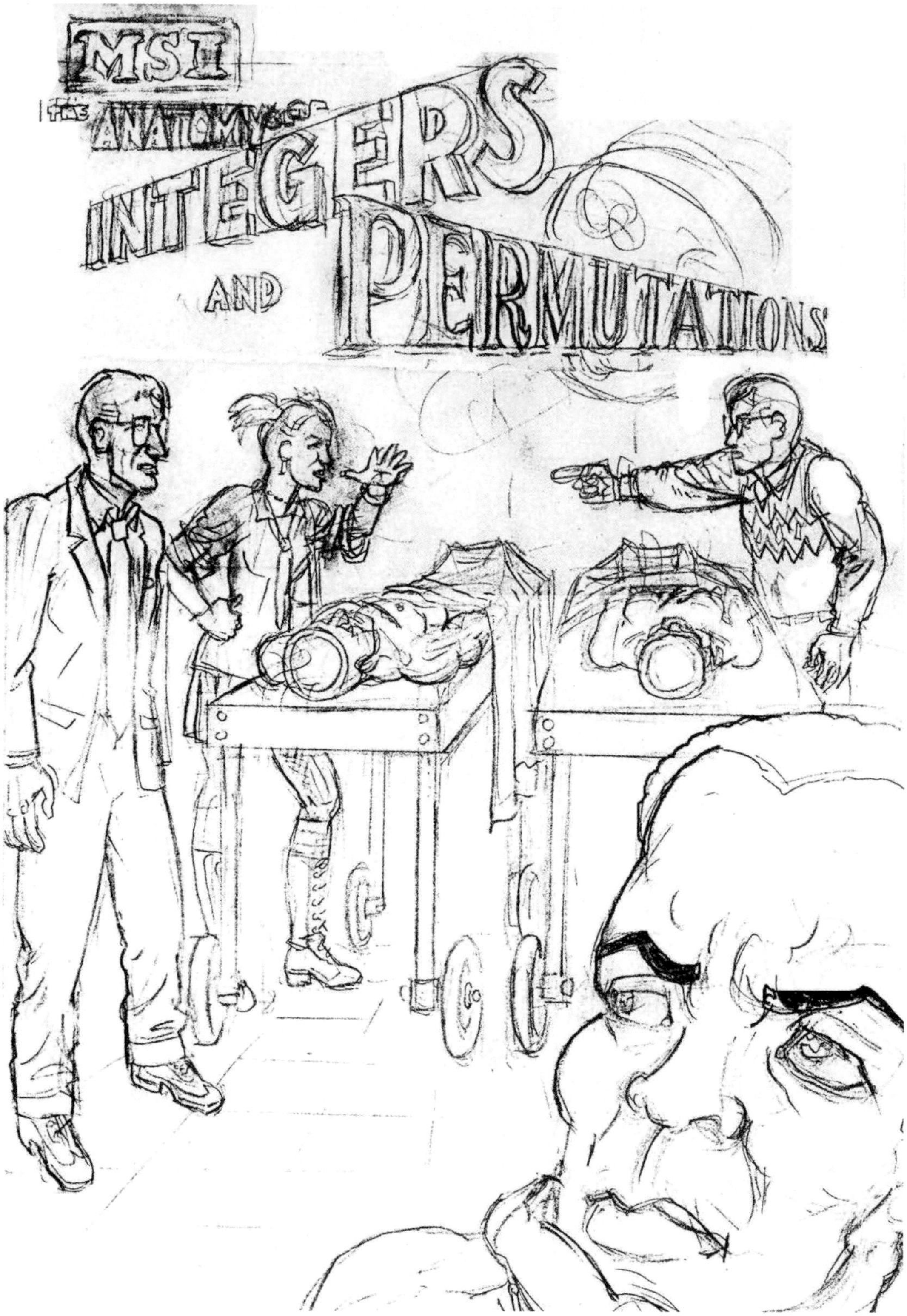
MSI
THE ANATOMY OF
INTEGERS
AND
PERMUTATIONS

# THE UNFINISHED COVER ART OF
# *SPAIN RODRIGUEZ*

The idea of developing a graphic novel arose after several live readings/performances of our original script. While staging the script at Berkeley, Jennifer stayed with a long-lost high-school friend, Susan Stern, not knowing that Susan was married to Spain Rodriguez. The Spain Rodriguez. Legendary underground cartoonist, creator of *Trashman* and, coincidentally, the artist for *Sherlock Holmes' Strangest Cases*. Hearing about our graphic novel, Spain offered to draw the cover. For Robert, our artist, this was his opportunity to "stand on the shoulders of a giant" but sadly, Spain died after only drawing a first draft. We debated whether to "finish it" for him in his style, but in the end, decided to include his original draft as a tribute to his generosity and creativity, which you can see on the page opposite.

# CREDITS AND ACKNOWLEDGMENTS

| | |
|---:|:---|
| Written by | Jennifer and Andrew Granville |
| Illustrated by | Robert J Lewis |
| Coloring by | Gabriel Cassata |
| Lettering by | John Trauscht |
| Additional inking by | Steve Wands |
| Variant cover by | Spain Rodriguez |
| Music composed and placed by | Robert Schneider |
| Documentary by | Tommy Britt |

We owe an enormous debt of gratitude to Princeton University Press for having the idea of this graphic novel and helping us see it through along a path littered with obstacles. Our hero and main point of contact is Vickie Kearn, executive editor for mathematics.

"Reverie in Prime Time Signatures," composed by Robert Schneider, ©2009 The Elephant 6 Publishing Co., score reprinted by permission. Credits for the linked music recordings: Musicians: Camilo Davila (clarinet), Christine Han (harpsichord), Chris James (flute), Emma Schmiedecke (cello), Robert Schneider (synthesizer). Producer: Robert Schneider. Recording engineers: Stuart Bremner, Pouya Hamidi (Banff Centre, Banff, Alberta, Canada).

We thank the reviewers of our original text for their helpful suggestions. We also thank the Institute for Advanced Study in Princeton, the Mathematical Sciences Research Institute at Berkeley, the Canadian Mathematical Society and the Centre Recherche mathematiques in Montreal for their support in staging readings of our original script. Moreover, Michael Spencer's varied stagings gave us the courage to embark on this comic book adventure. We also thank the Banff International Research Station for allowing us to record there.

We thank Kevin Smith for allowing us to use his "Silent Bob" character for our sound man.

Andrew expresses his thanks, for varied reasons, to Enrico Bombieri, Nassif Ghoussoub, Robert Hartshorne, Anna Kepes, Jason Lotay, Vincent Masciotra, Terry Tao, and the participants in Greene Center Wednesday.

Jennifer expresses her thanks to Sue Palmer, Andrew Fryer, and colleagues at the Northern Film School, Leeds Beckett University, for their generous support.

A massive debt is owed to the tolerance and encouragement of our families.